CHEMICAL AND BIOCHEMICAL PHYSICS

A Systematic Approach to Experiments, Evaluation, and Modeling

Edited by
David Schiraldi, PhD
Gennady E. Zaikov, DSc

APPLE ACADEMIC PRESS

Apple Academic Press Inc.
3333 Mistwell Crescent
Oakville, ON L6L 0A2
Canada

Apple Academic Press Inc.
9 Spinnaker Way
Waretown, NJ 08758
USA

Exclusive worldwide distribution by CRC Press, a member of Taylor & Francis Group
No claim to original U.S. Government works
Printed in the United States of America on acid-free paper

International Standard Book Number-13: 978-1-77188-302-3 (Hardcover)

International Standard Book Number-13: 978-1-77188-303-0 (eBook)

Library and Archives Canada Cataloguing in Publication

Chemical and biochemical physics : a systematic approach to experiments, evaluation, and modeling / edited by David Schiraldi, PhD, Gennady E. Zaikov, DSc.

Includes bibliographical references and index.
Issued in print and electronic formats.
ISBN 978-1-77188-302-3 (hardcover).--ISBN 978-1-77188-303-0 (pdf)

1. Chemistry, Physical and theoretical. 2. Physical biochemistry.
I. Schiraldi, David Anthony, 1956-, editor II. Zaikov, G. E. (Gennadiĭ Efremovich), 1935-, author, editor

QD453.3.C54 2016 541 C2016-901762-1 C2016-901763-X

CIP data on file with US Library of Congress

Apple Academic Press also publishes its books in a variety of electronic formats. Some content that appears in print may not be available in electronic format. For information about Apple Academic Press products, visit our website at **www.appleacademicpress.com** and the CRC Press website at **www.crcpress.com**

CONTENTS

LIST OF ABBREVIATIONS

AFM	atomic force microscopy
ALT	alanine aminotransferase
AN	acrylonitrile
APS	ammonium persulfate
AST	aspartate aminotransferase
BA	butyl acrylate
CAT	catalase
CMC	critical micelle-forming concentration
DEA	diethylacetamide
DHAT/IA	dichloranhydride off tere- and isophthaloyl acids
DMSO	dimethyl sulfoxide
DOP	dioctyl phthalate
DPPH	diphenylpicrylhydrazyl
DSC	differential scanning calorimetry
DTA	differential thermal analysis
EA	ethyl acrylate
EVA	ethylene with vinyl acetate
EVAMA	ethylene with vinyl acetate and maleic anhydride
FFA	free fatty acids
FL	fly larvae
GER	germatrane
GPx	glutathione peroxidase
HAETIK	n-carboxy phenylethylene
HDPE	high density polyethylene
HMPC	Herbal Medicinal Product Committee
HPE	hypericum perforatum extract
IL-1	inter leukin-1
JEFF	Journal of Engineered Fabrics and Fibers
LDH	lactate dehydrogenase
LDHs	layered double hydroxides
LDPE	low-density polyethylene
LLDPE	linear low-density polyethylene
LPO	lipid peroxidation

LS	lauryl sulfate
MA	methyl acrylate
MAA	methacryl amide
MALDI-TOF	matrix-assisted laser desorption/ionization
MC	moderate cooling
MDA	malondialdehyde malondialdehyde
MFI	melt flow index
MHTBs	active membranotropic homeostatic tissue specific bioregulators
MM	molecular mass
MMA	methyl methacrylate
MP	n-methylpyrrolidone
OI	oxygen index
PA	aromatic polyamides
PAE	polyamidoether
PAGE	polyacrylamide gel
PET/PBT	polyethylene terephthalate/polybutylene terephthalate
PE	polyethylene
PMP	polymeric and monomeric particles
PP	polypropylene
PSW	plastics solid waste
RAS	Russian Academy of Sciences
RCR	respiratory control rate
SAS	surface active substances
SOD	superoxide dismutase
SSU	Saratov State University
TEM	transmission electron microscope
TEMED	n,n,n,'n'-tetramethylethylene diamine
TEMPO-1	tetramethylpiperidine-1-oxyl
TGA	dynamic thermogravimetric analysis
TNF-α	tumor necrosis factor-α
ULD	ultra-low doses
UMD-70	universal measuring device
VA	vinyl acetate
WD	water deficit

LIST OF CONTRIBUTORS

R. M. Akhmetkhanov
Bashkir State University, 32 Zaki Validi St., 450076 Ufa, Russia, E-mail: rimasufa@rambler.ru

Z. M. Aleschenkova
Institute of Microbiology, NAS Belarus, Kuprevich Str. 2, 220141 Minsk, Belarus, E-mail: microbio@mbio.bas-net.by, Tel./Fax: +375-17-267-77-66

A. I. Bastrakov
A.N. Severtsov Institute of Ecology and Evolution, Russian Academy of Sciences (RAS), Moscow, Leninski Avenue, 119071 Russia, E-mail: naushakova@gmail.com

R. Ch. Bazheva
Kabardino-Balkarian State University, Chernishevskay St. 173, Nalchik, KBR 360004, Russia; E-mail: am_charaev@mail.ru

Yu. V. Berestneva
L.M. Litvinenko Institute of Physical Organic and Coal Chemistry. 70 R. Luxemburg Street, 83-114, Donetsk

Yuliya N. Biglova
Chemistry Department, Bashkir State University, Ufa, Russia, E-mail: bn.yulya@mail.ru

Raisa Z. Biglova
Chemistry Department, Bashkir State University, Ufa, Russia, E-mail: bn.yulya@mail.ru

V. I. Binyukov
Emanuel Institute of Biochemical Physics, Russian Academy of Sciences, Street Kosygina, 4, Moscow, 119334 Russia; Fax: +7 (499) 137-41-01; E-mail: elenamil2004@mail.ru, zhigacheva@mail.ru

T. A. Borukaev
Kabardino-Balkarian State University, 173 Chernishevsky Street, Nalchik, Russia, E-mail: borukchemical@mail.ru

D. V. Burko
Institute of Microbiology, National Academy of Sciences, Kuprevich Str. 2, 220141, Minsk, Belarus; Fax: +375(17)267-47-66; E-mail: zinch@mbio.bas-net.by

A. A. Efremova
Kazan National Research Technological University, Kazan, Russia

A. A. Fedorenchik
Institute of Microbiology, NAS Belarus, Kuprevich Str. 2, 220141 Minsk, Belarus, E-mail: microbio@mbio.bas-net.by, Tel./Fax: +375-17-267-77-66

I. T. Gabitov
Bashkir State University, 32 Zaki Validi St., 450076 Ufa, Russia, E-mail: rimasufa@rambler.ru

R. M. Garipov
Kazan National Research Technological University, Kazan, Russia

I. P. Generozova
Timiryazev Institute of Plant Physiology, Russian Academy of Sciences, ul. Botanicheskaya 35, Moscow, 127276, Russia, Tel.: (495)903-93-40, E-mail: igenerozova@mail.ru

M. D. Goldfein
Biological Faculty, Saratov State University (SSU), Saratov, Russia

A. P. Ilyina
Nesmeyanov Institute of Organoelement Compounds, Russian Academy of Sciences, St. Vavilova 28, Moscow, 119991 Russia

S. G. Karpova
Laboratory of Physical Chemistry of Synthetical and Natural Polymers, Emanuel Institute of Biochemical Physics of Russian Academy of Sciences, Russia; E-mail: kolesnikova@sky.chph.ras.ru, karpova@sky.chph.ras.ru

A. M. Kharaev
Kabardino-Balkarian State University, Chernishevskay St. 173, Nalchik, KBR 360004, Russia; E-mail: am_charaev@mail.ru

A. I. Khasanov
Kazan National Research Technological University, Kazan, Russia

E. L. Khmeleva
Moscow University of Fine Chemical Technology, pr. Vernadskogo 86, Moscow, 119571, Russia

N. N. Kolesnikova
Laboratory of Physical Chemistry of Synthetical and Natural Polymers, Emanuel Institute of Biochemical Physics of Russian Academy of Sciences, Russia; E-mail: kolesnikova@sky.chph.ras.ru, karpova@sky.chph.ras.ru

G. A. Korablev
Izhevsk State Agricultural Academy, Russia, E-mail: korablevga@mail.ru, devugen@mail.ru

Vladimir A. Kraikin
Institute of Organic Chemistry, URC RAS, Ufa, Russia

L. Z. Kravtsova
The "NTC BIO" LLC, 309292 Russia, Belgorod Region, Shebekino Town, Dokuchayev Str., Russia, E-mail: ntcbio@mail.ru

N. V. Kozhevnikov
Biological Faculty, Saratov State University (SSU), Saratov, Russia

A. A. Kozlov
Kazan National Research Technological University, Kazan, Russia

G. V. Kozlov
Kh.M. Berbekov Kabardino-Balkarian State University, Nal'chik – 360004, Chernyshevsky St., 173, Russian Federation, E-mail: i_dolbin@mail.ru

O. G. Kulikova
Nesmeyanov Institute of Organoelement Compounds, Russian Academy of Sciences, St. Vavilova 28, Moscow, 119991 Russia

V. K. Kumykov
Kabardino-Balkarian State University, Chernishevskay St. 173, Nalchik, KBR 360004, Russia; E-mail: am_charaev@mail.ru

S. V. Kvach
Institute of Microbiology, National Academy of Sciences, Kuprevich Str. 2, 220141, Minsk, Belarus; Fax: +375(17)267-47-66; E-mail: zinch@mbio.bas-net.by

L. R. Lyusova
Moscow University of Fine Chemical Technology, pr. Vernadskogo 86, Moscow, 119571, Russia

Sh. Maghsoodlou
University of Guilan, Rasht, Iran

D. I. Maltsev
Nesmeyanov Institute of Organoelement Compounds, Russian Academy of Sciences, ul. Vavilova 28, Moscow, 119991 Russia

E. E. Mastalygina
Laboratory of Advanced Composite Materials and Technologies, Department of Chemistry and Physics, Plekhanov Russian University of Economics, Russia; E-mail: elena.mastalygina@gmail.com, popov@sky.chph.ras.ru

Mansur S. Miftakhov
Institute of Organic Chemistry, URC RAS, Ufa, Russia

A. K. Mikitaev
Kh.M. Berbekov Kabardino-Balkarian State University, Nal'chik – 360004, Chernyshevsky St., 173, Russian Federation, E-mail: i_dolbin@mail.ru

E. M. Mil
Emanuel Institute of Biochemical Physics, Russian Academy of Sciences, Street Kosygina, 4, Moscow, 119334 Russia; Fax: +7 (499) 137-41-01; E-mail: elenamil2004@mail.ru, zhigacheva@mail.ru

A. G. Mustafin
Bashkir State University, 32 Zaki Validi St., 450076 Ufa, Russia, E-mail: rimasufa@rambler.ru

Yu. A. Naumova
Moscow University of Fine Chemical Technology, pr. Vernadskogo 86, Moscow, 119571, Russia

N. I. Naumovich
Institute of Microbiology, NAS Belarus, Kuprevich Str. 2, 220141 Minsk, Belarus, E-mail: microbio@mbio.bas-net.by, Tel./Fax: +375-17-267-77-66

R. V. Nekrasov
All Russian Research Institute of Animal Husbandry After Academy Member L.K. Ernst, 142132 Russia, Moscow Region, Podolsk District, Settlement Dubrovitsy, Russia, E-mail: nek_roman@mail.ru

R. Z. Oshroeva
Kabardino-Balkarian State University, 173 Chernishevsky Street, Nalchik, Russia, E-mail: boruk-chemical@mail.ru

L. A. Pashkova
All Russian Research Institute of Animal Husbandry After Academy Member L.K. Ernst, 142132 Russia, Moscow Region, Podolsk District, Settlement Dubrovitsy, Russia, E-mail: nek_roman@mail.ru

A. A. Popov
Laboratory of Advanced Composite Materials and Technologies, Department of Chemistry and Physics, Plekhanov Russian University of Economics, Russia; E-mail: elena.mastalygina@gmail.com, popov@sky.chph.ras.ru

S. Poreskandar
University of Guilan, Rasht, Iran

I. V. Pravdin
The "NTC BIO" LLC, 309292 Russia, Belgorod Region, Shebekino Town, Dokuchayev Str., Russia, E-mail: ntcbio@mail.ru

E. V. Raksha
L.M. Litvinenko Institute of Physical Organic and Coal Chemistry. 70 R. Luxemburg Street, 83-114, Donetsk

M.M. Rasulov
Research Institute of Chemistry and Technology Organoelement Compounds, Enthusiasts Highway 38, Moscow 111123, Russia, Tel.: +7(495)673-13-78; Fax: +7 (495) 783-6444; E-mail: maksud@bk.ru

A. O. Roshchin
Nesmeyanov Institute of Organoelement Compounds, Russian Academy of Sciences, ul. Vavilova 28, Moscow, 119991 Russia

S. N. Rusanova
Kazan National Research Technological University, 68 K. Marksa Str., Kazan 420015, Russia

A. N. Rymko
Institute of Microbiology, National Academy of Sciences, Kuprevich Str. 2, 220141, Minsk, Belarus; Fax: +375(17)267-47-66; E-mail: zinch@mbio.bas-net.by

L. M. Sakhtueva
Kabardino-Balkarian State University, Chernishevskay St. 173, Nalchik, KBR 360004, Russia; E-mail: am_charaev@mail.ru

N. I. Samoilyk
Kabardino-Balkarian State University, 173 Chernishevsky Street, Nalchik, Russia, E-mail: boruk-chemical@mail.ru

A. S. Shchokolova
Institute of Microbiology, National Academy of Sciences, Kuprevich Str. 2, 220141, Minsk, Belarus; Fax: +375(17)267-47-66; E-mail: zinch@mbio.bas-net.by

S. Yu. Sofina
Kazan National Research Technological University, 68 K. Marksa Str., Kazan 420015, Russia

O. V. Stoyanov
Kazan National Research Technological University, 68 K. Marksa Str., Kazan 420015, Russia

N. E. Temnikova
Kazan National Research Technological University, 68 K. Marksa Str., Kazan 420015, Russia

N. A. Turovskij
Donetsk National University, 24 Universitetskaya Street, 83–001 Donetsk, E-mail: elenaraksha411@gmail.com

N. A. Ushakova
A.N. Severtsov Institute of Ecology and Evolution, Russian Academy of Sciences (RAS), Moscow, Leninski Avenue, 119071 Russia, E-mail: naushakova@gmail.com

Yu. G. Vasiliev
Izhevsk State Agricultural Academy, Russia, E-mail: korablevga@mail.ru, devugen@mail.ru

I. A. Yamskov
Nesmeyanov Institute of Organoelement Compounds, Russian Academy of Sciences, St. Vavilova 28, Moscow, 119991 Russia

V. P. Yamskova
Koltsov Institute of Developmental Biology, Russian Academy of Sciences, St. Vavilova 26, Moscow, 119334 Russia, E-mail: yamskova-vp@yandex.ru

A. I. Zagidullin
Kazan National Research Technological University, Kazan, Russia

Vadim V. Zagitov
Chemistry Department, Bashkir State University, Ufa, Russia, E-mail: bn.yulya@mail.ru

G. E. Zaikov
N.M. Emanuel Institute of Biochemical Physics of Russian Academy of Sciences, Moscow 119334, Kosygin St., 4, Russian Federation, E-mail: chembio@sky.chph.ras.ru

V. P. Zakharov
Bashkir State University, 32 Zaki Validi St., 450076 Ufa, Russia, E-mail: rimasufa@rambler.ru

I. V. Zhigacheva
Emanuel Institute of Biochemical Physics, Russian Academy of Sciences, Street Kosygina, 4, Moscow, 119334 Russia; Fax: +7 (499) 137-41-01; E-mail: elenamil2004@mail.ru, zhigacheva@mail.ru

A. I. Zinchenko
Institute of Microbiology, National Academy of Sciences, Kuprevich Str. 2, 220141, Minsk, Belarus; Fax: +375(17)267-47-66; E-mail: zinch@mbio.bas-net.by

PREFACE

This valuable new book, *Chemical and Biochemical Physics: A Systematic Approach to Experiments, Evaluation, and Modeling*, focuses on a systematic approach to experiments, evaluation, and modeling. By providing an applied and modern approach, this volume will help readers to understand the value and relevance of studying chemical and biochemical physics and technology to all areas of applied chemical engineering. It gives readers the depth of coverage they need to develop a solid understanding of the key principles in the field. Presenting a wide-ranging view of current developments in applied methodologies in chemical and biochemical physics research, the papers in this volume, all written by highly regarded experts in the field, examine various aspects of chemical and biochemical physics and experimentation.

The book:

- highlights applications of chemical and biochemical physics for industry and research
- introduces the types of challenges and real problems that are encountered in industry and research
- presents biochemical examples and applications

The book is ideal for upper-level research students in chemistry, chemical engineering, and polymers. The book assumes a working knowledge of calculus, physics, and chemistry, but no prior knowledge of polymers.

This book focuses on the limitations, properties, and models in the chemistry and physics of engineering materials that have potential for applications in several disciplines of engineering and science. The contributions range from new methods to novel applications of existing methods.

The topics in this book reflect the diversity of recent advances in chemistry and physics of engineering materials with a broad perspective that will be useful for scientists as well as for graduate students and engineers. This new book presents leading-edge research from around the world.

This comprehensive anthology covers many of the major themes of chemical and biochemical physics, addressing many of the major issues, from concept to technology to implementation. It is an important reference publication that provides new research and updates on a variety of physical chemistry and biochemical physics uses through case studies and supporting technologies, and it also explains the conceptual thinking behind current uses and potential uses not yet implemented. International experts with countless years of experience lend this book credibility.

ABOUT THE EDITORS

David Schiraldi, PhD

Professor and Chair, Department of Macromolecular Science and Engineering, Case Western Reserve University, Cleveland, Ohio, USA

David Schiraldi, PhD, is Professor and Chair of the Department of Macromolecular Science and Engineering at Case Western Reserve University, Cleveland, Ohio, USA. He is an expert in the field of chemistry, physics and mechanics of polymers, and composites and nanocomposites. He has published several hundred scientific original papers as well as many reviews. His research interests focus on the transport properties of ionic and non-ionic solutes in multicomponent systems, such as host-guest compounds, as well as in the characterization of the transport properties in polymeric matrices, with particular emphasis to polyelectrolytes, gels, and functional blends and composites.

Gennady E. Zaikov, DSc

Head of the Polymer Division, N. M. Emanuel Institute of Biochemical Physics, Russian Academy of Sciences, Moscow, Russia; Professor, Moscow State Academy of Fine Chemical Technology, Russia; Professor, Kazan National Research Technological University

Gennady E. Zaikov, DSc, is Head of the Polymer Division at the N. M. Emanuel Institute of Biochemical Physics, Russian Academy of Sciences, Moscow, Russia, and Professor at Moscow State Academy of Fine Chemical Technology, Russia, as well as Professor at Kazan National Research Technological University, Kazan, Russia. He is also a prolific author, researcher, and lecturer. He has received several awards for his work, including the Russian Federation Scholarship for Outstanding Scientists. He has been a member of many professional organizations and is on the editorial boards of many international science journals. Dr. Zaikov has recently been honored with tributes in several journals and books on the occasion of his 80th birthday for his long and distinguished career and for his mentorship to many scientists over the years.

PART I

CHEMICAL PHYSICS

CHAPTER 1

HALOGEN CONTAINING SIMPLE AND COMPLICATED BLOCK COPOLYETHERS

A. M. KHARAEV, R. Z. OSHROEVA, G. E. ZAIKOV,
R. CH. BAZHEVA, L. M. SAKHTUEVA, and V. K. KUMYKOV

Kabardino-Balkarian State University, Chernishevskay St. 173, Nalchik, KBR 360004, Russia; E-mail: am_charaev@mail.ru

CONTENTS

ABSTRACT

Bifunctional halogen, containing oligomers of various composition and structure, are synthesized. Simple and complex aromatic block copolyethers of constructional and film purpose are obtained by various methods of polycondensation. Physical and chemical properties of block copolyethers are studied in this chapter.

1.1 INTRODUCTION

From numerous references it is known, that simple and complex polyethers have a number of unique properties. So, such a simple polyethers as polyethersulphones, polyetherketones and others, possess the high thermal resistance and plasticity in combination with some other operational characteristics. At the same time complex polyethers are characterized by high heat resistance and increased rigidity of the material.

The content of multiple links in a macrochain makes polymers capable to formation of spatially structured materials, differing by higher operational properties. By this way it is possible to increase considerably the thermal resistibility of materials that expands an operation temperature interval of its products, to raise the thermooxidizing destruction start temperature by tens degrees, to improve strength properties, to obtain polymeric materials, resistant even to the concentrated solutions of acids and alkalis.

Inserting the halogen atoms to the structure of polymer significantly increases the fire resistance of polymers. At their significant amounts the polymers become nonflammable and they are not secondary sources of ignition, because they do not burn and do not sustain burning. The content of atoms of halogen also positively influences mechanical, heat physical and other properties. Presence of volume polar atoms of bromine raises strength characteristics of polymers. Polar atoms of bromine also expand a temperature interval of operation of products due to increase of temperatures of glass transition and fluidity. Showing the shielding effect, atoms of bromine increase polyether's resistance to solutions of various acids and alkalis, including concentrated [1–14].

1.2 EXPERIMENTAL PART

For obtaining the non-saturated halogen containing block copolyethers by high-temperature polycondensation various new oligoethers [15, 16] were synthesized according to the following scheme (n = 1–20):

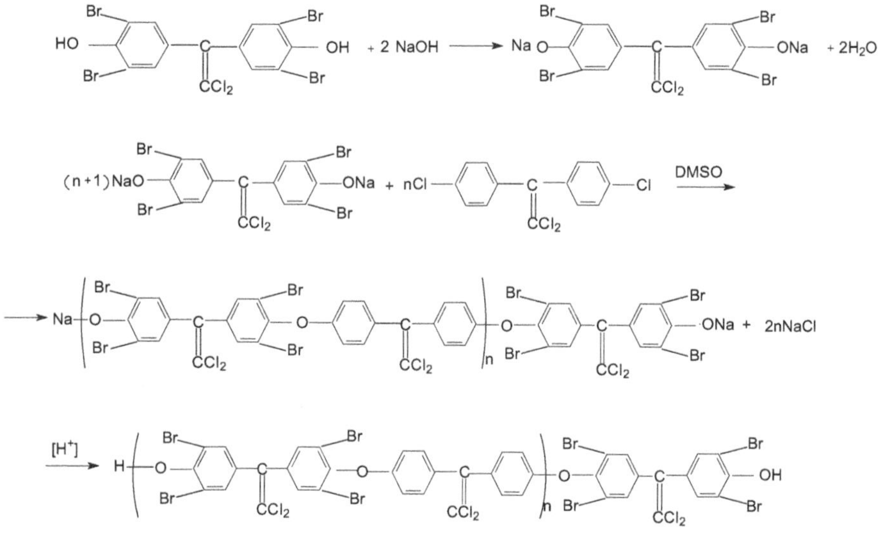

The reaction between 1,1-dichlor-2,2-di (3,5-dibrom-n-hydroxy-phenyl)ethylene and 1,1-dichlor-2,2-di(n-phenyl chloride)ethylene is carried out in the environment of aprotic dipolar solvent – dimethyl sulfoxide (DMSO) in the atmosphere of inert gas (nitrogen).

Simple polyethers are obtained by high-temperature polycondensation in N,N-dimethylacetamine at 170–180°C within 6 hours in inert gas atmosphere – nitrogen.

The scheme of polyether obtaining may be presented in a general view as follows:

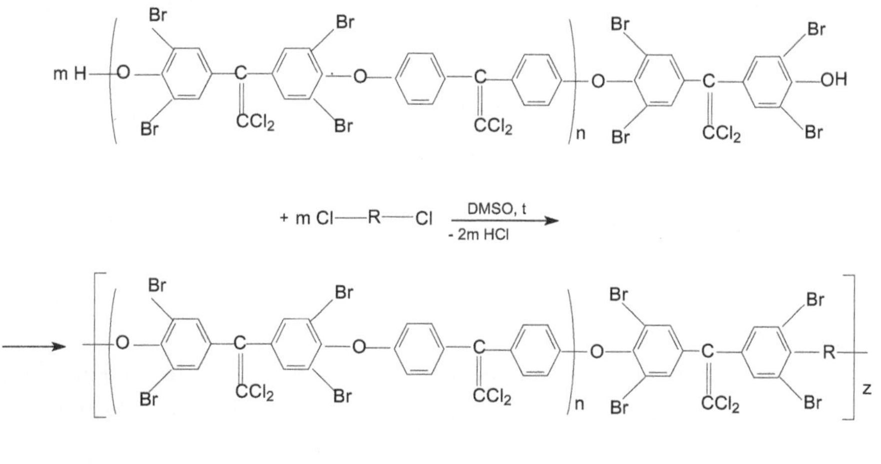

where:

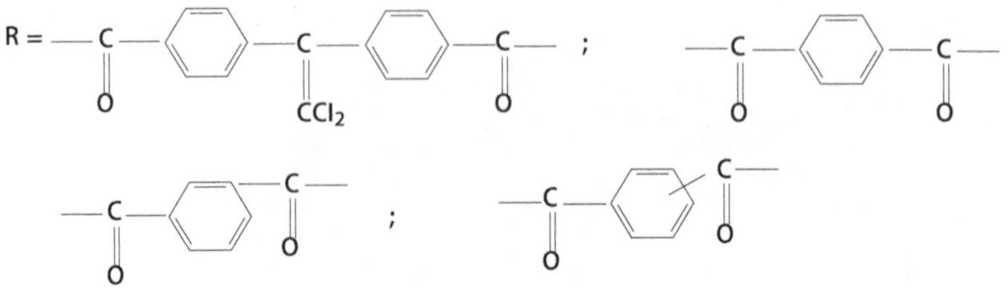

The viscosity measurements were made according to GOST 10028-81 (RF Standards) using an Ubbelohde viscometer with d = 0.56 mm. The viscosity was measured in 1,2-dichlorethane and the density of the solution was 0.5 g/dL.

Mechanical properties of film PAEK specimens (100 mm × 10 mm × 0.1 mm) were tested (GOST 11262, ASTM D638) on a MRS-500 with a constant strain rate of 40 mm/min at 20°C. The film specimens were obtained by pouring the polymeric solution into 1,2-dichlorethane.

Thermo-gravimetric analysis of poly(arylene ether ketone)s was performed on the derivatograph "MOM" under a temperature increase rate at 5 degree/min in the atmosphere.

The investigation of the polydispersity of the block copolymers was conducted by the turbidimetric titration method on an FEC-56M device. The principle of titration is that the diluted polymeric solution will become turbid if a precipitator is added and will have a different optical density from the original solution. The density of the solution was 0.01 g/mL. 1,2-dichlorethane was used as a solvent, and isopropyl alcohol – as precipitator.

The thermal-resistance was tested on the film specimens (strips) fixed vertically in a cylindrical camera, and laminar stream of nitrogen and oxygen mixture of the given correlation was put through it. The investigation was carried out under various structures of the gas mixture unless the optimal structure, which provided burning of the specimen was found. Thermal-resistance was evaluated by the percentage of oxide, contained in the specimen (GOST 21207–75).

The fire resistivity of the polymers was evaluated by the oxygen index method. The oxygen index test carried out on film samples (strips) fixed vertically in the cylindrical chamber through which passes a laminar stream of a mixture of nitrogen with oxygen. Tests are carried out at various ratios of a gas mixture until the optimum burning the sample is reached. The sample is set on fire from the top end with the help of a gas torch that is then withdrawn.

1.3 RESULTS AND DISCUSSION

Composition and structure of oligoethers are confirmed by IR-spectroscopy and element analysis. Some properties of oligoethers are given in Table 1.1.

Non-saturated halogen containing simple and complex polyethers with block structure was obtained with the use of synthesized oligoketone [4].

Complex polyethers are obtained by acceptor catalytic polycondensation with the use as acid components the equimole quantities of dichloranhydride off tere- and isophthaloyl acids (DHAT/IA), and also a dichloranhydride of 1,1-dichlor-2,2-di (n-carboxy phenyl)ethylene (HAETIK). Synthesis is carried out in ethylene dichloride medium at room conditions with the use of triethylamine as an acceptor catalyst.

Polyethers are obtained quantitatively and with high rates of given viscosity (Table 1.2). These indicators together with the data of IR-spectroscopy and turbidimetric titration confirm the formation of polyethers of supposed structure. On IR spectrums there are absorption strips, corresponding to simple and complex ethereous links, dichloroethylene group, Ar-Br – linkage and there are no absorption strips for OH-group.

Curves of molecular-mass distributions show a low polydispersity and good solubility of polyethers in chlorinated organic solvents. From

TABLE 1.1 Properties of Oligomers

Oligomers	n	Yield, %	T soft., °C	Calculated, MM	Hydroxyl group content	
					Calculated	Obtained
OE-1TBC-2	1	96	96–97	1446	2.35	2.34
OE-5TBC-2	5	95	102–105	4806	0.71	0.70
OE-10TBC-2	10	95	105–107	9015	0.38	0.37
OE-20TBC-2	20	94	108–110	17433	0.19	0.20

TABLE 1.2 Some Properties of Polyethers

№№	Polymer	Yield, %	Intrinsic Viscosity η., dL/g	T_{sl}., °C
1	OE-1TBC-2 + HAETIK	98.0	0.81	260
2	OE-10TBC-2 + HAETIK	98.0	0.80	263
3	OE-20TBC-2 + HAETIK	96.5	0.63	271
4	OE-1TBC-2 + DHAT/IA	97.5	0.90	236
5	OE-10TBC-2 + DHAT/IA	95.0	0.82	239
6	OE-20TBC-2 + DHAT/IA	95.0	0.66	242
7	OE-1TBC-2 + DPBP	96.0	0.71	247
8	OE-10TBC-2 + DPBP	97.5	0.66	251
9	OE-20TBC-2 + DPBP	96.5	0.60	250
10	OE-1TBC-2 + DHDPC	97.5	0.75	240
11	OE-10TBC-2 + DHDPC	96.0	0.70	243
12	OE-20TBC-2 + DHDPC	96.5	0.61	241

Figure 1.1 it is visible, that thresholds of coagulation of polyethers lie in the field of enough large volumes of solvents that confirms their good solubility. It is shown, that in the ranks of polyethers with the increase of condensation rate of initial oligomers the solubility in organic solvents increases, which is possible to connect with a loosening of structure of macromolecules because of volume atoms of bromine.

Glass transition temperatures of nonsaturated bromine, containing polyethers on the basis of various dihalogenides are given in Table 1.2. It is visible, that in the ranks of polyethers this characteristic doesn't change significantly. If some increase of it is observed, it can be explained with prevalence of process of structuring in macromolecules because of saturation of a macrochain by dichloroethylene group, over process of a structure loosening because of increase in a share of volume atoms of bromine at initial oligomer lengthening.

The highest rates of heat resistance are characteristic for polyethers on the basis of dichloride anhydride of HAETIK. Possibly, it is concerned to insertion to macromolecules the remains of dichloride anhydride $>C=CCl_2$ group. Double link promotes the formation of mesh structure, and atoms of chlorine increase polarity of macromolecules. In other three rows of

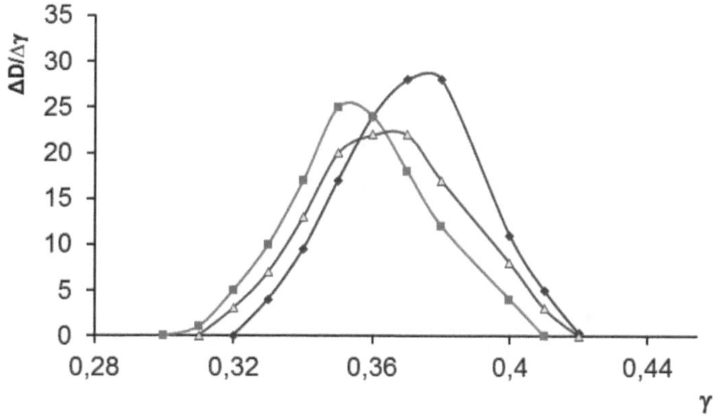

FIGURE 1.1 Integral (1, 2) and differential (1', 2') curves of turbidimetric titration of polyethers: 1, 1'-OE-1TBS-2 + HAETIC and 2, 2'-2OTBS-2 + HAETIC.

polymers, where in the remains of halogenides there is no dichloride ethylene group, lower indicators of temperatures of glass transition are observed.

Polyethers, obtained in present work, possess the increased thermal resistance. For the majority of polymers the beginning of thermooxidizing destruction is observed at 400°C and above. In the ranks of polyethers with the increase of condensation rate of initial oligomers the increase of thermal stability indicators is observed. In complex polyethers it is related to the reduction of content of thermally unstable complex ether groups. In simple polyethers, which are rather resistant to high temperatures, the decisive factor is the existence of polar and bulk atoms of bromine, and concentration of the last increases in ranks.

The manifestation of structuring effect is easily observed when comparing temperatures of 10% mass loss. In polyethers on the basis of HAETIK it courses a substantial increase of this indicator. Other polymers, which do not contain the $>C=CCl_2$ in acid components, loose the specified mass at significantly lower temperatures. Below in the Table 1.3 the results of thermal stability studies and other properties of the synthesized polyethers are presented.

All polyethers show high resistance to direct pull at rather small lengthening. The defining factor is the existence and concentration of bromine atoms. In polyether ranks concentration of the last increases, coursing the

TABLE 1.3 Thermal and Mechanical Properties of Polyethers

№	Polymer	Percentage		σ_{ts}, MPa	ε, %	OI, %
		2%	10%			
1	OE-1TBC-2 + HAETIK	406	526	96.8	12.6	53.0
2	OE-10TBC-2 + HAETIK	410	532	99.8	11.1	56.0
3	OE-20TBC-2 + HAETIK	418	537	101.2	11.0	56.5
4	OE-1TBC-2 + DHAT/IA	397	471	103.6	11.4	50.5
5	OE-10TBC-2 + DHAT/IA	402	479	106.7	11.0	52.0
6	OE-20TBC-2 + DHAT/IA	409	480	106.3	10.2	53.5
7	OE-1TBC-2 + DPBP	378	480	93.4	11.4	50.0
8	OE-10TBC-2 + DPBP	394	474	104.3	10.1	50.5
9	OE-20TBC-2 + DPBP	400	482	105.1	9.6	52.5
10	OE-1TBC-2 + DHDPC	393	491	97.7	11.8	50.5
11	OE-10TBC-2 + DHDPC	407	495	101.9	10.7	51.0
12	OE-20TBC-2 + DHDPC	410	497	102.4	10.8	52.0

increase of direct pull resistance. In the competing processes of a structure loosening and increase of macromolecules polarity, probably, the last prevails.

Same explanation has the low indicators of relative lengthening and falling of this indicator in ranks with the increase in length of initial oligomers. The tensile strength and relative lengthening of synthesized polyethers are not differ significantly and are in the range of 93.4–106.7 MPa and 9.6–12.6% respectively.

The characteristics of inflammation and combustibility of polymer materials are connected closely with the existence of macromolecules haloid containing groupings in the chain. The introduction of a macromolecule $>C=CCl_2$ groupings into the chain and the increase of their percentage in block-copolymers promote the increase of the index of the oxygen index (OI). The ramp OI of block-copolymers with the increase of the content of chlorinated components, is apparently connected with the changes of the amount of combustible products, exuded from unit of volume of block-copolymers when burning.

As one would expect, the real polyethers do not burn and do not sustain burning. The fire resistance, estimated in oxygen indexes is in the range of

50.0–56.5%, and it is the result of the maximum saturation of macrochains by the atoms of halogens, and particularly by bromine atoms. We came to such conclusion because of the increase of OI values in all synthesized ranks of polyethers.

The essential contribution to fire resistance is also made by the atoms of chlorine, which are available in dichloroethylene group of HAETIK dichloranhydride remains. For this reason OI indicators for a number of polyethers on the basis of given dihalogenid are higher, than in those, which structures do not have mentioned $>C=CCl_2$ group. Polyethers on naked flame just carbonizing and are not secondary sources of fire.

1.4 CONCLUSION

The results of studies show that the synthesized nonsaturated block copolymers possess a complex of valuable properties and can find applications in various branches of technique as thermo- and fire-resistant constructional and film materials.

KEYWORDS

- **bifunctional oligomer**
- **fire resistance**
- **polycondensation**
- **thermal resistance**

REFERENCES

1. Ozden, S., Charaev, A. M., Shaov, A. H. High impact thermally stable block copolyethers. J. Mater. Sci. 2001. v. 36. 4479–4484.
2. Ozden, S., Charayev, A. M., Shaov, A. H. Synthesis of block copolyether ketones and investigations of their properties. J. Appl. Polym. Sci. 2002. v. 85. 485–490.
3. Charayev, A. M., Shaov, A. H., Shustov, G. B. Mikitaev, A. K. Synthesis and Properties of Polyetherketones. Chemistry and chemical technology. 1998. Vol. 41. № 5. 78–81.

4. Ozden, S., Kharayev, A. M., Bazheva, R.Ch. Synthesis and modification of aromatic polyesters with chloroacetyl 3,5-dibromo-p-hydroxybenzoic acid. J. of Appl. Pol. Sci. V. 111, I. 4, 2009. 1755–1762.
5. Blencowe, A., Davidson, L., Hayes, W. European Polymer Journal, 39, 10, 2003. 1955–1963.
6. Mikitaev, A. K., Shustov, G. B., Kharaev, A. M. Synthesis and properties of block-copolysulfonarilates. Polymer Science. Series, A., 1984, 26, №1, 75–78 (in Rus.).
7. Kharaev, A. M., Mikitaev, A. K., Shustov, G. B. and others. Synthesis and properties of block-copolysulfonarilates on the basis of oligoarilensulfophenolphthalein. Polymer Science. Series, B., 1984, V.26 №14. 271–274 (in Rus.).
8. Mikitaev, A. K., Kharaev, A. M., Shustov, G. B. Unsaturated aromatic compound polyesters on the basis of chloral's derivatives as constructive and membraneous materials. Polymer Science. Series, A., 1998, v.39. №15, 228–236 (in Rus.).
9. Kharaev, A. M. Aromatic polyesters as thermostable constructive and membraneous materials. Thesis PhD, – Nalchik, 1993. 297 p.
10. Charayev, A. M., Shaov, A. H., Mikitaev, A. K., Matvelashvili, G. S., Khasbulatova, Z. S. Polymeric composite materials on a basis polyetherethers. International Polymer Science and Technology. V.3, №3. 1992.
11. Ozden, S., Kharayev, A. M., Shaov, A.Kh., Bazheva, R.Ch. The synthesis of blok copolyetheketones on 4,4′-dichlodiphenylketone, phenolphthaleine and bisphenol A and investigation of their properties. J. of Applied Polymer Science. 2008. V. 107, I. 4, 2459–2465.
12. Barokova, E. B., Bazheva, R.Ch., Haraev, A. M. Oligosulphones on the basis of 1,1-Dichlor-2,2-di(4-oxyphenyl)-Ethylene and 4,4′-dichlordiphenyl-sulphone obtained by high-temperature polycondensation. J. of the Tribological Association. V. 16, № 2. 2010. 284–287.
13. Kharaev, A., Shaov, A., Bazheva, R. The Synthesis and Stabilization of Polymers (Monograph). Saarbrucken, Deutschland. Palmarium Academic Publishing. ISBN-13: 2013. 978-3-65948590-9.
14. Kharaev, A. M., Bazheva, R.Ch., Chaika, A. A. et al. Aromatic block copolymers as heat-resistant structural materials and film. International Polymer Science and Technology. 2013. № 9, 22–26.
15. Patent 2445304 (RF), 2012. Bazheva, R.Ch., Kharaev, A. M., Kharaeva, R. A., Kazancheva, F. K., Khasbulatova, Z. S. Halogenated aromatic simple oligoethers.
16. Patent 2413713 (RF), 2011. Bazheva, R.Ch., Kharaev, A. M., Kharaeva, R. A., Istepanov, M. I., Begieva, M. B. Monomer for polycondensation.
17. Patent 2513757 (RF), 2014. Kharaev, A. M., Bazheva, R.Ch., Mikitaev, A. K. Halogenated aromatic polyesters.

CHAPTER 2

WELDING MODES AND THEIR INFLUENCE ON THE ADHESIONS

A. I. ZAGIDULLIN, R. M. GARIPOV, A. I. KHASANOV,
A. A. EFREMOVA, and A. A. KOZLOV

Kazan National Research Technological University, Kazan, Russia

CONTENTS

ABSTRACT

The chapter concerns the effect of the clamping force of welding jaws, time and temperature of welding on the barrier properties of the heat-shrinkable multilayer packets of polymer film materials and establishes optimal process parameters for polymer films of such brands as Barrier shrink tube LT50, Barrier shrink tube LT9 and PVB M-50.

2.1 INTRODUCTION

The purpose of food packaging is the preserving the quality of a product from the moment of production to the consumption time. The most common reason of spoiling are the water vapor and oxygen getting in the packaging by molecules of H_2O and O_2 migrating through the packaging material [1]. Therefore, from the technological point of view there is a great importance in studying of barrier properties of the polymer film materials [2–3]. The requirements for polymer packaging materials, concerning gas and vapor permeability, are very strict right now [4–7], which promotes the elongation of expiration of products in a package. However, high barrier properties of multilayer polymer films can be useless, unless the process of bag making and food packaging will provide continuity of the barrier layers of film material [8].

The interface surface of the welded material is the "weakest link" in the conjunction of two melt layers. Figure 2.1 shows schematically the intermolecular relation between the melts in the welding process of two polymer films.

When welding the films (sector 1) the oriented areas of macromolecules form the interface surface on the edges of the material. In the process of contacting two heated materials, the orientation during the period of relaxation of oriented macromolecules partially persists. As shown by the statistical study of macromolecules segment mobility, the cross-linking

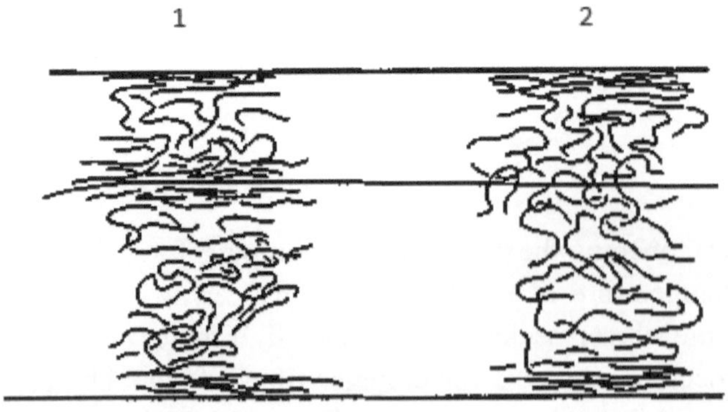

FIGURE 2.1 Pattern that describes adhesion on the border of two polymer melts [1] (1 – there are no weaves between molecular chains, the surface is oriented; 2 – molecular chains are mutually intertwined, the surface is no longer oriented).

(sector 2) takes much more time [9, 10]. If the welding time exceeds the relaxation period for the given terms of welding, then the orientation completely disappears on the interface surface and the cross-linking appears on the entire parting surface between the layers.

Since the interface surface between the layers is the weakest link in the interlayer connection [4] in polymer packaging for food products, such welding parameters as pressure, temperature and time of temperature impact (welding rate) must be correctly agreed (with consideration to packaged products) and correspond to the properties of welded polymers.

That is why the topics of great interest are the technological process of forming bags from shrink barrier films and the interrelation of adhesive reliability and barrier properties of the welded part of the packaging.

2.2 EXPERIMENTAL PART

We have chosen the multilayer barrier polymer film material – FHB M-50, FB C-47 and FB FP-47 by Tasma, Ltd. (Russia) – as a subject of study. Table 2.1 shows the specifics of these material.

The subjects of study were used in the production of various batches of thermal shrink barrier packages 200×200 mm^2 on the PACKNOVA EXTRA 700/108 bag-making machine. The following technological parameters of the bag making process was varied:

- pressure of welding clamps – 0.4–0.7 MPa.;
- welding temperature – 230–250°C;
- welding time – 0.1–0.5 s.

The welding temperature is chosen according to thickness of the original film.

We have studied the produced bags and estimated their weld strength, O_2, CO_2 and H_2O permeability of the weld.

We used the tearing machine XLW by Labthink (China) to value the weld strength with accordance to GOST 14236. The O_2 и CO_2 permeability at the weld was calculated according to ASTM D 1434, using the gas permeability rate estimation machine PERME VAC-V1 by Labthink (China). The H_2O permeability at the weld was estimated according to ASTM F-1249 on the H_2O permeability estimation machine PERMATRON-W 3/33 (USA).

TABLE 2.1 Film Specifics

Title	Unit of measurement	Value		
		FHB M-50	FB C-47	FB FP-2
Thickness	μm	50±5	47.5±2.5	27.5±7.5
Heat shrinkage index				
(after being in water at 90°C in 1 s), no less	%			
longitudinal direction		55	45	55
transverse direction		55	45	55
Oxygen permeability	sm³/(m²·24 hours·atm)	4	100	4
(at 23°C and 75% humidity)				
CO_2 permeability (at 23°C and 0% humidity)	sm³/(m²·24 hours·atm)	less than 50	Less than 50	Less than 50
Water vapor permeability (at 38°C and 90% humidity)	g/(m²·24 hours)	10	20	10
Tear resistance, no less	MPa			
longitudinal direction		70	60	70
transverse direction		60	50	60
Breaking elongation, no less	%			
longitudinal direction		110	110	110
transverse direction		160	160	160

2.3 RESULTS AND DISCUSSION

Figures 2.2–2.4 show the relation of the weld strength, O_2, CO_2 and H_2O permeability of the produced bags to technological parameters of the welding process.

Having analyzed the pictures we see that increasing of the welding clamps pressure (Figure 2.2a) leads to increased weld strength. However, when reaching certain pressure (6 atm) this index turns constant. The O_2 (Figure 2.2b), CO_2 (Figure 2.2c) and H_2O (Figure 2.2d) permeability of the bags at the weld depends on the welding clamps pressure in the formation process. However, starting from 6 atm. the weld permeability becomes defined by the permeability of the film that the bag is made of.

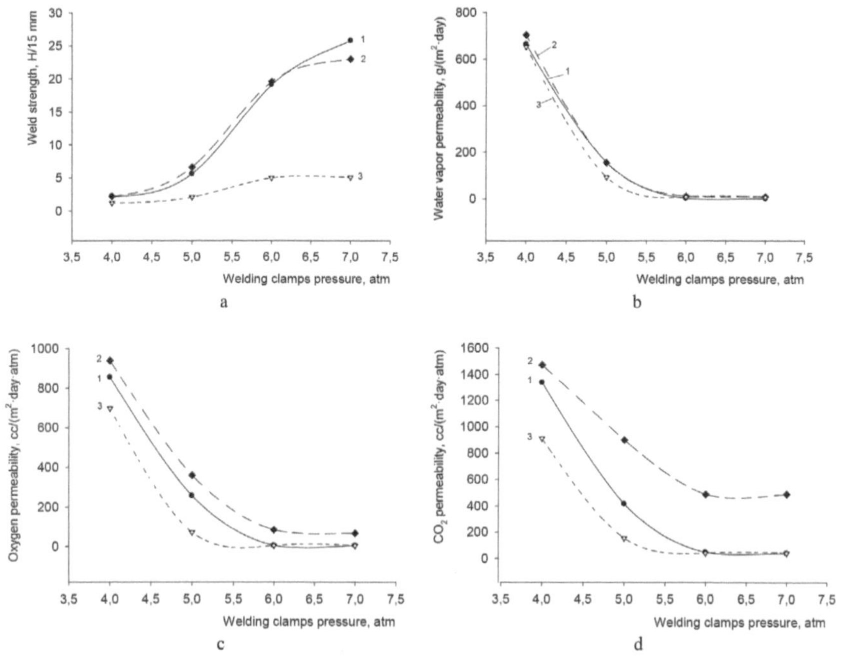

FIGURE 2.2 Weld strength (a), O_2 (b), CO_2 (c), H_2O (d) permeability dependence on welding clamps pressure.

The weld strength, O_2, CO_2 and H_2O permeability dependences on the welding clamps temperature (Figure 2.3) are extreme (except for the FB FP-27 film). Besides that, the weld strength depends on the film thickness. According to Figure 2.3, the appropriate temperature for welding bags of 50 μm thickness is 240–260°C. The film with thickness of 47 μm requires 230–250°C, whereas 27 μm thickness film needs 140–170°C. If the temperature goes lower, the weld strength decreases. When increasing this range, high welding temperature and high pressure lead to disrupted continuity of the barrier layer of the film material near the weld area, which is proven by the increased O_2, CO_2, H_2O permeability (Figures 2.3b–2.3d). When it comes to the FB FP-27 film, the "sparing" welding mode does not lead to emergence of extreme points in the curves of gas and vapor permeability dependencies on the welding temperature, as in this case the decreased welding temperature does not cause faults in barrier layers in the weld area.

If the welding time is not enough, the weld strength decreases and O_2, CO_2, H_2O permeability of the bags increases (Figure 2.4). If the welding

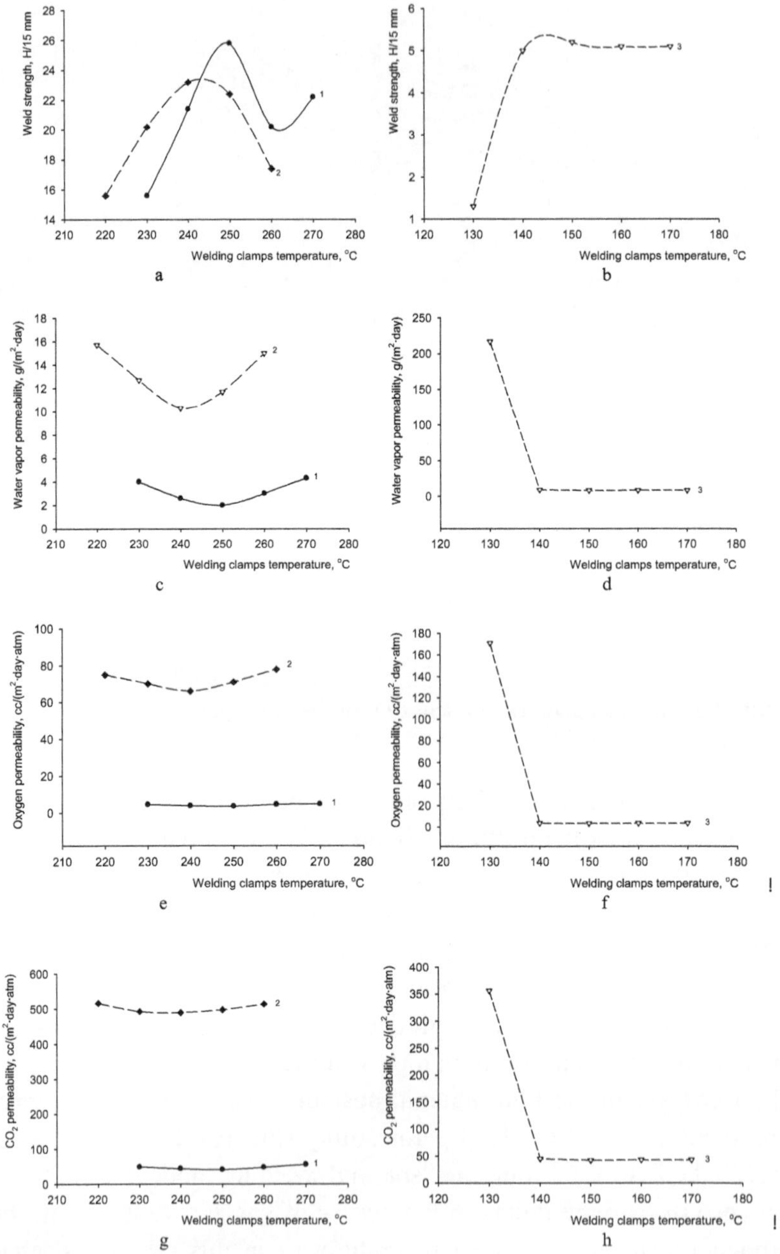

FIGURE 2.3 Weld strength (a, b), H_2O (c, d), O_2 (e, f), CO_2 (g, h) permeability dependence on welding clamps temperature.

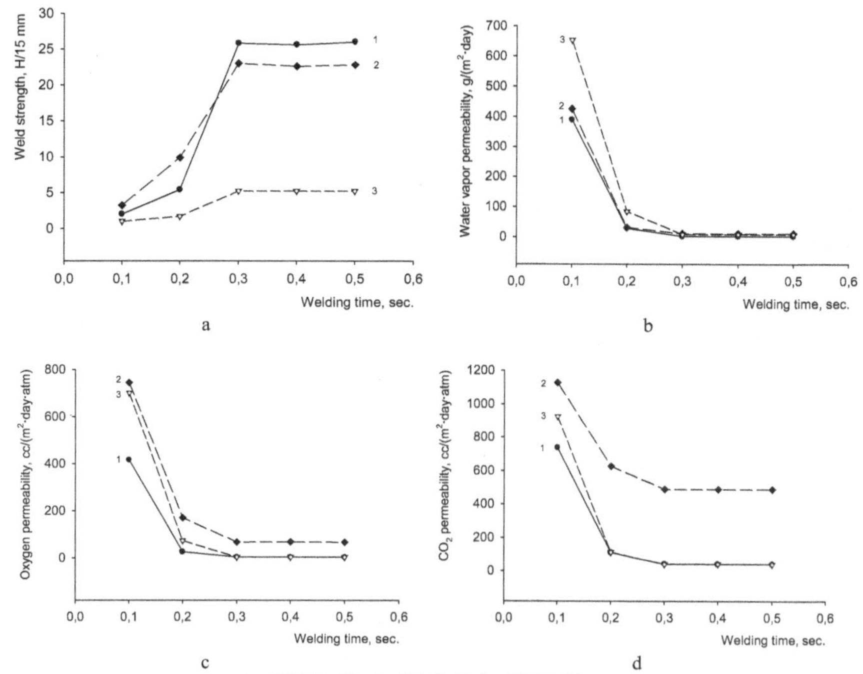

1 – FHB M-50; 2 – FB C-47; 3 - FB FP-27

FIGURE 2.4 Weld strength (a), H_2O (b), O_2 (c), CO_2 (d) permeability dependence on welding time.

time equals 0.3 s and more, the weld strength and O_2, CO_2, H_2O permeability become constant.

Thus, the analysis of the effect of welding clamps pressure on the properties of shrink bags made of tubular barrier films – FHB M-50, FB C-47 and FB FP-27 – shows that the most suitable clamp pressure is 6 atm. and higher.

The analysis of the effect of welding clamps temperature on the properties of shrink bags made of tubular barrier films FHB M-5, leads to the conclusion, that the appropriate welding temperature is 240–260°C, but in case of FB C-47, the temperature must be 230–250°C, and for the FB FP-27 film, the required temperature is 130–170°C.

The analysis of the effect of the welding time on the properties of shrink bags made of tubular barrier films FHB M-50, FB C-47 and FB FP-27 leads to the conclusion, that the appropriate welding time is 0.3 seconds.

ACKNOWLEDGMENTS

The work was supported by the Government of the Russian Federation (Russian Ministry of Education) as a part of the complex project of developing high-quality production under the contract № 02.G25.31.0037, according to the resolution № 218 of the Government of the Russian Federation from April, 9, 2010.

KEYWORDS

- **permeability**
- **polymeric film material**
- **shrinkable multilayer barrier film**
- **welding**

REFERENCES

1. Tihminlioglu, F., Atik, I., Ozen, B. *Journal of Food Engineering*, 96, 342 (2010).
2. Jang, W., Rawson, I., Grunlan, J. C. *Thin Solid Films* 516, 4819 (2008).
3. Minelli, M., De Angelis, M. G., Doghieri, F., Marini, M., Toselli, M., Pilati, F. *European Polymer Journal* 44, 2581 (2008).
4. Jagadish, R. S., Raj, B., Asha, M. R. *Journal of Applied Polymer Science* 113, 3732 (2009).
5. Steven, M. D., Hotchkiss, J. H. *Packaging Technology and Science* 15, 17 (2002).
6. Pajin, B., Lazic, V., Jovanovic, O., Gvozdenovic, J. *International Journal of Food Science & Technology* 41, 717 (2006).
7. Rhim, J. W., Hong, S. I., Ha, C. S. *LWT-Food Science and Technology*, 42, 612 (2009).
8. Zagidullin, A. I., Garipov, R. M., Khasanov, A. I., Efremova, A. A., Kozlov, A. A. Welding parameters and their influence on the barrier properties of thermal shrink multilayer bags. Vestnik Kazanskogo tekhnologicheskogo universiteta. 2013, 16, №20, 83–86.
9. Rauvendaal K. Polymer extrusion (Text) – St.P.: Professiya, 2006. p. 768.
10. Vlasov, S. V., Kandyrin, L. B., Kuleznev, V. N., Markov, A. V., Simonov- Yemelyanov, I. D., Surikov, P. V., Ushakova, O. B. Plastics processing technology: Study book. M: Khimiya, 2004. p. 600.
11. Kryzhanovskyi, V. K., Kerber, M. L., Burlov, V. V., Panimatchenko, A. D. Production from polymer materials: Tutorial (Text). St.P.: Professiya, 2004. p. 464.

CHAPTER 3

KINETICS AND MECHANISM OF POLYMER DISPERSION FORMATION ON BASED OF (METH) ACRYLATES

M. D. GOLDFEIN, N. V. KOZHEVNIKOV, and
N. I. KOZHEVNIKOVA

Biological Faculty, Saratov State University (SSU), Saratov, Russia

CONTENTS

3.1 CLASSICAL AND MODERN CONCEPTS

The kinetics and mechanism of emulsion polymerization of vinyl monomers differ from those of a homogeneous reaction in bulk and solution in a number of ways. In some cases these differences are of a fundamental character; sometimes they are related to technological aspects of the processes of polymer formation. The former include the mechanism, topochemistry and kinetic parameters of a reaction. The importance of polymerization in emulsion is connected with advantages associated with the process being conducted at a high rate to yield a polymer with a high molecular mass (MM), producing a highly concentrated latexes with a relatively low viscosity allowing further processing without separation of the polymer from the reaction mixture, considerably lowering of the fire-hazard properties of the product, etc.

The kinetics, mechanism and topochemistry of emulsion homo and copolymerization are determined by their solubility in water and polarity, diffusion, mass transfer, adsorption, interpenetration of phases, character of interaction of an emulsifier with a monomer and initiator. All this create difficulties in developing a quantitative theory of (co)polymerization in emulsion. The basic theories of the formation of latex particles are based on micellar, homogeneous nucleation, microemulsification and several coexisting mechanisms.

The classic theory [1–3] refers to polymerization of monomers, insoluble or weakly soluble in water, in the presence of water-soluble initiators and ionogenic micelle-forming emulsifiers. Since this theory has been repeatedly described [4–7], we will briefly discuss its main features. Polymerization proceeds within the bulk of the polymeric and monomeric particles (PMP) and is characterized by three main steps. The first step involves the PMP number growth occurs as a result of penetration of radicals from the water phase into micelles containing monomer. After the emulsifier has been completely depleted, the process proceeds with a constant number of latex particles and a stationary concentration of the monomer in the latter, which is maintained by diffusion of the monomer from drops into PMP – the second step. Disappearance of drops of the monomer in the course of polymerization leads to a decrease of its concentration in particles and termination of the reaction.

The basic equations describing the stationary step of the reaction are

$$W = k_p[M]\bar{n}N / N_A$$

$$N = K(W_i / \omega)^{0,4}(S_o[E])^{0,6}$$

where K is a constant ranging from 0.37 to 0.53 dependent on the efficiencies of radicals capture by micelles and PMP; k_p is the chain propagation reaction rate constant; *[M]* is the equilibrium concentration of the monomer in PMP; W_i is the penetration rate of radicals into all particles (the initiation rate); ω is the PMP volume increase rate; S_o is the surface area of polymeric particles which can be covered by 1 g of the emulsifier; *[E]* is the emulsifier concentration; $\bar{n}$ is the number of radicals in a particle; N_A is the Avogadro constant.

This equations are valid in the absence of coagulation of latex particles, at a quasi-stationary distribution of radicals between particles, constant concentration of the monomer in a particle at any degree of conversion, as well as at a constant rate of initiation and total surface area of micelles and latex particles at the first step of the process [7].

According to classic theory [1–3], the time required for the interaction of two radicals in a particle is small as compared to the period of time between consecutive penetration acts of radicals into a particle. Hence one half of PMP contains one radical, while the other contains none of it, e.g., $\bar{n} = 0.5$.

The common assumption about a uniform distribution of free radicals in latex particles is not valid in all cases. It is not observed if growing oligoradicals contain hydrophilic end groups, e.g., have the properties of surface active substances (SAS) and accumulate at the surface of particles [8].

Analysis of the quantitative theory allowed to describe the dependence of such kinetic parameters of the process as the number of particles, nucleation time; the conversion value at which nucleation is terminated; averaged MM of the polymer from the characteristics of the emulsion system [9].

According to the data of Ref. [10], the volume fraction of the monomer in a swollen particle are: for styrene 0.60; for vinyl acetate 0.85; for methyl methacrylate 0.73; for butyl acrylate 0.65 and for butadiene 0.56.

Monomers, which are somewhat more soluble in water than styrene, proceed by a mechanism of homogeneous nucleation in aqueous solution. The homogeneous nucleation theory [11, 12] is based on the assumption that oligomers which grow in water precipitate from the solution after having attained a critical size, giving rise to PMP. The particles formation rate is given by the equation

$$dN/dt = W_i - W_a$$

where the absorption rate of radicals by the latex particles W_a is proportional to the radicals formation rate, total PMP surface and the distance passed by a growing radical until it has attained the critical size. According to this theory, the particles formation rate is independent from the presence of an emulsifier, which, as a rule, does not agree with experimental data.

The effect of an emulsifier was taken into consideration within the framework of the homogeneous nucleation theory [13], where micelles were assumed to be the place in which the emulsifier is present. Polymer chains growing in water can form PMP only when adsorption of an emulsifier appears sufficient. If the oligomer is not capable of this, it is captured by a previously formed particle. Proceeding from the same assumptions about the radicals adsorption rate and the adsorption isotherm polymer-emulsifier that are made in the classic theory, an expression similar to that found in the miceller theory was obtained for the PMP formation rate.

The micellar and homogeneous mechanisms of particles nucleation taken separately cannot describe all the kinetic features of emulsion polymerization. In the opinion of the authors [14], both mechanisms have the same physical basis. Predominance of one of them is determined by the conditions under which the process is being conducted; in particular, the solubility of the monomer in water and amount of emulsifier.

The micellar theory, in contrast to the homogeneous nucleation theory, cannot explain the formation of particles when the content of an emulsifier in the system is below the critical micelle-forming concentration (CMC), for example, in emulsifier free processes, which are of great importance for emulsion polymerization involving monomers soluble in water. However the homogeneous nucleation theory does not take into account the effect of emulsifier on the final number of particles.

Some efforts to surmount the drawbacks of both theories have been made [15, 16]. The kinetics of capture of oligomers by polymer particles are analyzed in detail, taking into account the diffusion of oligomer radicals within particles, mass transfer through the external interface surrounding a particle, oligomer desorption and electrostatic repulsion between charged oligomers and similarly charged particles.

Nevertheless, it should be noted that these theories are complementary to one another, and, in fact, there is no explicit boundary between the micellar and homogeneous ways of nucleation of latex particles. The conditions under which this transformation may be represented by a single mechanism, as well as a more detailed discussion of such a mechanism, simulating the process of PMP creation and forming in a wide range of emulsifier concentrations and solubilities of monomers in water [14]. At very low emulsifier concentrations the homogeneous nucleation mechanism is appropriate, and the number of particles is determined mainly by the processes of their coagulation. If the amount of the emulsifier increases but has not yet reached the CMC, the processes of coagulation become less important, since primary particles begin to be stabilized by the emulsifier. Over the CMC the micellar mechanism begins to be appropriate which competes with the homogeneous nucleation and becomes prevalent at high emulsifier concentrations. For these reactions the solubility of the monomer in water is of great importance, since it increases considerably the homogeneous nucleation rate. Nevertheless, for such a weakly water-soluble monomer as styrene the homogeneous nucleation also plays a role [17]. This process leads to the formation of new PMPs practically at all degrees of conversion. As a result, latex particles size non-uniformity increases with a conversion growth, giving rise to a bimodality of the corresponding distribution.

A model of PMP formation, which is based on the coagulative nucleation theory and fits well the experimental data on emulsifier-free polymerization, has been developed [18].

The question of the number of latex particles has been thoroughly investigated. During micellar nucleation, N monotonously increases up to a certain value and after that remains constant (at the second step of the process). However, in the case of styrene as well under certain conditions at the step of stationary kinetics the number of particles may increase [19].

In order to evaluate the classic theory, dependences of N on the emulsifier and initiator concentrations are often used.

According to this microemulsification theory, [20–28] along with monomer drops and micelles, there are monomer microdroplets between 200 and 2000 A in size which are formed by spontaneous splitting of monomer drops. This results in a noticeable increase in the monomer-water interface. On initiation of polymerization in emulsion with such a high degree of dispersion, PMPs are formed mainly from monomer microdroplets. Monomer diffusion into a PMP is assumed to proceed through the microdroplet-PMP interface during direct contact, rather than through the water phase. From electron microscopic data, the basic site where polymerization occurs is the superficial layer of a PMP to which the monomer has access from microdroplets from the outside and from the inner part of the PMP. Thus a stationary concentration of the monomer has to be determined by the surface area of monomer drops, which is greater, the higher the degree of dispersion of the drops. After a certain critical value of the surface area of the drops has been attained, the monomer concentration and reaction rate approach their stationary values. A growing degree of dispersion of monomer drops leads to an increase in mass transfer and polymerization rates, as well as a decrease in the PMP dimensions.

In order to validate the microemulsification mechanism, a series of experiment have been performed. Comparison of the dimensions of PMP and monomer drops, estimated by means of electron microscopy has shown that an increase in the degree of splitting of monomer drops by 1.5 orders of magnitude results in a 4- or 5-fold decrease in the particles size. Hence a decrease in the PMP size must be accompanied by an increase in both the rate and length of the linear part of the kinetic curve. Upon addition of various amounts of a monomer to the same latex, an increase in the amount added (i.e., an increase in the total surface area of drops) leads to an increase in the rate up to a certain value. Monomer, addition shows no influence on the rate at the stationary part of the kinetic curve, while in the third part the latter grows again.

The use of microemulsification theory makes it possible to explain high rates of emulsion polymerization in the presence of oil-soluble initiators, as well as the dependence of the length of the linear part of the kinetic curve on physicochemical conditions under which the formation of latex particles proceeds.

The mechanism of particles nucleation exerts a determining influence on the dependence of the emulsion polymerization rate upon the emulsifier concentration. Upon micellar nucleation and an essential adsorption of polymer molecules (or radicals) by an emulsifier, the reaction order with respect to the emulsifier *ni* is equal to 0.6 when its concentrations exceeding the CMC. In the case of homogeneous nucleation, the emulsifier concentration must not influence the reaction rate, e.g., $n_e = 0$. Since during polymerization of a variety of monomers the formation of latex particles proceeds simultaneously with different mechanisms, the reaction order with respect to the emulsifier has an intermediate value ($0 < n_e < 0.6$) and may be regarded as a measure which determines the contribution of a certain nucleation mechanism to the overall process. The effect of emulsifier on PMP formation depends on the state of the interface and its adsorption efficiency, [5, 29], thus being determined by the nature of both monomer and emulsifier. A dependence of n_e on the structure of an emulsifier (i.e., the length of alkyl substituents, presence of double bonds et al.), as well as of an initiator has been observed [30, 31].

For acrylates, on account of an increasing contribution of homogeneous nucleation and a lesser number of emulsifier molecules at the surface of a particle as compared to some monomers, n_e must decrease [6, 32, 33]. The latter effect being the more pronounced, the higher is the solubility of the monomer in water [34]. Upon its "artificial" increasing the reaction order with respect to the emulsifier decreases as well. Addition of methanol (50%) to water the styrene polymerization reaction order with respect to the emulsifier decreases to 0.2 [35].

An increase in the solubility of a monomer in water not always results in an increase in the role of homogeneous nucleation and a decrease in n_e. During copolymerization of methyl acrylate (MA) or methyl methacrylate (MMA) with acryl amide and methacryl amide (MAA), which are soluble in water, the reaction order with respect to an emulsifier is bigger than in the case of MA or MMA homopolymerization [36–38]. This effect is connected with the possibility of oligomer radicals termination in the water phase and a decrease in the solution polymerization rate in the presence of amides owing to the high rates of square-law termination.

Since the polymerization reaction order with respect to emulsifier is determined by different competing mechanisms of PMP formation, n_e must also be dependent on conditions of the reaction which influence the relationship between the rates of homogeneous and micellar nucleation. A dependence has been found between the reaction order with respect to emulsifier and the temperature and initiator concentration [37, 38]. If a bimolecular termination of a chain is possible in the water phase, then a decrease in the initiation rate leads to prolonging radicals lifetime and provides conditions for the formation of longer oligomers and an increase of the role of homogeneous nucleation.

Like any polymerization process, polymerization in emulsion proceeds through several basic steps (initiation, propagation and termination of a chain). Considering the topochemical features of this reaction, all substances used as polymerization initiators may be divided into two classes depending on their solubility in water. Water-soluble initiators (e.g., some peroxides and azo-derivatives, redox systems) are most frequently used. Radicals formed in the water phase penetrate into micelles with monomer and into PMP where they take part in the reactions of chain propagation and termination. Due to the fact that anionic SAS are widely used as emulsifiers, and primary radicals formed on decomposition of such initiators as potassium or ammonium persulfates are also negatively charged, their penetration into micelles and PMP is retarded owing to electrostatic repulsion. Some authors assume transformation of anion-radicals $SO_4^{-\bullet}$ into hydroxyl radicals which initiate polymerization [5, 39]. It is more likely that the addition of primary radicals $SO_4^{-\bullet}$ to monomer molecules dissolved in water with the formation of oligomer radicals capable of penetrating into a micelle [6]. The rate constant of oligomer ion-radical penetration increases with an increasing number of monomer elements up to a diffusion-controlled limit at a certain critical chain length which depends on monomer solubility in water [40]. The critical chain length varies from 3 in the case of styrene, upto 40 and 200 during polymerization of MMA and acrylonitrile, respectively.

The rate of initiation by persulfates often increases with an increasing emulsifier concentration [5, 41–43].

The dependence of potassium persulfate decomposition rate on emulsifier concentration (sodium lauryl sulfate) has an extreme character [44].

The maximum of this dependence is in the CMC region and is stipulated, by a balance of quantities of molecular and micellar forms of the initiator-emulsifier interaction product.

With a rise in the pH value of the medium, persulfate decomposition rate decreases. Application of organic peroxides is hindered owing to their limited solubility in water. Some data have been reported on application of water-soluble azo-compounds as initiators [45]. For emulsion polymerization at relatively low temperatures it is expedient to use redox systems, one of the components of which is soluble in monomer, and the other – in the water phase [5, 6, 46].

The question of the mechanism of initiation of emulsion polymerization by oil-soluable initiators is still debatable. In a number of cases polymerization caused by such initiators is described by the same kinetic relationships as for water-soluble initiators [43, 47]. However, according to the micellar theory, when oil-soluble initiators are used, polymerization must proceed at a very low rate with the formation of low-molecular products or even not be observed at all [23]. This is connected with the fact that radicals, formed during decomposition of an initiator which is present in a particle of a small size, may rapidly react with one another. In order to overcome this contradiction, it should be assumed that one of primary radicals comes out of a particle into the water phase [6]. This process may occur on account of diffusion or as a result of transfer of the free valence to an emulsifier molecule with subsequent replacement of the resulting emulsifier radical by its molecule from the water phase [48, 49]. Since at each act of initiator decomposition, one of its radicals may have time to come out of a particle before their recombination, a low efficiency of initiation by an oil-soluble initiator should be expected. Initiation of emulsion polymerization by AIBN give the value of 0.04 [50].

Within the framework of the microemulsification theory it is believed that radicals do not leave a particle; an effective emulsion polymerization with an oil-soluble initiator is possible if radicals formed are rapidly fixed in the polymer matrix [23], polymerization proceeds within a superficial layer of PMP where radicals are fixed.

It should be noted that some oil-soluble initiators are hydrophilic to a certain extent. Initiation of emulsion polymerization of butyl methacrylates by AIBN is caused by the part of the initiator dissolved in water 6.9% of its initial concentrations [48].

An oil-soluble initiator is present not only in monomer-containing micelles, or in PMP, but also in monomer drops where polymerization is possible [51]. AIBN molecules in the presence of some monomers, MMA in particular, are practically not transferred from drops to micelles, and therefore there is no chain propagation in micelles. They believe that in the case of oil-soluble initiators, independently from the nature of an emulsifier and monomer solubility in water, the process is initiated and polymerization proceeds mainly in monomer drops. However, this conclusion contradicts generally the accepted view in which polymerization in drops plays a negligible role, since the rate of processes proceeding through an emulsion mechanism is higher than for polymerization in bulk [52]. The conclusion about a negligible contribution of polymerization in monomer drops in the case of systems with a weakly soluble initiator has been drawn on the basis of analysis of the chain termination reaction mechanism [53]. By means of radiochemical technique it has been found that upon emulsion polymerization of butyl methacrylate initiated by AIBN, termination of a chain occurs via disproportionation of growing and primary radicals in a PMP, while under homogeneous conditions and in the reaction of monomer drops partial recombination of macroradicals would be possible.

The reaction order with respect to initiator ni is an important characteristic of an emulsion system. According to the micellar theory, this value is equal to 0.4, being independent of the nature of the components of the system. Such a dependence of the rate and number of latex particles on initiator concentration is usually observed during emulsion polymerization of nonpolar monomers with poor solubility in water, for example, styrene. In the case of more soluble monomers n_i increases, being within the interval 0.4–1.0 [54]. The reaction order with respect to initiator, polymerization of acrylates is close to 0.5 [6, 32, 46], which is attributed to the reaction proceeding simultaneously in the water and micelle phase. More pronounced influences of initiators on emulsion polymerization are also known, and an increase in the rate with a growing initiator concentration is often observed without a corresponding increase in the number of latex particles [55]. This provides evidence for a growing amount of radicals in the particles. In accordance with classic ideas, PMP may contain either one growing radical or none. This follows from the postulate of immediate termination in a latex particle on penetration of another radical. On the other hand, if one assumes that under certain conditions the termination rate is not so high, then two or more radicals may coexist in a particle [56].

Termination may be retarded by a high viscosity of the reaction medium in a PMP containing a percent of polymer formed by at earlier steps, as well as by the large dimensions (over 100 nm) of latex particles which also contribute to a decrease in the probability of chain termination [6, 56]. A slow termination is observed under certain conditions which lead to a decrease in the mobility of radicals in particles. When initiators with surface-active properties were used, as well as when radicals appeared to be localized within adsorption layers of the emulsifier, increased $\bar{n}$ values were found growing with an increasing initiator concentration and accounting for a high reaction rate order with respect to initiator [57] and n_i has been shown to take different values in different regions of concentrations of the initiator [37, 38].

Emulsion polymerization rate of MA as a function of APS concentration in logarithmic coordinates is represented by a broken line. In the region of low APS concentrations ($<10^{-4}$ wt.%; 60°C) $n_i \approx 0.4$. Then n_i increases up to 0.64, and at APS concentrations exceeding 0.025 wt.% it decreases again. The high n_i value in the intermediate part of initiator concentrations was explained by an increase in the number of radicals in a particle and are a manifestation of the gel-effect. Coexistence of several radicals is possible only in a PMP of sufficiently large dimensions which may be formed at low initiation rates. However, the n_i value depends on the frequency of penetration of radicals into PMP and grows with an increasing initiation rate. Thus, since the probabilities of coexistence of several macroradicals in a latex particle and appearance of the gel-effect are determined by factors which depend in an opposite way on initiator concentration, the increased kinetic order values with respect to initiator are observed only in some certain range of APS concentrations.

At present it is generally accepted that the chain propagation during emulsion polymerization proceeds mainly within PMP. Hence the rate of this reaction depends on the numbers of latex particles and free radicals in them, chain propagation reaction rate constant and monomer concentration in PMP. It is commonly assumed that the propagation reaction rate constant during emulsion polymerization is the same as for the reaction in bulk or in solution. Polymerization in emulsion is regarded as one of the means for experimental estimation of k_p [58, 59]. In a number of cases significant discrepancies have been found, which are probably connected with uncertainties in the $\bar{n}$ value and other parameters of the emulsion system involved in determining the chain propagation rate constant [52]. Possible deviations of the average number of radicals in latex particles from 0.5 have been mentioned above.

The monomer concentration in PMP depends on the relationship between the rates of polymerization and monomer feed into particles from drops, as well as on the interphase tension value at the monomer–water interface which decreases with increase in monomer polarity [5, 29]. During polymerization in PMP in the presence of drops of monomer, an equilibrium concentration is reached which, in the case of polar monomers, is higher than during styrene polymerization, and increases with growing hydrophilic properties of the monomer [5]. However the monomer concentration in PMP cannot be considered constant and independent of the polymerization conditions [25]. This conclusion appears reasonable for polymerization of acrylic monomers [6, 54], and deviations from equilibrium arise from high chain propagation reaction rates.

The monomer concentration in PMP is closely connected with the mass transfer mechanism. Penetration of monomer into latex particles is possible not only as a result of its diffusion through the water phase from monomer drops, but also when PMPs directly contact with drops [25, 60]. In the literature there are different opinions concerning the question on the zone where the chain propagation reaction occurs. According to the micellar theory, the reaction proceeds within the bulk PMP During emulsion polymerization of isoprene [61] the PMP surface area remains constant. This fact and a constancy of the polymerization rate over a broad range of conversions [61] have led the to the conclusion that polymerization occurs within the adsorption layer of emulsifier at the PMP surface. As PMPs grow, the adsorption layer moves towards the water phase, permanently remaining at the surface of the particles. At present the majority of researchers believe that this mechanism takes place only in some special cases [6].

A heterogeneous model of emulsion polymerization considers a growing PMP consists of a nucleus enriched with polymer and a shell saturated by monomer which is the main place where the reaction occurs [62, 63].

During emulsion polymerization the reaction, zone contains an equilibrium monomer concentration which is constant throughout the second step of the process. The polymerization rate should be independent of the total monomer content [6]. In a number of cases such a dependence has been found to exist [37, 38, 64]. Within the framework of the microemulsification theory this is explained by the formation of a large number of macro- and microdrops when monomer concentration increases, which may become the place of the formation of latex particles. The amount of the emulsifier used appears insufficient for their effective stabilization, which leads to coagulation resulting in a decrease in the PMP number and polymerization rate.

During emulsion polymerization the chain termination reaction occurs on penetration of a radical from the water phase into a PMP where a growing radical is already present. In the course of this process the determining factor is not the radicals interaction rate, but the frequency of their penetration into the latex particles. Under such conditions the average number of radicals in a PMP is 0.5. In emulsion systems certain conditions may be reached under which chain termination proceeds at a relatively low rate. Under these conditions growing radicals may be simultaneously present in a latex particle, interaction between which proceeds in a highly viscous medium giving rise to the appearance of the gel-effect [36–38, 65, 66]. In the case of emulsion polymerization with high polymer content a corresponding viscosity is reached in PMP in the initial reaction steps. Owing to the gel-effect the degree of conversion does not depend on the extent of reaction, as in the case of polymerization in bulk [67]. The gel-effect grows with an increase in the volume of the latex particles and initiation rate.

During emulsion polymerization primary radicals are formed in water, where they undergo several stages of growth before they penetrate into a micelle or a PMP. Therefore chain termination is possible during interaction of oligomer radicals in the water phase, which is usually not taken into account. An evaluation of the probability of such termination have shown [40] that it is sufficiently high for high initiation rates of monomers with a low solubility in water and a small chain propagation rate constant of value. It was also assumed that the rate constant of penetration of a radical into a particle is equal to zero for primary $SO_4^{-\bullet}$ radicals and increases with a growing degree of polymerization. On the basis of the analysis it was concluded that among the monomers under consideration, styrene, MMA, vinyl acetate, vinyl chloride, acrylonitriles a square-law termination in the water phase should be taken into account only during polymerization of styrene and, to a considerably lesser extent, MMA. In the case of acrylic monomers, characterized by a higher chain propagation rate constant and a better solubility in water, it evidently does not take place. Nevertheless, a semiquantitative evaluation of the probability radicals termination in water [40] seems insufficiently strict to provide a basis for unambiguous conclusions. An assumption about the existence of such termination allowed a number of kinetic relationships, found during emulsion polymerization of MA, MMA and butyl acrylate (BA) [36–38], to be explained. An increasing solubility of monomers in water must lead to an increase in the oligomer radicals lifetime, since a loss of solubility upon the fomation of a PMP primer is reached by them at higher degrees of polymerization.

Square-law termination in the water phase plays an essential role during copolymerization of a hydrophobic and a hydrophilic monomer. In this case oligomer radicals formed in water are enriched with the hydrophilic monomer. Hence they have a better solubility in water, and their penetration into micelles filled mainly with the hydrophobic monomer is more hindered. As a result, the period of existence of a radical in the water phase is prolonged and the role of square-law termination of radicals increases, as is assumed for copolymerization of (methyl) acrylates with acryl amide or MAA [33,36–38].

Investigations of emulsion copolymerization, which allows the process kinetics and properties of polymer emulsions to be controlled, are of special interest. In spite of the fact that emulsion copolymerization is rather frequently used in practice, there are relatively few reports on the study of its kinetics and mechanism.

During emulsion copolymerization of hydrophobic monomers with partially water-soluble monomers the kinetics and mechanism become more complicated owing to formation of particles by homogeneous mechanism. Under these conditions chain propagation may occur at different sites.

Analysis of the reaction mechanism, require a knowledge of copolymerization constants. If mass transfer in the system proceeds by diffusion, then, due to different solubilities of monomers in water, the ratio of their concentrations in the latex particles, diffusion flow and monomer consumption may vary, leading, to a difference between copolymerization constants in emulsion, bulk and solution. However the interphase resistance to diffusional flow at the micelle-water interface would be lower than that at the PMP–water interface [68]. Therefore the ratio of monomers in micelles may correspond to that in drops, while in PMP, formed by flocculation of oligomer radicals in water, it is essentially different. Thus if PMP formation proceeds mainly via micelles, then the copolymerization constants may be the same as in the bulk. Consequently, the coincidence of copolymerization constants in bulk and emulsion found in styrene–AN [69, 70] points either to a direct-contact mass transfer or to a negligible role of homogeneous nucleation, if a monomer penetrates into particles by diffusion. These conclusions are valid only for mutually soluble monomers.

Analysis of literature data on emulsion copolymerization of monomers differing in their solubility in water leads to the conclusion that despite a supposedly possible change in relative activities of monomers owing to their diffusion-controlled mass transfer, in most cases no changes in copolymerization constants has been found [68]. Nevertheless, in a number of studies differences in these values, obtained for the reaction in bulk (solution) and in emulsion, have been reported [5, 29]. The copolymerization constants in emulsion observed are effective values. Their comparison with the true values, estimated during copolymerization under homogeneous conditions, enables some specific features of the emulsion process topochemistry to be elucidated. An essential difference was found between copolymerization constants of MA with MAA in emulsion and in solution [38]. It follows that the MAA content in the reaction zone under emulsion conditions is higher than average in the reaction medium. Considering that the MAA is localized in the water phase (the distribution constant of MAA between water and MA is 2.5) and that its solubility in MA is very limited, the conclusion has been drawn that copolymerization occurs in the superficial layer of a latex particle where a high local concentration of MAA may be reached.

In the case of acryl amide, which is characterized by a significantly higher distribution constant (9.4), the relative content of the amide in the reaction zone is lower than in the initial mixture.

An important role of the water phase was confirmed by the results of investigations on the kinetics, particle size and copolymer composition during periodic emulsion copolymerization of MMA with ethyl acrylate (EA) [71] and of vinyl acetate (VA) with BA [72, 73]. For MMA–EA extreme dependences of particles number and emulsion conductivity on conversion have been found. A simulation considering monomers distribution between phases and latex particles swelling in monomers made it possible to determine an averaged propagation rate constant and evaluate the change of the average number of radicals in a particle. Under the conditions of the study the prevalent mechanism of particles formation during copolymerization of VA with BA is homogeneous or coagulation nucleation. Specific interaction between the hydrophilic parts of the surface of latex particles and an emulsifier leads to an extremal dependence of copolymerization rate of BA with acrylic acid on the degree of oxyethylation of the emulsifier [74], the reaction rate with respect to emulsifier being a measure of its adsorption at the surface of particles.

A series of studies have been published [75–80] on emulsion copolymerization of monomers strongly differing in their solubility in water: AN and BA. Copolymerization kinetics depends on the distribution of monomers between oil and water phases, the [AN]:[BA] ratio. The composition of a copolymer depends on the degree of conversion at <15% and over 90% of conversion the copolymer is enriched with AN, while in the intermediate region – with BA. The basic reason for such a dependence is believed to be different mechanisms of nucleation of PMP. In the beginning of the process the basic role is played by homogeneous nucleation involving AN molecules dissolved in water. Then the process proceeds in PMP, and, on account of a high concentration of BA in the oil phase, its content in the copolymer increases. At the end of the reaction, owing to a very low monomer content in PMPs, the polymerization process practically ceases. Homopolymerization of free AN dissolved in water leads to the formation of a polymer which after that is absorbed by latex particles. The outer PMP shell will therefore consist mainly of PAN. The copolymer composition is a function of the monomers ratio, copolymerization constants and concentrations of monomers in water.

In the presence of a mixture of an anionogenic and non-ionogenic emulsifiers [79] two types of micelles were found depending on the concentrations ratio of the emulsifiers. With a growing concentration of the anionic emulsifier, an increase in the rate of copolymerization of AN with BA and decrease in PMP dimensions are most noticeable in certain regions of emulsifier concentrations. Note that non-ionogenic emulsifier does not determine the reaction mechanism, but only improves stabilization of particles.

Oxygen has been used as an inhibitor [58, 68, 81], as it inhibits radical reactions. The influence of oxygen on the polymerization kinetics in emulsion depends on its location in the reaction system. If oxygen is present in both phases or only in the water phase, induction periods are observed. It is removed from aqueous solution but is present in the monomer, then it is rapidly solubilized in micelles, nucleation ceases, and kinetics and copolymer composition change drastically. The process moves to the water phase, and monomer solubility in water becomes the determining factor affecting diffusion and monomers composition in the reaction. The copolymerization constant values found in such a system differs from those for polymerization in bulk. One of the reasons for this is believed to be a low adsorptional saturation of the surface of PMP formed according to the homogeneous nucleation mechanism. With an increase in emulsifier concentration its adsorption rate grows, which may lead to a change in copolymerization constants up to values corresponding to the true relative activities of the monomers. Thus by varying the emulsifier concentration in the presence of the micellar oxygen one can vary the values of effective copolymerization constants.

The use of benzoquinone and diphenylpicrylhydrazyl (DPPH) in the emulsion copolymerization of AN and BA [75] enables conclusions about the reaction mechanism to be drawn. At the second and third polymerization steps the kinetics is influenced not only by disappearance of monomer, but also by the reactivity of the radicals and their concentration. In the presence of oil-soluble benzoquinone the reaction rate is greatly diminished, latex particles size decreases and the contribution of homopolymerization of AN in water increases, which affects the copolymer composition. The presence of DPPH in the system noticeably increases the PMP nucleation time and particles number, decreasing their dimensions.

A mathematical model of the behavior of the emulsion system MMA–APS–sodium lauryl sulfate in the presence of oxygen has been developed [82, 83]. The model takes into account the mechanism of particles formation, kinetics of polymerization in emulsion, heat and mass transfer. A method for limiting the rate of heat release and controlling the rate of polymerization by means of introducing oxygen into the system without an essential decrease in the polymer MM has been prepared [82, 83].

Inhibition of the reaction in the emulsion has its specific features connected with distribution of inhibitors between different phases of a system owing to their different solubilities in water and in monomer. A known radical inhibitor 2,2,6,6-tetramethy1-4-stearyloxypyperidine-1-oxyl insoluble in water does not retard emulsion polymerization of MMA [84], but leads to a noticeable induction period in AN polymerization, since nitroxyl is soluble in aqueous solution of this monomer. During polymerization of MMA and BMA in the presence of AIBN a stable radical present mainly in the organic phase [85] retards polymerization in the latter, while initiation occurs in water.

The effect of an inhibitor insoluble in water on the emulsion polymerization kinetics depends also on the relationship between the lifetime of a growing chain prior to its interaction with the inhibitor and period of time between penetration of radicals in a PMP [86]. If an inhibitor is partially soluble in water, its activity in emulsion polymerization may be considerably higher than during polymerization in bulk [87], since the initial formation of active radicals occurs in the water phase.

3.2 MATERIALS AND BASIC PHYSICAL-CHEMICAL RESEARCH METHODS

The emulsion polymerization of the following monomers was studied: methyl acrylate (MA), ethyl acrylate (EA), butyl acrylate (BA), methyl methacrylate (MMA), and their copolymerization with acrylonitrile (AN) and methacrylic acid (MAA). These monomers were thoroughly purified (releasing from the stabilizer, drying, distilling under reduced pressure in an inert gas, and recondensation in vacuum).

The polymerization initiator, ammonium persulfate (APS), was purified by recrystallization from its water-alcohol solution.

The following surfactants were used as emulsifiers:

- sulfated oxyethylated alkylphenol S-10, which is a reaction product of the nonionic wetting agent OP-10 (monoalkylphenyl ether of polyethylene glycol C_nH_{2n+1}-C_6H_4-$(OC_2H_4)_m$-OH, where n = 8–10, m = 10–12) with concentrated sulfuric acid followed by neutralization with ammonia [5];
- Neonol $AP_{9-12}S$, an analog of S-10, oxyethylated nonylphenol sulfate with an ethoxylation degree of 12;
- sodium lauryl sulfate (LS).

Kinetic studies of emulsion polymerization mean estimating the reaction rate and main characteristics of the resulting polymer dispersion (the amount and size of latex particles) in different conditions of its implementation and at each stage of the process. The methods to be used should have high sensitivity, provide continuous control over the process, and, at the same time, feature availability and simplicity.

Since polymerization is accompanied by volume effects, dilatometry fully conforms to the specified requirements to rate estimation [86]. There exists proportionality between the volumetric changes of the polymerizate during the reaction and the weight monomer-to-polymer conversion degree, determined by the densities of the monomer and polymer:

$$\frac{\Delta V}{V_o} = \frac{\rho_{\text{ï}} - \rho_{\text{ì}}}{\rho_{\text{ï}}} \cdot \frac{\Delta P}{P_o}$$

where ΔV and ΔP are the changes in volume of the polymerizate and the weight of the monomer during polymerization, V_o and P_o the initial volume and weight of the monomer, ρ_m and ρ_p the densities of the monomer and polymer at the polymerization temperature.

Polymerization was carried out in special glass dilatometers of original design. The main elements of such a dilatometer are a measuring capillary, a reaction vessel, and a filling system providing the opportunity of releasing the monomer and dispersion medium from dissolved gases before polymerization. To avoid the influence of oxygen on the processes studied, the water-soluble and oil-soluble components of the reaction system were separately freed from dissolved air by multiple freezing (liquid nitrogen), high vacuum pumping, and thawing in vacuo with subsequent transfusions through the measuring capillary into the reaction vessel of the dilatometer. The tool filled with inert gas was then placed into a water bath over an electrical magnetic stirrer. The desired temperature was maintained via an ultrathermostat.

The estimation of the number and size of the resulting latex particles formed in the reaction is another task of experimental studies of emulsion polymerization. The classical theory of emulsion polymerization [2] establishes the nature of the influence of the concentration of initiator *In* and emulsifier *E* just on the number of particles in the emulsion *N*, which, starting from the second (fixed) stage of the reaction, was considered to be independent of the conversion degree and, for micellar nucleation, is described by: $N = K[In]^{0.4}[E]^{0.6}$. The emulsion polymerization rate is determined by the number of particles, the average number $\bar{n}$ of radicals, and the monomer concentration $[M]_p$ therein:

$$W = k_p\bar{n}[M]_p N/N_A$$

There is much experimental evidence of failure of these classical ideas, especially for describing the emulsion polymerization of monomers with increased hydrophilicity [6, 29]. In order to clarify and develop the theory of emulsion polymerization it is necessary to establish the dependence of the number of particles on the conversion degree and the synthesis conditions of polymer dispersions.

The method of latex property evaluation should be, on the one hand, sufficiently accurate, and on the other hand, simple and relatively low laborious as measurements at these kinetic studies are massive. The turbidity spectrum method [88] meets these requirements, it is based on the determination of the wavelength exponent in Ångstrøm's equation $\tau = const\ \lambda^{-u}$, describing the spectral dependence of the turbidity τ of colloidal solutions within a relatively narrow wavelength range. Turbidity is the coefficient in the exponent of Bouguer's law $I = I_o e^{-\tau \cdot l}$, where l is the scattering layer thickness (the cuvette length) characterizing the ability of the disperse system to reduce the intensity of the incident light due to scattering. Turbidity depends on the number *N* of scattering centers, their sizes and optical properties:

$$\tau = N\pi r^2 K(\alpha, m)$$

where $K(\alpha, m)$ is the scattering efficiency factor (or scattering coefficient), $\alpha = 2\pi r n_0/\lambda$ the relative size of a particle, $m = n/n_0$ their relative refractive index (*r* the radius of a particle, λ the wavelength of light, *n* and n_0 the refractive indices of a particle and the dispersion medium). The wavelength exponent *u* is a function of α and *m* and can be found by measuring the slope of the straight portion of the graphic dependence of turbidity (or absorbance) $D = \lg(I_o/I) = \tau \cdot l/2.303$ on λ in the log-log coordinates.

It is possible to determine the parameters of the dispersion particles from the found wavelength exponent with the known dependences $u(\alpha, m)$ and $K(\alpha, m)$ on the relative size and the relative refractive index. These relationships for a discrete set of *m* according to the formulae of G. Mie's light scattering theory [89] were calculated [88]. However, these dependencies have an oscillating character for monodisperse systems. The allowance for the actual polydispersity of colloidal systems is typically provided by a graphical smoothing on the principle of symmetry of the oscillating curves plotted for specific values of the relative refractive index *m*. It is more convenient to use approximate analytic expressions for the characteristic functions of light scattering [90]. This approach allows computer processing of experimental data. Besides, due to the fact that the relative refractive index of the particles may continuously vary during the experiment (e.g., as the polymerization depth increases, the monomer-polymer ratio in the particles varies and, therefore, their spectral characteristics do as well), the table data calculated for a discrete set of *m* turn out to be inapplicable.

The mean radius and number concentration of particles in dispersed systems were calculated by the formulae [88]:

$$\bar{r} = \alpha\lambda_m / 2\pi n_o, \; N = 4\pi\tau(\lambda_m) n_0^2 / \lambda_m^2 K(\alpha, m)\alpha^2$$

where λ_m is the middle of the wavelength range in the logarithmic scale ($\lambda_m = \sqrt{\lambda_{max}\lambda_{min}}$).

One of the problems in the turbidity spectrum method is the question of the need to consider the spectral dependence of the refractive indices of the particles $n = n(\lambda)$ and medium $n_0 = n_0(\lambda)$. In many papers devoted to the development and application of this method, the refractive index dispersion is neglected [88]. However, Ref. [91] concludes on the unreasonableness of such an approach, especially in systems with small particles. It has been shown that the results obtained by the turbidity spectrum method match those given by any absolute method (such as electron microscopy) only when considering refractive index dispersion. Otherwise, the turbidity spectrum method gives much underestimated values of particle sizes. The correction value associated with the refractive index dispersion depends on the parameters of the particles, and hence the emulsion polymerization conditions. Therefore, neglecting the spectral dependence of the refractive indices may lead to deviation of the values of the kinetic characteristics of the process determined from data on the number of particles in the polymer dispersion from their true values.

The allowance for refractive index dispersion was made in accordance with the approaches to this problem developed in Ref. [91]. In this paper it is shown that the theoretical value of the wavelength exponent u_0, which is calculated by the approximate equations with $\bar{r}$ and N, and which corresponds to systems with negligible dispersion is related to the experimental parameter u by a relation like: $u_0 = u + \Delta u$, where $\Delta u = k_0 u + 2m(k - k_0)/(m - 1)$ when $u \geq 2$ and $\Delta u = k_0 u + um(k - k_0)/(m - 1)$ when $u \leq 2$.

The quantities k and k_o describe the refractive index dispersion of the dispersed phase $n(\lambda)$ and the dispersion medium $n_o(\lambda)$ and represent their logarithmic derivatives with respect to wavelength. The values of the parameter k are proposed [93] to be evaluated from the inverse relative dispersion δ and the refractive index at the wavelength corresponding to the yellow line of sodium n_D. Approximating $n(\lambda)$ by Cauchy's binomial formula, we obtain:

$$k(\bar{\lambda}) = (n_D - 1)(\lambda_F^{\,2} + \lambda_C^{\,2}) / n_D\delta(\lambda_F^{\,2} - \lambda_C^{\,2}) = -3.43(n_D - 1) / n_D\delta$$

where $\bar{\lambda} = [(\lambda_F^{-2} + \lambda_C^{-2})/2]^{-1/2} = 552.4$ nm is the middle of the spectral interval in the $1/\lambda^2$ coordinates between the F and C lines.

The relation for $k(\bar{\lambda})$ was used for the final polymer dispersion when the conversion q approached 100%, and the latex particles were composed of polymer entirely. However, this approach is inapplicable in analysis of the properties of latex particles depending on the polymerization depth, as the particles consist of the polymer and monomer with their varying ratio depending on q. In the absence of monomer droplets, when almost all the monomer and polymer are in the latex particles, the refractive index of the scattering centers can be estimated through the corresponding values for the monomer n_m and polymer n_p, in view of the volume ratio φ of the monomer and polymer in a particle, calculated, in turn, from the conversion degree:

$$n = (\varphi n_m + n_p)/(1 + \varphi)$$

But the values of δ for such particles are unknown. It has been shown [91] that the parameter $k(\bar{\lambda})$ can also be estimated based only on n_D, since for more than 100 samples of polymers it is described by a general dependence approximated as: $-k(\bar{\lambda}) = B_o + B_1(n_D - 1.4) + B_2(n_D - 1.4)^2$, where $B_o = 0.01675$, $B_1 = -0.026858$, and $B_2 = 0.780829$. The use of this approximation requiring the knowledge of n_D, within the range $n_m \div n_p$ (in the case of MA polymerization it is 1.4040–1.4725) to assess the value of k, apparently gives wrong results, since when $n_D = 1.4 - B_o/2B_2 = 1.4172$, this function has a minimum, which contradicts our experimental data. The dependence of k on the refractive index in this range of its values can be more accurately represented by linear interpolation between the values calculated for the monomer and polymer.

The parameter k_o for the medium in the case of aqueous dispersions can be obtained from direct spectral measurements of $n_o(\lambda)$. Approximation of the data from Ref. [92] for $n_o(\lambda)$ of water at 20°C by Cauchy's tripartite formula gives $k_o = -0.0155$ ($\lambda = 552.4$ nm).

Turbidity spectra were recorded on a SF-26 spectrophotometer. The value $\lambda_{cp} = \bar{\lambda} = 552.4$ nm was selected as the wavelength mid-range, and measurements were performed with a constant logarithmic step $\Delta \lg\lambda = 0.02$.

The polymer dispersion formed by the emulsion polymerization of acrylates (the monomer concentration is 10–20%, the initiator is ammonium persulfate, APS) have relatively high optical density; therefore, they were diluted with water before measuring turbidity. However, it turned out that the results of measurements, and, in particular, the wavelength exponential in Ångstrøm's equation and the reduced turbidity (the turbidity multiplied by the dilution *R*) depended on the dilution (Figure 3.1), especially at its relatively small values. These dependencies are not observed at high dilutions. These data indicate that in sufficiently concentrated dispersions there is multiple secondary scattering due to which more (as compared with single scattering) light passes through the cuvette and reaches the receiver as if to reduce the turbidity of the system. The theory underlying the turbidity spectrum method does not account for multiple light scattering. Therefore, to estimate the parameters of the dispersion it is necessary to dilute it to such an extent that the reduced turbidity becomes independent of *R* and remains constant at any dilution, which serves as the criterion of no multiple scattering [93]. The obtained data have allowed us to estimate the contribution of multiple light scattering at different dilution degrees, since the proportion of multiply scattered light at the exit of the cell T_2 can be expressed [93] through the reduced turbidity at the given dilution (τR) and infinite dilution $(\tau R)_\infty = \lim_{R\to\infty} (\tau R)$:

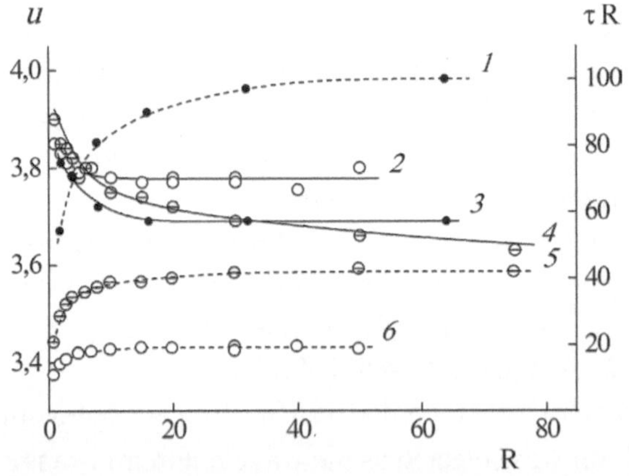

FIGURE 3.1 Dependence of the wavelength exponent (*2–4*) and the reduced turbidity of the dispersions (*1, 5, 6*) obtained by MA emulsion polymerization (50°C) with the monomer concentration of 10 (*2, 4–6*) and 20% (*1, 3*) on the dilution prior to measuring turbidity. The emulsifier (1%) is LS (*1–3, 6*) and Neonol (*4, 5*); [APS] × 10^3 = 2 (*2, 4–6*) and 36 mol/L (*1, 3*).

$$T_2 = I_2/I = 1 - 10^{D-D_1}, \text{ where } D - D_1 = \frac{l}{2.303} \cdot \frac{(\tau R) - (\tau R)_\infty}{R}$$

where, I being the intensity of light transmitted through the dispersion, which consists of the intensity I_1 weakened due to single scattering, and the additional intensity I_2 arising due to multiple scattering of radiation in the direction of the receiver.

The effect of multiple light scattering quickly decreases with increasing dilution (Figure 3.2).

The necessary dilution degree depends on the dispersion properties, in particular, it is higher when sulfated oxyethylated alkylphenols (C-10, nonoxynol-9-12) are used as an emulsifier in comparison with sodium lauryl sulfate (LS), and it also depends on the monomer content in the initial emulsion.

Dilution simultaneously facilitates washing of the latex particle surface from the emulsifier molecules stabilizing them, whose refractive index differs from the corresponding quantities of the monomer or polymer, which, in the absence of dilution, introduces an additional uncertainty into the results. Therefore, polymer dispersions are pre-diluted with water by 100–200 times and measurements are usually performed in 0.3-cm cuvettes with the value of transmittance maintaining within the range of 0.2–0.8.

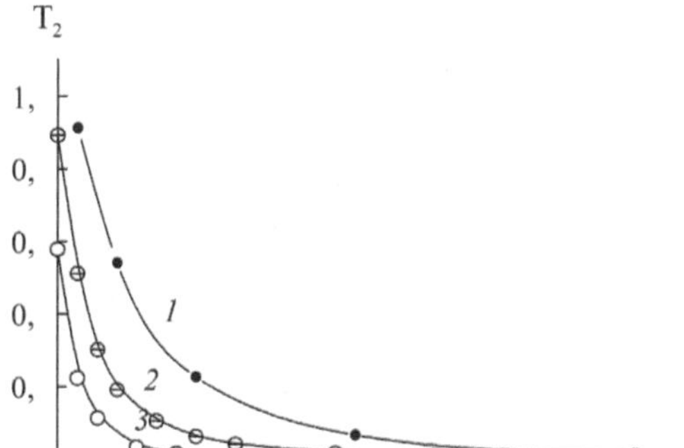

FIGURE 3.2 Dependence of the fraction of multiple light scattering at λ = 552 nm at the cell outlet with l = 0.1 cm on the dilution degree of the dispersion obtained by MA polymerization of 20% (*1*) and 10% (*2, 3*) with LS (*1, 3*) and Neonol (*2*) as the emulsifier. [APS] × 10^3 = 2 (*2, 3*) and 36 mol/L (*1*); 50°C.

Thus, our studies have shown that the turbidity spectrum method can be applied not only to the final polymer dispersions resulting from the emulsion polymerization of acrylic monomers but also to the emulsion systems arising in the course of the reaction at various conversions. It is necessary to take into account multiple (secondary) light scattering, and the dependences of the scattering properties of particles on the polymerization depth, those of the refractive indices of the particle and medium on wavelength. Under these conditions, the turbidity spectrum method can be used in kinetic studies of this reaction.

3.3 NEW RESULTS ABOUT STUDY OF KINETIC SPECIFICS OF EMULSION POLYMERIZATION OF (METH) ACRYLATES

Emulsion polymerization remains one of the most problematic sections of radical polymerization as a whole. Development of its general theory meets principal difficulties. The reasons are multi-phase structure of emulsion system, variety of parameters determining kinetics and mechanism of the process. These parameters depend not only upon reagents' reactivity, but also on their phase distribution, reaction topochemistry, mechanism of particles' nucleation and stabilization.

The present paper studies the emulsion polymerization of methylacrylate (MA) and methylmethacrylate (MMA).

As it noted early, emulsion polymerization goes through three main stages – PMP formation (when free emulsifier is available), stationary process (when monomer droplets in water phase are present, and monomer equilibrium concentration is established in PMP), and reaction completion (when monomer in PMP is depleted). For many monomer–polymer systems the parameters of equilibrium swelling of latex particles are well-known [29]. Such parameters indicate that in the case of polar monomers droplets' disappearance and completion of constant rate sections on kinetic curves must occur at relatively low conversions. For example, in the case of MA polymerization – at 16% conversion, MMA – 34% [29].

Our research showed that constant rate stage continues to the higher conversions *q*, depending on monomer structure and reaction conditions. In some cases polymerization rate does not slow down at high conversions, but, on the contrary, increases. Such effects, occurring notwithstanding the fact that monomer concentration in PMP is decreased, indicate that average radicals' amount is growing up, which is possible when large enough particles are formed and when high viscosity takes place in PMP [6].

Simultaneous growth of several radicals in the particle leads to polymerization acceleration and initiation of the so-called "gel-effect." The probability of gel-effect, according to [67], can be judged by the value of parameter α, which is defined as: $\alpha = \vartheta V / k_o N$, where ϑ – the total rate of radical entrance to the particles, N and V – their number and average volume, k_o – rate constant of bimolecular chain termination. Polymerization rate growth, caused by increasing of k_o, starts to be noticeable when $\alpha > 10^{-4}$, at $\alpha > 10^{-2}$ gel effect is pronounced. If, as a first approximation, to assume that ϑ is equal to the rate of radicals formation in water phase (initiation rate), and use the final dispersion data, than α parameter value can be estimated. The results are shown in the Figure 3.3. It indicates the existence of gel-effect, which is especially strongly pronounced when Neonol $AP_{9-12}S$ and S-10 are used as emulsifiers. In the presence of these emulsifiers larger latex particles are formed, which promotes the origination of gel-effect. It weakens while emulsifier concentration is increased, but becomes stronger with the growth of initiation rate.

In the case of MA polymerization gel-effect starts already at initial reaction stages. MMA is characterized by lower propagation rate constant,

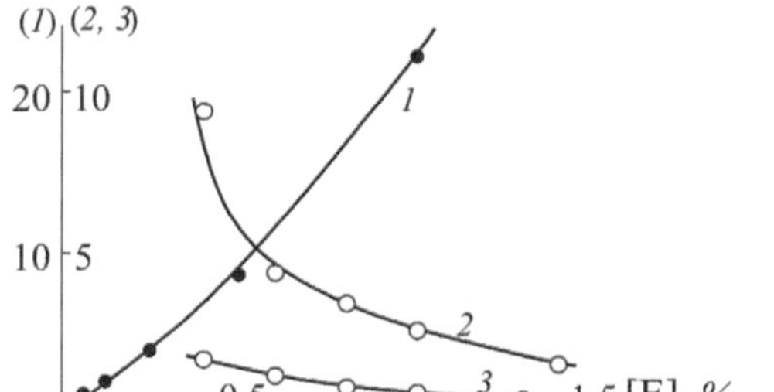

FIGURE 3.3 Parameter α value against initiator concentration (*1*), emulsifier Neonol $AP_{9-12}S$ (*2*) and LS (*3*) concentrations, emulsion polymerization of MA. (Parameter α is the measure of gel-effect probability). [Neonol] = 1%, 60°C (*1*); [APS] = 5 × 10^{-3} mol/L, 50°C (*2, 3*).

and because of that conditions for gel-effect arise only at high enough conversion, when monomer concentration decreases and viscosity in PMP grows up. This leads to more pronounced influence on the shape of kinetic curves.

Thus, for the emulsion polymerization of studied monomers classical conceptions of instantaneous chain termination in PMP in the moment of second radical entrance are not valid.

The growth of emulsion polymerization kinetic order with respect to emulsifier (LS) with the increase of initiator concentration is reported in Ref. [38]. This is connected to the interaction of oligomeric radicals and chain termination in the water phase. In the present work, studying reaction rates, we also found similar relationships (Figure 3.4). Moreover, kinetic order with respect to emulsifier n_e is greater in the case of S-10 or Neonol, than with LS. This is the result of partial solubility of oxyethylated alkylphenols in the monomer. We found [66] (using electronic spectroscopy) that with the growth of these emulsifiers' concentration the part of them staying in water also increases. This part is promoting PMP formation, which increases the influence of Neonol and S-10 concentration on polymerization rate comparing with LS.

Established dependences of kinetic order with respect to emulsifier on reaction conditions, emulsifier properties indicate that n_e, contrary to the prevailing view, cannot characterize the emulsion polymerization of certain monomer.

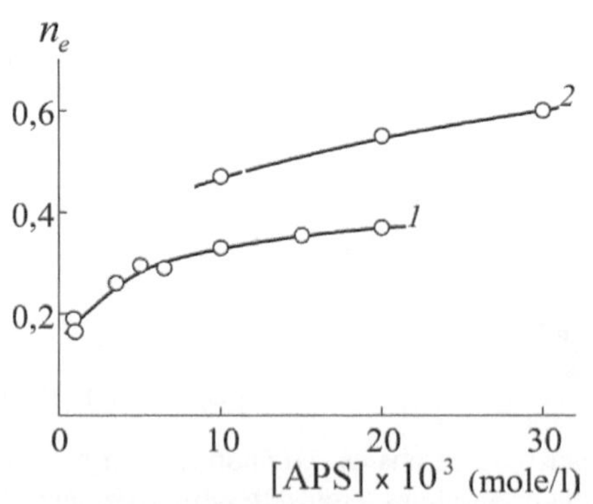

FIGURE 3.4 Kinetic order of MA emulsion polymerization with respect to emulsifier LS (*1*) and Neonol $AP_{9–12}S$ (*2*) versus initiator concentration. [MA] = 20%; 50°C.

The classic emulsion polymerization theory quantitatively describes the influence of emulsifier and initiator on the number of formed latex particles. Including the reaction rate into the same description pattern presumes that it is proportional to the number of particles. However, the reaction rate is determined not only by number of particles, but also by the rate of monomer conversion inside them, which can change due to the conditions and depth of polymerization (for example, due to different degree of gel-effect).

Our studies of emulsifier concentration effect on latex particles number in the final dispersion N_{100} (q=100%) showed, that kinetic order with respect to emulsifier, obtained using this data, is much larger than calculated using rate values (especially in the case of Neonol and S-10).

It's important to keep in mind that emulsion polymerization rate and N_{100} value characterize the process and forming dispersion at different conversions and comparison between them is justified only when one of the basic postulate of classical theory is working – number of PMP during the reaction after completion of its first stage is constant. However, our research showed that number of particles in the emulsion depends on polymerization depth. The most common effect – decreasing of particles number during the third stage of the reaction (Figures 3.5 and 3.6).

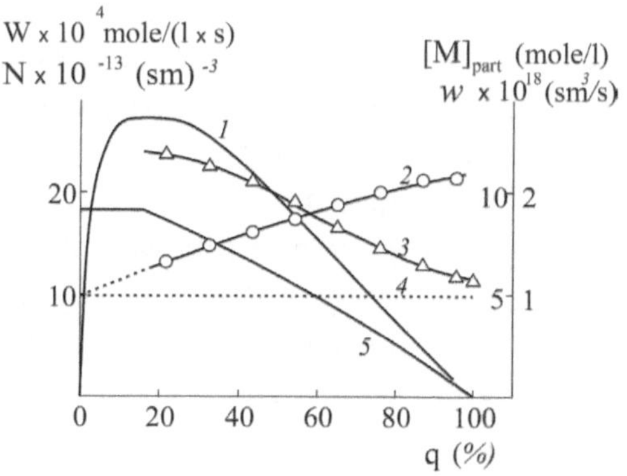

FIGURE 3.5 Polymerization rate (*1*), number of particles in emulsion (*3*), monomer concentration and rate in the particle (*5*, *2*) against monomer-polymer conversion depth, emulsion polymerization of MA. *4* – Specific rate in the particle at $\bar{n}$ = 0.5. [APS] = 0.25 × 10^{-3} mol/L; [LS] = 1%; 60°C.

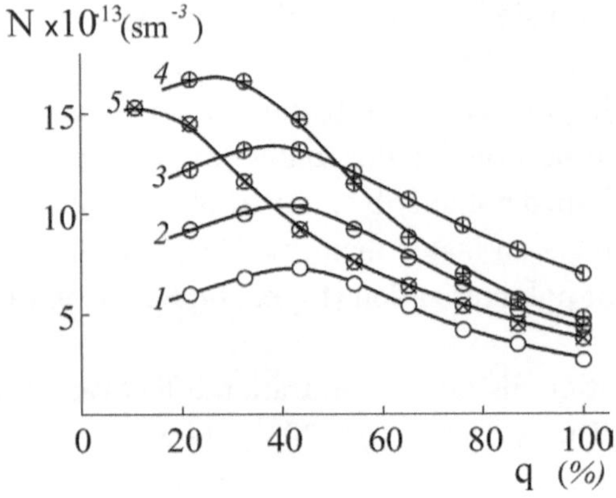

FIGURE 3.6 Number of particles in the emulsion versus monomer–polymer conversion depth, polymerization of MA. [APS] × 10^3 = 0.1 (*1*), 0.25 (*2*), 1 (*3*), 4 (*4*), 8 mol/L (*5*); [Neonol] = 1%; 60°C.

This indicates the changing of interphase surface condition and decreasing of its stabilization degree with the conversion growth. Such effects can be caused by various reasons, for example by changing the ratio of monomer/polymer concentrations in the particle. In addition, all the time during polymerization new and new charged oligomer radicals are entering PMP from water phase and sometimes start the reaction in the particle, sometimes terminate it. Such radicals are increasing the charge on the particles' surface and strengthening the electrostatic repulsion of surface active substances molecules (SAS), that hinders their absorption and can cause even desorption of emulsifier. As a result, flocculation of particles occurs, not only at the initial conversions, as assumed in [29], but also at more advanced stages of the reaction.

When partially soluble in monomer S-10 and Neonol $AP_{9-12}S$ are used as stabilizers, the conversion growth is followed at first by increase of PMP number, and only after that by its decrease (Figure 3.6). In this case during polymerization the emulsifier, dissolved in monomer, is released. This emulsifier is stabilizing new particles. At the same time the emulsifier replenishment promotes flocculation due to the growth of interphase surface, and also because new small particles are less stable [94].

Flocculation is the cause of abnormal high kinetic orders with respect to emulsifier (>0.6), found from particles number data in final dispersion, because at high emulsifier concentrations number of particles decreases due to the flocculation at third reaction stage not so heavily as when emulsifier concentration is low.

We determined specific reaction rate in the particles, comparing reaction rate and number of particles at the same conversion level. Specific reaction rate was found to change with the conversion. This is related to a considerable degree to the decreasing of monomer concentration in the particle $[M]_{part}$. At the initial polymerization stages, when monomer droplets are still present in the reaction system, monomer concentration in the particle is determined by equilibrium degree of dissolving in polymer (equilibrium swelling degree) and remains constant. At higher conversions it decreases. If we consider all monomer and polymer to sit in PMP, the dependence of $[M]_{part}$ (mol/L) on polymerization depth at third reaction stage will be given by:

$$[M]_{part} = \frac{1000 \cdot \rho_M}{m_o} \cdot \frac{1-q}{1-q \cdot (1-\rho_M/\rho_p)}$$

where ρ_M, ρ_P – densities of monomer and polymer, m_o – monomer molecular mass.

Specific rate in the particle $w = W/(N\,[M]_{part})$ grows with monomer–polymer conversion degree (Figures 3.5 and 3.7). This means that radicals' number in PMP is increasing. Moreover, the extrapolation of specific rate to the initial conversion gives the value corresponding to the specific rate at $\bar{n} = 0.5$ (dotted line *4* in Figure 3.5 is calculated assuming k_p for MA is equal to 1190 l/mole **x** sec [58]). Reaction rate in the particles grows with the conversion depth depending on initiator concentration – higher concentrations lead to higher growth (Figure 3.7), indicating the increase of gel-effect. Consequently, our data confirm the previous conclusion that in the case of MA emulsion polymerization gel-effect emerges already at low conversions and than grows gradually as particle size in increasing at second polymerization stage, and as viscosity is growing at third stage due to increasing of polymer concentration. When MMA is polymerized, gel-effect starts to be noticeable only at high conversions and is much more manifested than with MA (Figure 3.7).

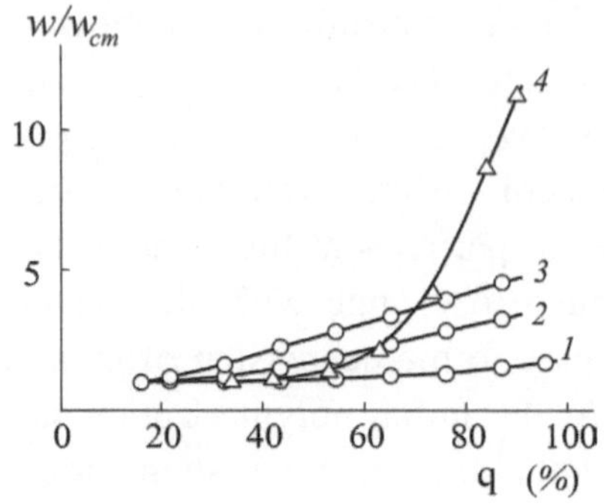

FIGURE 3.7 Behavior of specific reaction rate in the particle at the third stage of MA (*1–3*) and MMA (*4*) emulsion polymerization. [APS] × 10^3 = 1 (*1*), 1.5 (*4*), 4 (*2*), 8 mol/L (*3*); [Neonol] = 1 (*1–3*) и 2% (*4*); 60°C.

Initiation rate also influences emulsion polymerization of (meth)acrylates in an unusual way. The parameters of this process and polymer dispersion differently depend on initiator concentration. The maximal rate value W_{max}, as a rule, increased with the growth of APS, but differently in the various ranges of PSA concentration change. As a result, the graph of initiation rate versus initiator concentration in logarithmic scale (used to calculate the order with respect to initiator n_i) has the form of broken line with the point of inflection (Figure 3.8). Value of n_i at the relatively low initiator concentrations turned out to be higher than expected theoretical value of 0,4 (n_i = 0.6–0.7), but in the relatively high concentration region – lower

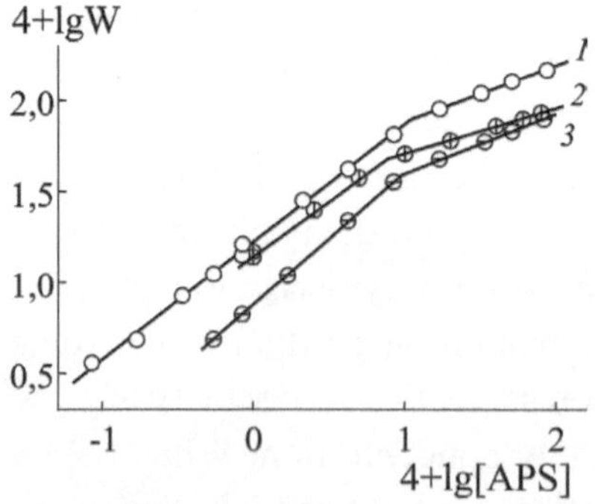

FIGURE 3.8 Emulsion polymerization of MA: rate against initiator concentration. Emulsifier – LS (0.4%) (*1*), Neonol (1%) (*2*), S-10 (1%) (*3*). 60°C.

(n_i = 0.25–0.35). Thus, kinetic order of the emulsion polymerization with respect to initiator depends not only on monomer properties, nature and concentration of emulsifier, but also on the region of initiator concentrations where it is measured.

Number of particles in the polymer dispersion N_{100} is growing initially, but later starts to decrease with the growth of PSA concentration in the presence of Neonol or S-10. The stationary rate curve also passes maximum. The observed external curves are related to the flocculation processes, taking place in the various polymerization stages. For instance, dependence of N_{100} on initiator concentration is determined by flocculation at all stages of the process, while stationary rate is influenced by flocculation only at initial stages. Behavior of maximum rate versus [APS] is influenced both by flocculation growth and by initiator concentration increase, which promotes acceleration of the reaction in the particles. Therefore, at relatively low initiator concentrations, when flocculation is still weak, the gel-effect intensification causes more drastic dependence of reaction rate on initiator concentration than is expected based on classic concepts (n_i > 0.4). At the high initiator content, depending on flocculation – gel-effect relative efficiency correlation, reaction rate continue to grow (though with low n_i value), or even starts to decrease.

Number of particles in emulsion at the conversion corresponding to maximum polymerization rate also depends externally on initiator concentration. This fact, together with the stationary rate passing maximum, indicates that flocculation takes place at the initial reaction stages. At high [APS] flocculation completely compensates the growth of particles' number due to the initiation rate increase; at low [APS] this is not the case. Thus, flocculation at initial polymerization stages increases with the initiation rate growth and reduces the n_i value. Flocculation at the third stage, as a rule, almost does not influence W_{max} and, consequently, n_i, because the rate reaches maximum at lower conversions. However, it reduces the number of particles in final dispersion. The reducing effect is more striking with emulsifiers S-10 or Neonol, than with LS, because the first two emulsifiers cause stronger flocculation at the third reaction stage, due to emulsifier replenishment from monomer droplets or PMP. In the case of LS N_{100} value is increasing in all studied initiator concentration range and does not pass maximum.

It is shown that kinetics behavior of the reaction is determined by: different nucleation mechanisms, initiation of gel-effect, bimolecular chain termination in water phase, flocculation of polymer-monomer particles during all polymerization stages, partial emulsifier solubility in the monomer. These effects lead to the dependence of particles number and reaction rate in particles on conversion and to the influence of polymerization conditions on kinetics orders with respect to initiator and emulsifier concentrations [95].

3.4 EMULSION COPOLYMERIZATION METHYL ACRYLATE WITH SOME HYDROPHILIC MONOMERS

Emulsion polymerization of vinyl monomers is always characterized by a rather sophisticated mechanism, because the emulsion system contains several phases and elementary reactions are localized in its various zones. The process is more complicated while joint polymerization of several monomers. In contrast to a reaction in block or in solution, emulsion copolymerization depends not only on the character of chemical interaction of comonomers with growing radicals but also on factors influencing the topochemistry of the process, the number of PMP formed, and their stabilization. At copolymerization of monomers with different solubility in water, the nucleation mechanism may switch. The influence of comonomers on emulsion polymerization is determined by their distribution in the phases and their affecting the processes in each of them. Besides, the parameters of reactivity of some monomers are medium-dependent [96]. The aim of the present work was to establish the kinetics and mechanism of joint emulsion polymerization of methyl acrylate with metacrylic acid (MAA) or acrylonitrile, which underlies production of many commercial acrylic polymeric dispersions.

At emulsion copolymerization of MA with AN or MAA, the composition of the comonomers influences not only the properties of the latexes formed and the reaction rate but also the character of the dependence of rate on time or conversion depth [97]. For example, increasing the MAA content retards polymerization and increases the value of conversion q_{max}, at which the maximum (for the given conditions) reaction rate W_{max} is achieved (Figure 3.9). Meanwhile, the data on equilibrium solubility of

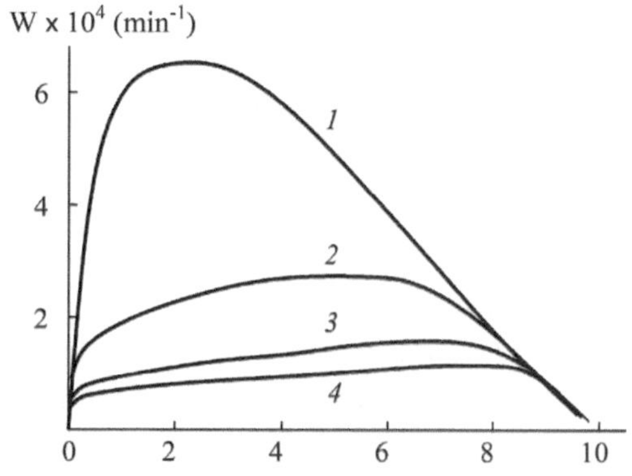

FIGURE 3.9 Rate of emulsion copolymerization of MA with MAA as a function of conversion depth. [MAA] = 0 (*1*), 4 (*2*), 8 (*3*) and 14% of [M] (*4*); [APS] = 2 × 10^{-3} mol/L; [Neonol] = 1%; 60°C.

MA in its monomer [29] point to that disappearance of monomer drops and the end of the constant rate plateau on the kinetic curves (finishing the second stage of emulsion polymerization) should be observed as early as at q = 16%.

The dependence of the rate of emulsion copolymerization on the composition of the monomers is due, on the one hand, to changes in the reaction rate in PMP, and, on the other hand, to changes of the amount of latex particles in the emulsion. The latter fact is determined by the influence of the monomer composition of the processes of nucleation and flocculation.

In the presence of MAA, in spite of the low coefficient of its water-monomer phase distribution [5], the concentration of the monomer dissolved in water increases. Judging by the values of MA–MAA copolymerization constants [98], the reaction in water proceeds with the predominating participation of acid molecules, and with a relatively low rate characteristic of MAA polymerization in low-acidic aqueous solutions [99]. The high content of MAA units in the oligomeric radicals formed in water elongates the chain length at which they can generate particles by the homogeneous nucleation mechanism, and hinders micellar nucleation (due to the poor solubility of oligomeric MAA-containing radicals in the monomers). As a result, MAA increases the probability of radical interaction in the aqueous phase, which leads to termination of reaction chains, a decrease in the number of PMP and an increase of their sizes.

The high rate of polymerization in the particles can also be maintained after exhaustion of monomer drops due to the creation of favorable conditions for coexistence of several growing polymeric radicals in PMP and the appearance of gel effect, which becomes stronger with the growing initiation rate. This affects the shape of the kinetic curves, being a cause of the observed increase of q_{max} (Figure 3.10). At MA–AN copolymerization, the reduction of the reaction rate is also observed, however, the conversion degree at which the maximum rate is achieved decreases in comparison with homopolymerization (Figure 3.10), which is evidence of decreasing the gel effect.

As well as at homopolymerization, the number of particles in the emulsion depends on the conversion degree. The character of this dependence is influenced by the emulsifier as well as the comonomer. Earlier [66] it was noted that in the case of Neonol, which is partially dissolved in acrylates, a gradual rise of the number of particles occurs when the conversion degree increases, which is caused by extraction of the emulsifier dissolved in the monomer drops and PMP. But at high conversion degrees, the amount of particles begins to decrease due to flocculation proceeding at the third stage of the reaction. When sodium lauryl sulfate (almost insoluble in the monomer) is used, no additional feed of the reaction solution with the emulsifier in the course of polymerization takes place and no increase in the number of PMP is observed (Figure 3.11, curve 4).

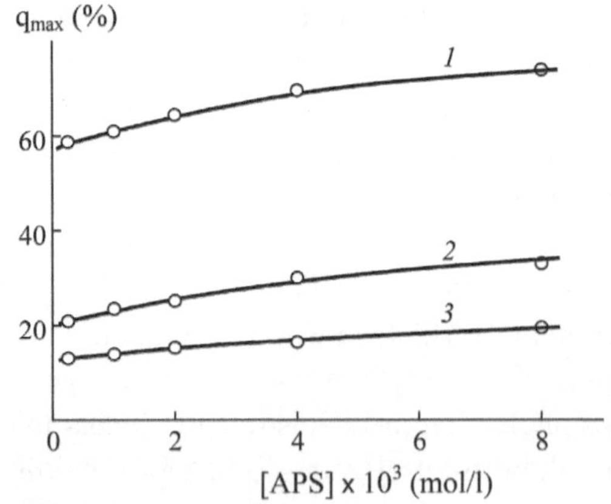

FIGURE 3.10 Dependence of the conversion depth, at which the maximum rate of emulsion homopolymerization (*2*) and copolymerization of MA with MAA (8%) (*1*) and with AN (8%) (*3*) is achieved, on the initiator concentration. [Neonol] = 1%; 60°C.

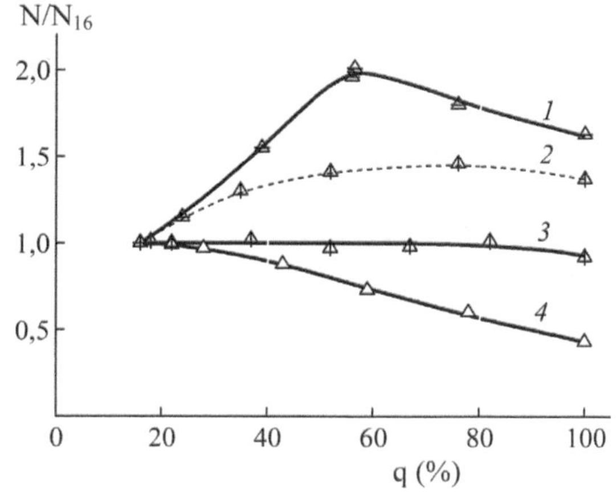

FIGURE 3.11 Changes in the number of particles in emulsion at the third stage of emulsion homopolymerization (*4*) and copolymerization of MA with MAA (8%) (*1*) and AN (8%) (*2, 3*). Emulsifier (1%): LS (*1, 3, 4*) and Neonol (*2*); [APS] = 0.25 × 10^{-3} mol/L; 60°C.

Similar regularities have also been found while studying MA–AN copolymerization. The reduction of the number of PMP occurs as late as at the very end of the reaction at high conversion degrees, while during almost the whole third stage of the process the number of particles gradually increases (Neonol as emulsifier, curve 2) or almost does not change (LS, curve 3). Comparison of these data allows us to conclude that AN reduces flocculation at the third polymerization stage. Moreover, at copolymerization with AN, more stable dispersions are formed and the amount of coagulum decreases in comparison with MAA-containing monomer systems.

Flocculation of particles at the third stage of polymerization plays a more important role while usage of monomer-soluble emulsifiers (Neonol), which is due to the formation of new particles in such conditions in the course of reaction. This flocculation entails an extremal dependence of the number of particles N_{100} in the final polymeric dispersion (when q tends to 100%) on the APS concentration, which is characteristic of polymerization of (met)acrylates in the presence of this emulsifier. At homopolymerization with LS, only gradual increase in N_{100} with the initiator content was observed [95]. However, at MA–MAA copolymerization, this extremal dependence of the number of particles was found with LS as well (Figure 3.12). Therefore, this comonomer attaches LS-containing systems such kinetic properties which are characteristic of Neonol-containing systems. They feature an increase in the number of PMP not only at the first stage of the process but also at higher conversion degrees (Figure 3.3, curve 1). However, the cause of the appearance of such an effect at copolymerization with LS (emulsifier) differs from that proposed for the case with Neonol. At MA–MAA copolymerization, the termination probability of reaction chains in the aqueous phase increases, which leads to the formation of water-soluble surfactant molecules with micelle-forming properties and promoting nucleation [100]. That is, oligomers appear which play the role of their "own" emulsifier. As a result, one of the possible mechanisms of particle stabilization is realized, characteristic of emulsifier free emulsion polymerization [100]. Owing to the bimolecular chain termination, the formation of such oligomeric radicals continues during the whole reaction, which results in the formation of new PMP. However, the appearance of new particles in the course of polymerization (due to extraction

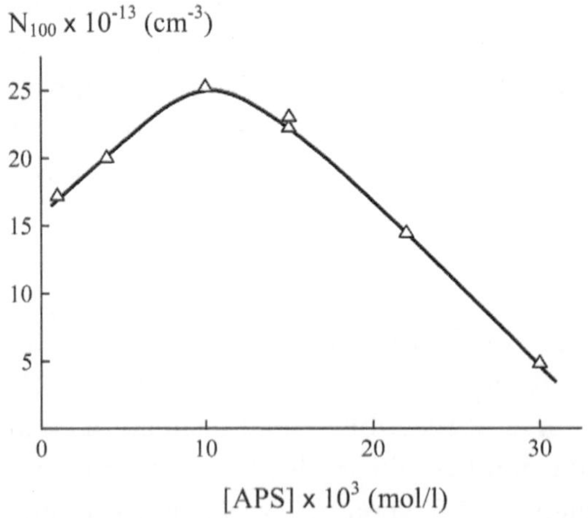

FIGURE 3.12 Number of particles in the polymeric dispersion formed at emulsion copolymerization of MA with MAA (5%) as a function of the initiator concentration. [LS] = 1%; 50°C.

of the emulsifier dissolved in the monomer to the aqueous phase from the drops and PMP, or due to the appearance of an "own" emulsifier at interaction of oligomeric radicals in water), which have a poorly protected surface, promotes flocculation.

Possibly, the oligomers with surfactant properties, playing the role of an "own" emulsifier, are formed at copolymerization with AN as well. But the effect of increasing the number of particles in this case is weaker than at copolymerization with MAA, and it is compensated by weak/poor flocculation, which results in the independence on conversion degree (Figure 3.11).

The specific reaction rate in the particle $w=W/(N[M]_{part})$ at the conversion degree corresponding to the attainment of the total maximum rate of polymerization, in the MA–AN system is lower than at homopolymerization of MA. Providing for the polymerization constants k_p of MA and AN having close values [58], one can conclude on a reduction of the average number of radicals in PMP and a poorer gel effect at copolymerization with AN. This is possible at faster chain termination in the particles in comparison with homopolymerization. Really, AN raises the termination rate since the constant of this reaction k_t for AN is almost two orders of magnitude higher than that for acrylic esters [58]. The influence of the gel effect reduces at decreasing the concentration of the initiator, and at its low content in the systems with AN the gel effect is almost not observed. This is evidenced by the independence of the specific rate in the particles on the conversion depth in these conditions (Figure 3.13, curve 3), whilst in the case of homopolymerization (curve 2), this value grows at the third stage of the reaction almost twice.

The increase in the specific rate of polymerization in the particles with the conversion depth at copolymerization with MAA speaks for a strong gel effect (Figure 3.13, curve 1). However, in contrast to MA homopolymerization, it appears at high conversion degrees only. This is reflected on the shape of the kinetic curves: the reaction rate rises at higher conversion degrees and its maximum values are attained at the final stage of polymerization. As was already noted, MAA is more actively, in comparison with acrylic esters, participates in the reaction of copolymerization in both aqueous and organic medium (judging by the values of copolymerization

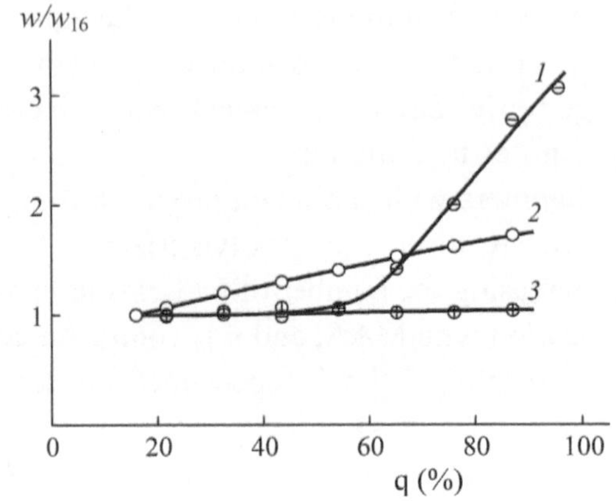

FIGURE 3.13 Variation of the specific reaction rate in particles at the third stage of emulsion homopolymerization (*2*) and copolymerization of MA with MAA (8%) (*1*) and AN (8%) (*3*). [APS] = 0.25 × 10^{-3} mol/L; [LS] = 1%; 60°C.

constants). However, the growth constant of MAA polymerization in block or in an organic solvent is much lower than in water, and lower that for acrylic esters [101]. This must entail a reduction of the polymerization rate, a decrease in the molecular mass of the polymer and viscosity in the particles. Hence, the conditions for the appearance of gel effect at copolymerization with MAA are created only at high conversion degrees, when the reaction in PMP accelerates, and much more stronger than in the case of homopolymerization (Figure 3.13).

The increase in the maximum rate of MA homopolymerization with the APS concentration occurs to some degree within various ranges of the initiator concentration. As a result, the plot of the dependence of the rate on the initiator concentration in the log coordinates (which is used for determination of the order of the reaction by initiator n_i) looks as a broken line with inflection point (Figure 3.14). The value n_i found at relatively small initiator concentrations (lower than at the point of inflection) has turned out to be higher than the theoretically expected value 0.4, and in the range of its relatively high content – lower than 0.4. This dependence is determined by the influence of the initiator on the number of particles in the emulsion, subject to flocculation processes, and also by influencing the reaction rate in PMP, which grows owing to the strengthened gel effect. That is why at relatively low initiator concentrations, when flocculation

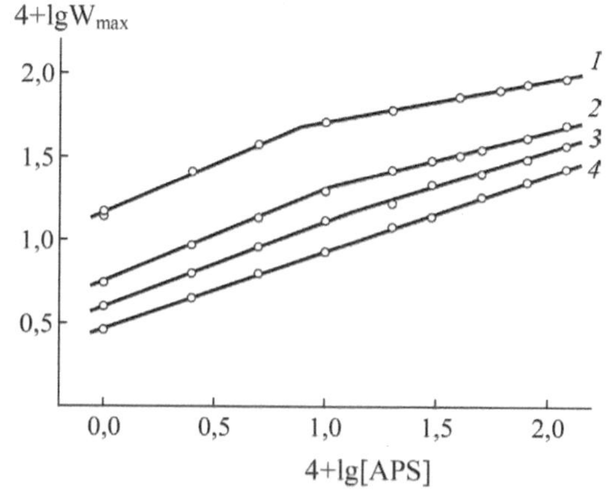

FIGURE 3.14 Dependence of the rate of emulsion homopolymerization and copolymerization of MA with MAA on the initiator concentration. [MAA] = 0 (*1*), 4 (*2*), 8 (*3*), and 14% of [M] (*4*); [Neonol] = 1%; 60°C.

is still weak, the strengthening of the gel effect with [APS] determines a sharper (than following the classical concept) dependence of the rate on the initiator concentration. At a high initiator content, the strengthening of flocculation reduces the value of n_i.

The presence of comonomers affects the character of this dependence. The reaction order by initiator n_i estimated in the range of relatively low APS concentrations reduces at increasing the content of MAA but increases with the AN concentration. Within the range of the initiator concentrations, higher than at the kink, the comonomers raise n_i in comparison with that observed at homopolymerization. As a result, at copolymerization with MAA the differences in n_i values before and after the point of inflection become weaker and at relatively high MAA concentrations (14%) the kink on the log W_{max} – log [APS] curve disappears (Figure 3.14).

The slight decrease in n_i at copolymerization with MAA (low [APS]) points to the weaker dependence of the gel effect on the initiation rate in the presence of this monomer. On the contrary, AN reduces the gel effect but raises the degree of its dependence on [APS], which entails a growth of n_i. The increase in the order by initiator at its relatively high content under the action of the studied comonomers speaks for flocculation moderation at the initial stages of polymerization. This leads to a growth of the APS concentration corresponding to the point of inflection.

The studies made have enabled a number of regularities of emulsion copolymerization of methyl acrylate with some hydrophilic monomers to be established. The effects revealed are explained by a concept of several nucleation mechanisms, gel effect, bimolecular breakage of oligomeric radicals in the aqueous phase, the appearance of surfactant oligomers acting as emulsifiers, flocculation of polymer-monomer particles and the influence of polymerization conditions (initiator concentration, emulsifier nature) and the composition of monomers on the said processes.

3.5 SOME ASPECTS OF KINETICS AND MECHANISM OF PROCESSES OF FORMATION POLYMERIC LATEXES IN ABSENCE EMULSIFIER

Polymer dispersions are widely used in industry, agriculture, building, and home. They are typically prepared by emulsion polymerization in the presence of stabilizers for polymer particles (emulsifiers) [29]. However, there is a possibility of producing some synthetic latexes by emulsion polymerization without specially added surfactants, which allows one to obtain an environmentally friendly product [102]. The synthesis of polymer dispersions is complicated by the need to ensure their colloidal stability, which is achieved by introducing an emulsifier in conventional emulsion polymerization [100].

This paper presents the results of our study of the emulsifier-free polymerization of the alkyl esters of acrylic acid and their copolymerization with methacrylic acid and acrylonitrile.

When emulsifier-free polymerization with persulfate-type initiators, the polymer particles are stabilized by the initiator's charged end groups. The water-dissolved oligomeric radicals $^{\bullet}M_nSO_4^-$ grow until some critical chain length is reached, whereby they lose their solubility and are extracted from solution to form nucleus PMP, leading to homogeneous nucleation [9]. Stabilization of the particles can be enhanced by copolymerization of monomers with ionizing or highly hydrophilic monomers.

Moreover, the "micellar" nucleation mechanism can be realized in the emulsifier-free conditions, which, with visible similarity, substantially differs from the particle formation occurring during conventional emulsion polymerization. In the absence of emulsifier micelles, the primary charged radicals resulting in the aqueous phase, when decomposition of a water-soluble persulfate-type initiator, after several acts of growth (interacting with the water-dissolved monomer) react with each other to form oligomeric molecules with surface activity and being able to create micelle-like structures, which play the role of an "own" emulsifier. Subsequently, the monomer and oligomer radicals are absorbed by these "micelles," where chains can grow [100].

Therefore, preconditions are created for PMP stabilizing and the implementation of polymerization by the emulsion mechanism and without the participation of a specially introduced emulsifier.

In the absence of emulsifier, the polymerization of the monomer systems discussed proceeds by the emulsion mechanism, primarily in monomer-polymer particles. The basic regularities typical for the reaction in the presence of emulsifier are observed. This is indicated by the formation of an emulsion just at the initial stage of polymerization and relatively stable polymer dispersion upon its completion, the similarity of the kinetic curves (Figure 3.15) and the kinetic regularities of both emulsifier-free and traditional emulsion polymerization in the presence of specially added emulsifiers (see Chapter 4). The most important differences between these two processes are the ways of particle generation and stabilization.

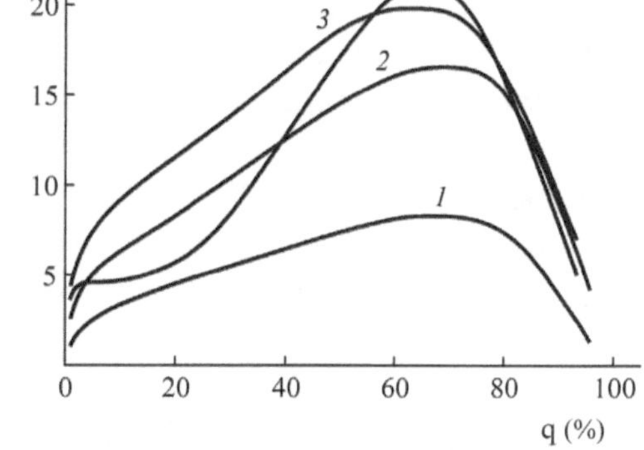

FIGURE 3.15 Dependence of the rate of the emulsifier-free emulsion copolymerization of MA with AN (4%) and MAA (14%) on the monomer conversion degree. [APS] × 10^3 = 1 (*1*), 4 (*2*), 10 (*3*), and 20 mol/L (*4*). 70°C.

The kinetics of classical emulsion polymerization is characterized by three main stages, namely: a rapid increase in the rate (due to the formation of primary latex particles), the stationary reaction (while monomer droplets are present in the aqueous phase and its equilibrium concentration is set in the particles), and the completion of the process (as the monomer in the particles is exhausted). The same stages are also peculiar to emulsifier-free polymerization, but their course and duration depend on individual monomers and reaction conditions.

The relatively low stability of the polymer dispersions synthesized in the absence of a special emulsifier is expressed in the fact that, under certain conditions, a coagulum could be formed during synthesis, as well as a precipitate and a clarified upper layer in the dispersion during its storage [103]. This indicates the existence of particles of different sizes: the larger ones form a precipitate while the smaller ones remain in a dispersed state. The considerable variation of the latex particles by size is possible if the step of their formation has a long duration. In the presence of emulsifier, more uniform dispersions are formed since PMP are formed with a high rate when radicals entering from the aqueous phase into the monomer-filled emulsifier micelles, where polymerization then proceeds with an equal probability. The polymerization in the presence of certain emulsifiers soluble in the monomer is an exception, which the formation of new particles during the reaction is of also characteristic [99]. The appearance of a clarified layer without precipitate formation indicates that in the absence of a specially added emulsifier, though large but size-uniform particles may occur which, under insufficiently effective stabilization, gradually settle to form two layers with a high and low content of latex particles.

Emulsifier-free polymerization proceeds at a slower rate than in the presence of emulsifier, and an increase in the rate, longer by time and deeper by conversion degree, is observed, e.g., the first stage of the process (latex particle formation) lasts longer. The further variation of the reaction rate with increasing the degree of conversion, as well as in the case of conventional emulsion polymerization, is caused by changes in the number of particles in the emulsion because of their flocculation at high conversions or additional nucleation in the course of the reaction, and peculiarities of the reaction proceeding inside PMP: the formation of coarse particles during emulsifier-free polymerization contributes to the gel effect and an increased rate at high conversions, despite of the lower monomer concentration therein.

It is shown that an increased temperature leads to an increased polymerization rate and an increased number of latex particles in the final product (*N*), their reduced size (*r*), and a lower amount of coagulum formed (*P*) with the dispersion stability improved. Increasing temperature accelerates the decomposition of the initiator and leads to more frequent penetration of oligomeric radicals into particles to increase their charge and stabilization. The amount of radicals in the aqueous phase increases and the probability of their bimolecular interaction to form associates of water-soluble surfactant oligomers having micelle-forming properties and promoting nucleation rises. The probability of oligomeric radicals to reach a critical chain length at which they would lose their solubility and form new particles increases as well. Therefore, changing temperature affects not only the rate of generation of primary radicals and their growth in PMP but also other processes involved in nucleation. Besides, increasing temperature influences the value of gel effect, which also affects the effective activation energy of polymerization. The ratio of the maximum rate to that at 10% conversion (corresponding to the second stationary stage of the reaction) can be regarded as a measure of the gel effect. In the copolymerization of butyl acrylate with AN and MAA in the absence of emulsifier, it increases with temperature up to 70–75°C, and decreases at higher values (Figure 3.16), reflecting opposite tendencies in the effect of the reaction temperature on the probability of coexistence of several radicals in PMP. This probability increases with the initiation rate but decreases when the particle size reduces, which occurs under these conditions. As a result, the maximum rate of polymerization (W_{max}) does not obey the Arrhenius dependence, and the experimental data in the lg W_{max} *vs.* 1/T coordinates do not fit a straight line (Figure 3.2), indicating a changed reaction mechanism in various temperature ranges and the role of individual factors which determine the rate of the process.

The initiator concentration also renders significant impact on emulsifier-free polymerization. With its increasing, the amount of latex particles in the final dispersion increases but decreases at higher concentrations, passing through a maximum (Figure 3.17). When BA is copolymerized with AN and MAA, this maximum is observed at [APS] = 15×10^{-3} mol/L (85°C).

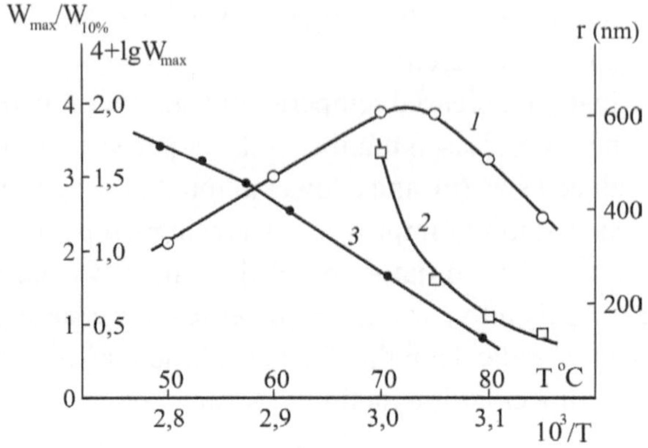

FIGURE 3.16 Temperature dependence of the size of latex particles (*2*) and the maximum rate–steady rate differences (*1*) of the emulsion copolymerization of BA with AN (4%) and MAA (14%). *3* – Arrhenius' dependence for W_{max}. [APS] = 40 × 10^{-3} mol/L.

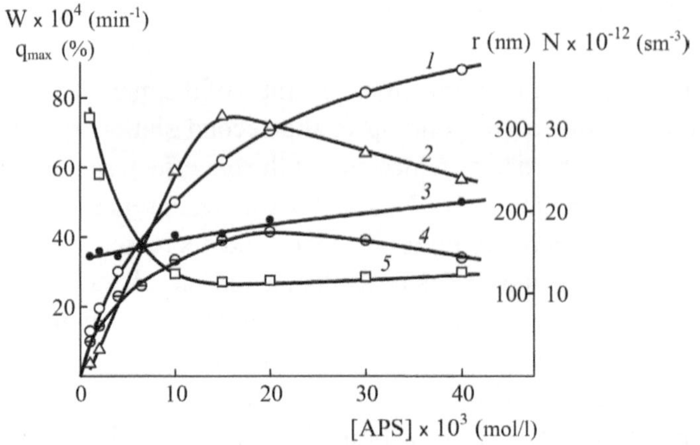

FIGURE 3.17 Dependence of the maximum rate (*1*) of the emulsifier-free emulsion copolymerization of BA with AN (4%) and MAA (14%), the corresponding degree of conversion (*3*), the number of particles in the resulting polymer dispersion (*2*), their average radius (*5*), and the rate at a 10% conversion (*4*) on the initiator concentration. 85°C.

The reaction rate also changes, to various degrees, depending on the monomer-to-polymer conversion degree. At low conversions (10%) the rate, like the number of particles, has an extreme dependence on the APS concentration.

When the initiator concentration increases, the depth of polymerization, at which the maximum rate is achieved for given conditions of the process, shifts towards higher values, and the rate itself rises. But its logarithmic dependence on the initiator concentration is nonlinear. In a BA-based monomer system the order by initiator (n_i) decreases from 0.6 when [APS] < 10^{-2} mol/L down to 0.3 at higher concentrations. (According to the classical theory, $n_i = 0.4$ [2]). In the case of the copolymerization of methyl acrylate (MA) at a sufficiently high content of the initiator (>10^{-2} mol/L) W_{max} almost no longer depends on the initiation rate as well. Dilatometric measurements in these systems were carried out at temperatures not exceeding 70°C, because the studied monomers are low-boiling (80°C for MA, 77°C for AN). However, the synthesis of dispersions without registering rate (in sealed ampoules) was performed at higher temperatures as well.

The initiation rate also affects the stability of the resulting product. BA-based dispersions which form neither precipitate nor clarified layer during storage were obtained at high temperatures (80–85°C) only in a narrow range of APS concentration (10–20) × 10^{-3} mol/L and simultaneous copolymerization with MAA and AN. In the presence of only one of the said comonomers within a concentration range of 0–14% we have failed to avoid the appearance of a clarified layer, but its volume and rate of formation decrease when the comonomer concentration increases [103].

Delamination to form a clarified layer (even more pronounced) occurs during storage of MA-based dispersions as well. This process is more probable in the case of the synthesis of dispersions with low initiation rates, and the amount of the coagulum produced has a minimum value in the range of APS concentrations (1–10) × 10^{-3} mol/L (90°C) (Figure 3.18).

Therefore, stable polymer dispersions were obtained only at a high temperature of synthesis, a relatively high content of hydrophilic comonomers, within a narrow concentration range of the initiator.

It should be noted that in the emulsifier-free copolymerization of alkyl acrylates the stability of emulsions and final dispersions decreases in the row butylacrylate > ethylacrylate > methylacrylate, e.g., the stability deteriorates with an increased water solubility (polarity) of the primary monomer. BA-based dispersions consist of a much larger number of smaller particles than those on the basis of MA. Their better stabilization, with surfactant oligomers involved, may be associated with the fact that the surface activity of the "own" surfactants increases with an increased hydrophobicity of the particle surface. This is confirmed by the increase in the interaction energy of the emulsifier with the organic phase as the polarity of alkyl acrylate decreases [29].

The extreme dependences on the APS concentration observed for the number of particles in the final dispersion and the reaction rate at low conversions (Figures 3.17 and 3.18), just at polymerization in the presence of emulsifier, are associated with insufficient stabilization of latex particles and their flocculation which occurs at various degrees of conversion, and not only at the initial stage of polymerization, as follows from the classical theory [2]. The reduction of the number of particles in the polymer emulsion when increasing the conversion at the third stage of emulsifier-free polymerization has been found experimentally. In the range of APS concentration corresponding to the maximum number of particles in the dispersion and their minimum size, the lowest number of the coagulum formed and the highest stability of the dispersion during storage are observed. Moreover, the content of the initiator affects the balance among various nucleation mechanisms.

At polymerization with low initiation rates, the aqueous phase of the emulsion system contains a low concentration of oligomeric radicals and their recombination probability is low. Under these conditions, many

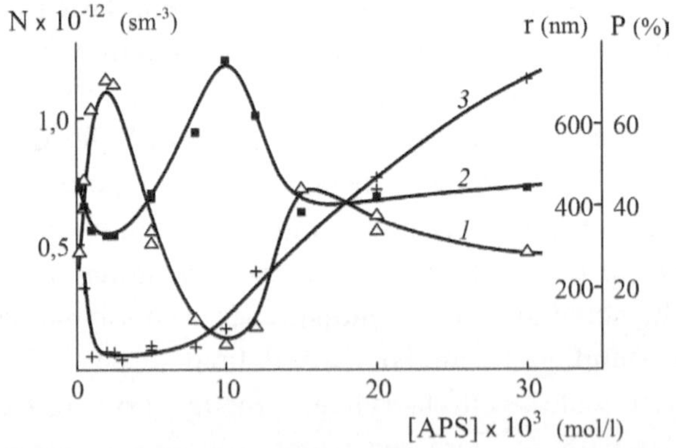

FIGURE 3.18 Dependence of the number of particles (*1*) in the polymer dispersion obtained by the emulsifier-free polymerization of MA with AN (4%) and MAA (14%), their average radius (*2*), and the amount of the coagulum formed (*3*) on the initiator concentration. 90°C.

of them would have time to reach a critical size and to form primary particles by the homogeneous mechanism. Their stabilization occurs due to the polar groups of the monomers and the surface charge arising due to the presence of end groups in the persulfate initiator. To create a sufficient charge density, the primary particles flocculate at the stage of their formation. During subsequent polymerization, together with an increase in the volume (and surface) of the particles, their charge increases due to the periodic introduction of charged oligomeric radicals $^{\bullet}M_nSO_4^-$ to PMP. An increased initiation rate leads to an increased number of radicals turning into primary particles, and the total number of PMP increases. More frequent penetration of oligomer radicals to the particles occurs as well, which improves their stabilization and prevents flocculation at later stages of the reaction.

At higher initiator concentrations, the role of the bimolecular chain breakage increases due to recombination of oligomeric radicals in the aqueous phase. As a result, water-soluble surfactant oligomers appear to play the role of an emulsifier. If their concentration in water exceeds the critical micelle concentration (which is possible at a sufficiently high initiation rate) then the associates of these oligomers form micelles which the monomer diffuses into and wherein polymerization can proceed. Under these conditions, the contribution of the micellar mechanism of PMP nucleation rises. Their surface is protected from the very beginning with the "own" emulsifier, which inhibits flocculation at the first stage of the reaction and promotes increasing N. However, if the "own" emulsifier is formed and adsorbed more slowly than the surface requiring stabilization grows (large rate constants of chain propagation and the concentration of the monomer in particles but a relatively small initiation rate), flocculation takes place at the second stage of polymerization.

At the third step of the reaction corresponding to the disappearance of monomer droplets and the reduction of the monomer concentration in PMP, the protective effect of the "own" emulsifier deteriorates due to an increased surface charge of the particles due to the introduction of new charged oligomeric radicals therein. As a result, flocculation of the particles occurs until the effect of another stabilization mechanism (due to the surface charge as in the case of homogeneous nucleation) gets sufficiently effective. The transition from stabilization by surfactant oligomers to stabilization due to electrostatic forces occurring at the third stage of polymerization is accompanied by reduction in the PMP number. As a result, N passes through a maximum with increasing APS concentration.

In the case of MA-based dispersions, the appearance of another peak on the *N vs.* [APS] dependence was detected at high initiator concentrations (Figure 3.18, curve 1). Under these conditions the rate of formation of the "own" emulsifier by recombination of oligomeric radicals in water may turn out to be so high that flocculation at the second step of polymerization begins to play a less important role, which leads to an increased number of particles. However, flocculation at the third stage of the reaction enhances with increasing the APS concentration, because oligomeric radicals enter the particle more often and the surface charge grows faster to worsen the operating conditions for the emulsifier. Therefore, the flocculation increase with an increased content of the initiator reduces *N* again.

The surface activity of the "own" emulsifier depends on the structure and composition of the oligomers, and, consequently, on the properties of the comonomers. In the case of BA no second peak has been detected, but it may, like the first maximum, occur at higher initiation rates than that of methyl acrylate latexes, outside of the range investigated. It can be assumed that the transition to the micellar nucleation mechanism as well as the termination of flocculation at the second stage of the reaction, at BA copolymerization occurs at higher rate initiation than in the case of MA. That is, the surfactant oligomers resulting from radical recombination in the aqueous phase are produced slowly. These oligomers must have a different composition at copolymerization of MA or BA due to their different reactivity and the solubility in water (5 and 0.2%, respectively [29]).

Taking into account that the MAA distribution coefficient between the aqueous and organic phases is less than unity, we can assume that within the investigated concentration range of MAA in the monomer system ($\leq$14%) the content of MA in water is higher than for MAA. However, MAA is more actively involved in the copolymerization reaction [98]. Therefore, the oligomeric radicals formed from MA and MAA copolymerization apparently consist of units of both monomers. In contrast, the concentration of BA in water is less than that of MAA, and the activity of MAA in copolymerization with BA is also high ($r_{MAA} = 1.3$, $r_{BA} = 0.3$ [104]). In this case, the oligomeric radicals are mainly composed of MAA. It is known that at copolymerization of monomers of different polarity a tendency to cross breakage is observed due to the effects of electron transfer [58]. The rate constant of cross break is higher than at homopolymerization of the corresponding monomers. In this connection, breakage in the aqueous phase to form surfactant oligomers at copolymerization of MA is more probable, which explains the lower (than for BA) APS concentrations, which the maximum number of particles in the dispersion is achieved at.

At copolymerization in the MA-based ternary system, all the monomers are (partially) dissolved in the aqueous phase, namely: MA, MAA, and AN. The copolymerization of AN with MA is characterized by the constants: $r_{AN} = 1.5$, $r_{MA} = 0.84$ [105]. Involving AN to oligomeric radicals leads to acceleration of their bimolecular interaction since the rate constant k_o of chain termination of AN is almost two orders of magnitude higher than that of MA [58]. This promotes strengthening of the role of the micellar mechanism of particle formation.

In the case of ternary BA-based systems, copolymerization of MAA and AN should mainly occur in the aqueous phase. And AN less actively participates in the reaction, as evidenced by its much slower consumption when the copolymerization reaction (by chromatographic data) and the resulting oligomeric radicals are mainly composed of MAA units. The breakage constant at polymerization of this monomer in water [101] is close to k_o of acrylic esters.

Flocculation of particles at different stages of the polymerization process and its dependence on the concentration of the initiator influence not only the number of particles in the final dispersion but, together with the gel effect, also the value of the reaction rate at different conversions and, consequently, the shape of the kinetic curves. Apparently, just the flocculation at the second stage, growing within some range of the APS concentration, causes an end of the rate growth with the polymerization depth (or even its slight decrease) at relatively small depths of conversion observed under these conditions (Figure 3.15, curve 4).

Unlike conventional emulsion polymerization with a specially introduced emulsifier which is spent for the formation and stabilization of the resulting latex particles, at emulsifier-free polymerization during the reaction an "own" emulsifier is formed, which participates in the origination of new particles. The gradual emergence of new surfactant oligomers increases the nucleation stage duration. Therefore, with the growth of the initiation rate, the instant of reaching the maximum rate of polymerization shifts towards higher conversion degrees (Figure 3.17). Besides, this increases the heterogeneity degree of particles by size, promotes enhancing flocculation, coagulum and precipitate formation during storage of the dispersion. On the other hand, an increased APS concentration promotes a better stabilization of the particles and the clarified layer decreases.

The formation of latex particles during emulsifier-free polymerization occurs by various mechanisms whose relative contributions depend on the reaction conditions. It has turned out that changes in some of them can influence the nature of the effects of other factors. E.g. an increased MAA concentration at relatively low initiation rates ([APS] = 5×10^{-3} mol/L, 70°C) leads to a decreased rate of the emulsifier-free polymerization of methyl acrylate, ethyl acrylate, and butyl acrylate and the number of latex particles, an increased conversion corresponding to the maximum rate. The particle size and the amount of coagulum formed increase as well (Figure 3.19). On the contrary, at high initiation rates (85–90°C) MAA improves the stability of the dispersion, increases the number of particles and the reaction rate (Figure 3.20).

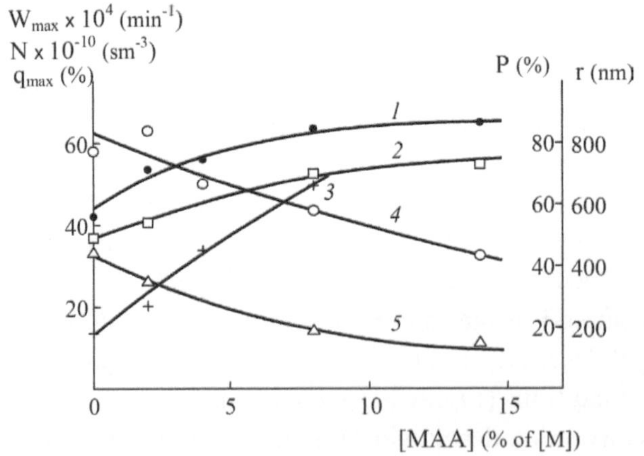

FIGURE 3.19 Dependence of the maximum rate of the emulsifier-free emulsion copolymerization of MA with MAA (*4*), the corresponding depth of polymerization (*1*), the number of particles in the polymer dispersion (*5*) and their average radius (*2*), and the amount of the coagulum formed (*3*) on the MAA concentration. [APS] = 5×10^{-3} mol/L; 70°C.

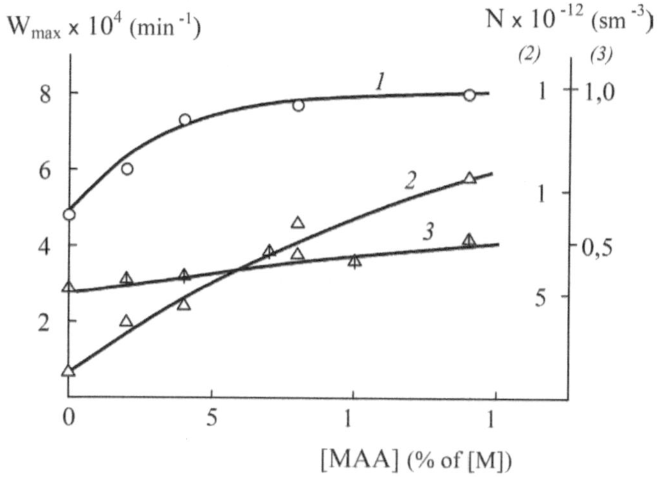

FIGURE 3.20 Effect of the MAA concentration on the maximum rate of reaction (*1*) and the number of particles in the dispersion (*2*, *3*) formed by the emulsifier-free emulsion copolymerization of BA (*1*, *2*) and MA (*3*). [APS] × 10^3 = 5 (*3*) and 15 mol/L (*1*, *2*); T = 85 (*1*, *2*) and 90°C (*3*).

In the presence of MAA, the probability of recombination of oligomeric radicals in the aqueous phase increases due to their hindered entry into PMP due to the poor solubility in the monomers. In the presence of a specifically introduced emulsifier this leads to reduction in the number of PMP and the polymerization rate and to strengthening of the gel effect in the resulting larger particles. The same phenomena are also observed while emulsifier-free polymerization at relatively low initiation rates (low temperatures) when the homogeneous nucleation mechanism predominates. However, at higher temperatures (higher initiation rates), PMP are mainly formed by the micelle mechanism as a result of the formation of their "own" emulsifier. Strengthening of bimolecular breakage in water under the influence of MAA promotes the appearance of water-soluble surfactant oligomers, playing the role of an emulsifier, the improved particle stabilization, an increase in their number and polymerization rate. Therefore, some of the processes proceeding in the reaction system may result in various effects at conventional and emulsifier-free emulsion polymerizations.

Similar differences in the effect on emulsifier-free emulsion polymerization at different initiation rates have been detected for AN as well, which retards the polymerization of MA and reduces N at 70°C (the predominance of the homogeneous nucleation mechanism) but increases it at 90°C, e.g., under those conditions in which the micellar nucleation mechanism involving surfactant oligomers is better expressed. Moreover, it turns out that the effect of one of the comonomers on the nucleation process may depend on the availability of other comonomers. For example, an increased AN concentration in the ternary monomer system with a high MAA content (14%), when the micellar nucleation mechanism seems more probable, leads to an increased number of particles at relatively low temperatures (70°C) as well, although in the absence of MAA or at its low concentrations, a decrease in N is observed (Figure 3.21).

An increased AN content in BA-based systems leads to an increased number of particles in the final polymer dispersion but the polymerization rate lowers. Increasing N in emulsifier-free conditions is associated with the improved particle stabilization during copolymerization with a hydrophilic monomer having polar CN-groups. However, just as in the presence of a specific emulsifier, AN weakens the gel effect, thereby reducing the reaction rate in the particle. Moreover, it suppresses flocculation at the third polymerization stage, whereby the number of particles reduces by the end of the reaction to a lesser extent than in the absence of AN.

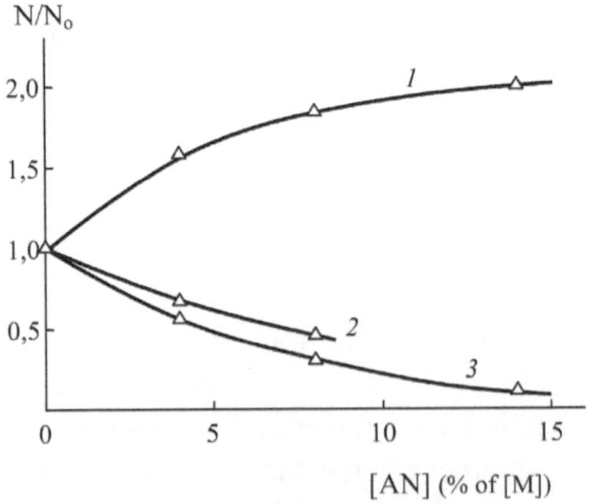

FIGURE 3.21 Effect of AN on the number of particles in the polymer dispersions produced during the emulsifier-free emulsion copolymerization of MA with AN (*3*) with MAA (4%) and AN (*2*), and with MAA (14%) and AN (*1*). [APS] = 4×10^{-3} mol/L; 70°C.

This leads to the observed differences in the effect of comonomer on the number of particles and the rate of emulsion polymerization.

Thus, this study shows the possibility of the synthesis of polymer dispersions in the absence of any emulsifying agent, but only under specially selected conditions (temperature, initiator concentration, the presence and nature of comonomers).

3.6 CONCLUSION

A review of the literature data obtained in studies of the kinetics, mechanism and topochemistry of polymer dispersion formation at emulsion homopolymerization and copolymerization of (meth)acrylic and other vinyl monomers is presented. The influence of the nature and concentration of the monomers, initiator, emulsifier on the nucleation mechanism of latex particles, on the reactions of initiation, chain growth (including copolymerization) and chain termination is discussed. Features of emulsion polymerization inhibition associated with the distribution of inhibitors in the phases of the reaction system are considered.

The authors' experimental study was conducted by dilatometry and the turbidity spectrum method. The possibility and conditions of applying the turbidity spectrum method in kinetic studies of emulsion polymerization are discussed. It is necessary to take into account the spectral dependence of the refractive index of the polymer-monomer particles and dispersion medium, secondary light scattering, as well as the use of approximating analytic expressions for the characteristic functions of light scattering, since the refractive index of the scattering centers depends on the depth of polymerization.

The kinetics and mechanism of the emulsion polymerization of (meth) acrylates and their copolymerization with hydrophilic monomers (methacrylic acid, acrylonitrile) in the presence of various emulsifiers (sodium lauryl sulfate, sulfated oxyethylated alkylphenols) were studied. It has been shown that the kinetics and mechanism of emulsion polymerization contradict to the classical concepts of this reaction due to a variety of nucleation mechanisms, the presence of several growing radicals in the polymer-monomer particles, the manifestation of gel effect, flocculation of the particles at different stages of polymerization, the partial solubility of some of the tested emulsifiers in the monomer, interactions between radicals in the aqueous phase, resulting in the formation of surfactant oligomers that act as a "self" emulsifier as well as chain termination. These effects lead to the number of particles and the reaction rate therein depending on the conversion degree, the influence of polymerization conditions on the kinetic orders by the emulsifier and initiator concentrations.

The feasibility of synthesis of stable polymeric dispersions by polymerization in the absence of emulsifying agent under specially selected conditions by temperature, the concentration of the initiator (ammonium persulfate), the presence and nature of comonomers is shown. The mechanisms of formation and stabilization of latex particles in the emulsifier-free conditions are discussed [106].

KEYWORDS

- **acrylonitrile**
- **alkyl acrylates**
- **ammonium persulfate**
- **emulsion polymerization**
- **kinetics**
- **mechanism**
- **metacrylic acid**
- **presence and absence of emulsifier**
- **stabilization of dispersions**

REFERENCES

1. Harkins, W. D. *J. Amer. Chem. Soc.*, 69(6), 1428–1444(1947).
2. Smith, W. V. and Ewart, R. H. *J. Chem. Phys.*, 16(6), 592–599(1948).
3. Yurzenko, A. I. and Mints, S. M. *Dokl. ANSSR,* 47(2), 106–108(1945).
4. Oudian, J. Bases of Polymer Chemistry (Moscow, 1974). Russian, ed., p. 256.
5. Eliseeva, V. I., Ivanchev, S. S., Kuchanov S. I. and Lebedev, A. V. Emulsion Polymerization and its Application in Industry (Moscow, 1976). 240 pp. (in Russian).
6. Ivanchev, S. S. Radical Polymerization (Leningrad, 1985). 280 pp. (in Russian).

7. Kuchanov, S. I. Advances of Science and Technology. Chemistry and Technology of High-Molecular Compounds (Moscow, 1975). Vol. 7, p. 167–225.
8. Chern, C. S. and Poehlein, G. W. *J. Polymer Sci. Polymer Chem. Ed.* 25(2), 617(1987).
9. Gardon, J. L. *Rubber Chem. and Technol.*, 43(1), 74(1970).
10. Gardon, J. L. *J. Polymer Sci.*, A-1, 6(3), 643–664(1968).
11. Fitch, R. M. and *Paint, J. Technol. Engng.*, 37(1), 32(1965).
12. Fitch, R. M. and Tsai, C. H. Polymer Colloids (N. Y., 1971), p. 73.
13. Roe, C. P. *Industr. and Engng Chem.*, 60(9), 20–23(1968).
14. Carra, S., Morbidelle, M. and Storti, G. Proc. Intern. School Phys. Enrico Fermi (Bologna, 1985), p. 483–512.
15. Ugelstad, J. and Hansen, F. K. *Rubber Chem. and Technol.*, 49(3), 536(1976).
16. Hansen, F. K. and Ugelstad, J. *J. Polymer Sci. Polymer Chem. Ed.* 16(8), 1953–1979(1978).
17. Bataille, P., Van, B. T. and Pham, Q. B. *J. Polymer Sci.*, 20(3), 795(1982).
18. Feeney, P. J., Napper, D. H. and Gilbert, R. G. *Macromolecules*, 20(11), 2992 (1987).
19. Robb, I. D. *J. Polymer Sci.*, A-1, 7(9), 2665(1969).
20. Ugelestad, J., El-Aasser, M. S. and Vanderhoff, J. M. *J. Polymer Sci.*, A-1.11(8), 503–513(1973).
21. Ugelstad, J., Hansen, F. K. and Lange, S. *Makromolek. Chem.*, 175(3), 507–521(1974).
22. Nikitina, S. A., Gritskova, I. A., Spiridonova, V. A., Sedakova, L. I., Malyukova, E. B. and Pavlov, A. V. *Vysokomolek, Soed. A.*, 17(3), 582(1975).
23. Gritskova, L. I. Sedakova, D. S. Muradyan, B. M. Sinekaev, A. V. Pavlov and A. N. Pravednikov, *Dokl. AN SSSR*, 243(2), 403–406(1978).
24. Gritskova, I. A. Dr. Chem. Sci. Thesis (Moscow Institute of Fine Chemal Technology, Moscow, 1979) 305 pp. (in Russian).
25. Gritskova, I. A., Sedakova, L. I., Muradyan, D. S. and Pravednikov, A. N. *Dokl. AN SSSR*, 238(3), 607–612(1978).
26. Malyukova, E. B., Egorov, V. V., Zubov, V. P., Gritskova, I. A., Pravednikov, A. N. and Kabanov, V. A. *Dokl. AN SSSR*, 265(2), 375(1982).
27. Simakova, G. A., Kaminskii, V. A., Gritskova, I. A. and Pravednikov, A. N. *Dokl. AN SSSR*, 276(1), 151–153(1984).
28. Tsar'kova, M. S., Gritskova, I. A., Nikitina, T. S. and Pravednikov, A. N. *Vysokomolek. Soed. A.*, 31(12), 2609–2613(1989).
29. Eliseeva, V. I. Polymeric Dispersions (Moscow, 1980). 296 pp. (in Russian).
30. Lee, P. I. *Plast. and Polymer.*, 39(100), 111–114(1971).
31. Lee, P. I. and Longrotten, R. M. *J. Appl. Polymer Sci.*, 14(5), 1377–1379(1970).
32. Banerjee, M. and Konar, R. S. *Polymer*, 27(1), 147–157(1986).
33. Bulkin, Yu.I., Usachyova, N. N., Lomonosova, G. A., Trubnikov, A. V., Gol'dfein, M. D., Kozhevnikov, N. V. and Zyubin, B. A. Plast. Massy (No.12), 16–20(1989).
34. Gershberg, D. Proc. Amer. Inst. Chem. Engng – Inst. Chem. Engng Joint Meet. (London, 1965), No.3, p.3.
35. Okamura, S. and Motogama, T. *J. Polymer Sci.*, 58(166), 221(1962).
36. Kozhevnikov, N. V. and Goldfein, M. D. *Vysokomolek. Soed. A.*, 33(11), 2398–2404(1991).

37. Goldfein, M. D., Kozhevnikov, N. V., Trubnikov, A. V., Tsyganova, T. V. and Bugreeva, L. R. Some Problems of Chemical Physics (Saratov, 1990). Part 2, p. 13–42 (in Russian).
38. Kozhevnikov, N. V., Goldfein, M. D., Zyubin, B. A. and Trubnikov, A. V. *Vysokomolek. Soed. A.*, 33(6), 1272–1280(1991).
39. Blackley, D. C. Emulsion Polymerizations. Theory and Practice (London, 1975), p. 163.
40. Hawkett, B. S., Napper, D. H. and Gilbert, R. G. *J. Polymer Sci. Polymer Chem. Ed.* 19(12), 3173–3179(1981).
41. Vinogradov, P. A., Odintsova, L. P. and Shitova, A. A. *Vysokomolek. Soed.*, 4(1), 98(1962).
42. Volkov, V. A. and Kalyuda, T. V. *Vysokomolek. Soed. B.*, 20(11), 862(1978).
43. Capek, J., Barton, J. and Karpatova, A. *Makromolek. Chem.*, 188(4), 703(1987).
44. Georgescu, C. S., Butucca, V., Sarbu, A., Sonescu, A. and Denconescu, J. Prague Meet. Macromolec. 31st Microsymp. Polyvinylchloride (Prague, 1988), p. 32.
45. Brietenbach, J. W., Kuchnar, K., Fritze, H. and Tarnavieski, H. *Brit. Polymer J.*, 2, 13(1970).
46. Cooper, W. Reactivity, Reaction Mechanisms and Structure in Polymer Chemistry; A. Jenkins and A. M. Ledwis, Editors (Moscow, 1977). Rus. ed., p. 208.
47. Medvedev, S. S. Kinetics and Mechanism of Formation of Macromolecules (Leningrad, 1968), p. 5 (in Russian).
48. Barton, J. and Karpatova, A. *Makromolek. Chem.*, 188(4), 693(1987).
49. Almgren, M., Grieser, F. and Thomas, J. K. *J. Amer. Chem. Soc.*, 101, 279(1979).
50. Al-Shahib, W. A.-G. and Dunn, A. S. *Polymer*, 21(4), 429(1980).
51. Arutyunyan, R. S. and Beileryan, N. M. *Arm. Khim. Zhurn.*, 40(1), 10(1987).
52. Pavlyuchenko, V. N. and Ivanchev, S. S. Reactions in Polymeric Systems; S. S. Ivanchev, Editor (Leningrad, 1987), p. 84.
53. Barton, J., Capek, J., Juranicova, V. and Riedel, S. *Makromolek. Chem., Rapid Commun.*, 7, 521(1986).
54. Pavlyuchenko, V. N. and Ivanchev, S. S. *Acta Polymerica*, 34(39), 521(1983).
55. Lukhovitski, V. I. and Polikarpov, V. V. Technology of Radiation Emulsion Polymerization (Moscow, 1980). 60 pp. (in Russian).
56. Gardonn, J. L. *J. Polymer Sci.*, A-1, 6(3), 665(1968).
57. Pavlyunchenko, V. N., Ivanchev, S. S., Byrdina, N. A., Alekseeva, Z. M. and Lesnikova, N. V. *Dokl. AN SSSR*, 259(3), 641(1981).
58. Bagdasariyan, H. S. Theory of Radical Polymerization (Moscow, 1966). 300.
59. Paoletti, K. P. and Billmeyer, F. W. *J. Polymer Sci.*, A-2(5), 2049(1964).
60. Malyukova, E. B., Gritskova, I. A. and Pavlov, A. V. *Trans. Moscow Inst. Fine Chem. Technol.* (in Russian) 3(1), 111(1973).
61. Sheinker, A. P. and Medvedev, S. S. *Dokl. AN SSSR*, 97(1), 111(1954).
62. Grancio, M. and Williams, D. *J. Polymer Sci.*, A-1, 8(9), 2617(1970).
63. Kensch, P., Price, J. and Williams, D. *J. Macromolec. Sci. Chem.*, 7, 623(1973).
64. Gritskova, I. A., Malyukova, E. B., Simakova, G. A. and Zubov, V. P. *Vysokomolek. Soed.*, A. 32(1), 14(1990).
65. Kozhevnikov, N. V., Goldfein, M. D. and Terekhina, N. V. *Proceedings of Russian Higher Educational Establishments. Chemistry and chemical technology*, 40(3), 78–83(1997).

66. Kozhevnikov, N. V., Terekhina, N. V. and Goldfein, M. D. *Proceedings of Russian Higher Educational Establishments. Chemistry and chemical technology,* 41(4), 83–87(1998).
67. Friis, N. and Hamielec, A. E. *Amer. Chem. Soc. Polymer Preprints,* 16(1), 192(1975).
68. Bagdasaryan, G. P. C., Chem. Sci. Thesis (Leningrad, Regional Scientific & Production Association "Plastopolymer", 1984). 154 pp. (in Russian).
69. Guillet, J. *Makromolek. Chem. Rapid Commun.,* 1(11), 697(1980).
70. Guillet, J. *Acta Polymerica,* 32(10), 593(1981).
71. Capek, J. *Acta Polymerica,* 39(5), 221(1988).
72. Kong, X. Z., Pichot, C. and Guillet, J. *Europ. Polymer J.,* 24(5), 485(1988).
73. Bataille, P. and Bourassa, H. *J. Polymer Sci. Polymer Chem., Ed.* 27(1), 357(1989).
74. Romantov, Al., Georgescu, E. and Dumiterscu, G. IUPAC Macro'83 (Bucharest, 1983). Abstrs. Sect. I, p. 616.
75. Capek, J. and Barton, J. *Chem. Papers,* 40(1), 45(1986).
76. Capek, J., Barton, B., Tuan Le Guang, Svoboda, V. and Novotky, V. *Makromolek. Chem.,* 188(7), 1723(1987).
77. Capek, J. and Than Le Guang, *Makromolek. Chem.,* 187(9), 2063(1986).
78. Capek, J., Mlynarova, M. and Barton, J. *Makromolek. Chem.,* 189(2), 341(1988).
79. Capek, J., Mlynarova, M. and Barton, J. *Chem. Papers,* 42(6), 763(1988).
80. Capek, J. *Makromolek. Chem.,* 190(4), 789(1989).
81. Ivanchev, S. S., Bagdasaryan, G. P. and Pavlyuchenko, V. N. *Dokl. AN SSSR,* 275(3), 653(1984).
82. Louic, B. M., Wen Yen Chiu and Soong, D. S. *J. Appl. Polymer Sci.,* 30(8), 3189(1985).
83. Louic, B. M., Franaszek, T., Pho, T., Wen Yen Chiu and Soong, D. S. *J. Appl. Polymer Sci.,* 30(9), 3841(1985).
84. Barton, J. and Juranicova, V. *Makromolek. Chem.,* 190(4), 763(1989).
85. Barton, J. and Juranicova, V. *Makromolek. Chem.,* 190(4), 769(1989).
86. Gladyshev, G. P. and Popov, V. A. Radical Polymerization at High Degrees of Conversion (Moscow, 1974), 244 (in Russian).
87. Gladyshev, G. P., Popov, V. A. and Penkov, E. I. *Vysokomolek. Soed.,* A.16(9), 1945(1974).
88. Klenin, V. I., Shchegolev, S.Yu. and Lavrushin, V. I. *Characteristic functions of light scattering from disperse systems.* Saratov: Saratov Univ. Press, 1977. 177 p.
89. Bohren, C. F. and Huffman, D. R. *Absorption and Scattering of Light by Small Particles* (Wiley-Interscience, 1983).
90. Ramazanov, K. R., Khlebtsov, N. G., Shchegolev, S.Yu. and Klenin, V. I. *Colloid Journal,* 45(3), 473–479(1983).
91. Khlebtsov, N. G., Melnikov, A. G. and Shchegolev, S.Yu. *Colloid Journal,* 53(5), 928–933(1991).
92. Joffe, B. V. *Refractometric methods in chemistry.* Leningrad: Khimiya. 1983. 352 p.
93. Klenin, V. I., Shchegolev, S.Yu. and Lebedeva, L. G. *Optics and Spectroscopy,* 35(6), 1161–1166(1973).
94. Dunn, A. S. and Chong, L.C-H. *Brit. Polym. Journal.,* 2(1–2), 49–59(1970).
95. Kozhevnikov, N. V., Kozhevnikova, N. I. and Goldfein, M. D., *Proceedings of Russian Higher Educational Establishments. Chemistry and chemical technology,* 53(2), 64–68(2010).

96. Gromov, V. F., Bune, E. V. and Teleshov, E. N., *Uspekhi Khimii* 63(6), 530–541(1994).
97. Kozhevnikov, N. V., Zyubin, B. A. and Simontsev, D. V., *Polymer Science,* A.37(5), 758–763(1995).
98. Nazarova, I. V. and Eliseeva, V. I., *Mendeleyev Soc. Journal,* 12(4), 587–588(1967).
99. Kabanov, V. A., Kurilova, A. I. and Topchiev, D. A. *Vysokomolek. Soed.*, B-15(9), 712–717(1973).
100. Eliseeva, V. I. and Aslamazova, T. R., *Uspekhi Khimii,* 60(2), 398–429(1991).
101. Kabanov, V. A., Zubov, V. P. and Semchikov Yu.D. Complex-Radical Polymerization (Moscow, 1987). 256 pp.
102. Aslamazova, T. R., Bogdanova, S. V. and Movchan, T. G., *Ross. Chemical Journal,* 37(4), 112–114(1993).
103. Kozhevnikov, N. V., Goldfein, M. D. and Trubnikov, A. V., Intern. Journal Polymeric Mater., 46, 95–105(2000).
104. Kulikov, S. A., Yablokova, N. V. and Nikolaeva, T. V. *Vysokomol. Soed.* A-31(11), 2322–2326(1989).
105. Encyclopedia of Polymers. Vol. 1. Moscow, *Sovetskaya Encyclopaedia,* 1972.
106. Kozhevnikov, N. V., Goldfein, M. D. and Kozhevnikova N. I., Proceeding of Saratov University. *New Series. Chemistry, Biology, Ecology*, 14(1), 38–47(2014).

CHAPTER 4

FILMS AND NONWOVEN MATERIALS BASED ON POLYURETHANE, THE STYRENE-ACRYLONITRILE COPOLYMER, AND THEIR BLENDS

S. G. KARPOVA,[1] YU. A. NAUMOVA,[2] L. R. LYUSOVA,[2] E. L. KHMELEVA,[2] and A. A. POPOV[1]

[1]*Emanuel Institute of Biochemical Physics, Russian Academy of Sciences, ul. Kosygina 4, Moscow, 119991, Russia*

[2]*Moscow University of Fine Chemical Technology, pr. Vernadskogo 86, Moscow, 119571, Russia*

CONTENTS

ABSTRACT

Films and nonwoven materials were investigated by the ESR method and thermophysical measurements. The effect of solvents on the structure and molecular dynamics of films and nonwoven based on PU and styrene-acrylonitrile co-polymers was studied. It is shown that a solvent weakly affects the molecular dynamics of chains in the film and nonwoven polyurethane materials and has a strong effect on the molecular mobility in the films based on the styrene-acrylonitrile co-polymers and blends with a high content of the copolymer. For the nonwoven material, this effect is insignificant. The dependence of correlation time on temperature and polymer ratio reveals the breaks at the melting temperatures of mesomorphic structures in PU-styrene-acrylonitrile blends. It was shown that ozone used as oxidizer influences the amorphous phase of investigated polymers. The spin-probe method demonstrade that for PU-based films and nonwoven materials ozone has no effect on the molecular dynamics, where as for blends and styrene-acrylonitrile co-polymer considerable changes in τ take place.

4.1 INTRODUCTION

An efficient way to solve problems related to the synthesis of materials with an improved complex of manufacturing and performance properties is to use blends of existing polymers. Choosing the type and ratio of components in the polymer blend makes it possible to vary the structures and properties of blend composites in a wide range. Therefore, the preparing polymer blends and designing new materials on their basis is a mainstream direction in the modern technology of polymer processing [1, 2].

The method of polymer processing and the choice of parameters of manufacturing processes are important aspects that determine the structures and properties of polymeric materials. The transfer of polymers to the viscous-flow state via preparation of their solutions is widely used in the manufacture of fibers and threads, paintwork materials, glues, sealants, etc. [3–6].

It is known [1, 3–6] that the choice of solvent in the technology of processing of polymer solutions is determined not only by its dissolving power but also by its effect on the structures and properties of the resulting materials. The structuring in solution depends on the nature of the used solvent. Note that the differences in the structures of the materials are largely associated with the differences in the thermodynamic qualities of solvents; their interaction with polymer macro-molecules; and, as a consequence, the differences in interactions between macro-molecules [1, 5–10]. For the ternary systems solvent–polymer 1–polymer 2, the role of the solvent is even much more complicated.

Along with the thermodynamic quality of a solvent, the structure of polymeric materials obtained from solutions is strongly affected by the volatility of a solvent and the conditions of its removal.

As a continuation of our studies [11] devoted to the effect of solvents on the structures and properties of polymeric materials obtained via processing of polymer solutions, comparative studies of films and fibrous nonwoven materials based on thermoplastic PU, the styrene–acrylo-nitrile copolymer, and their blends are presented below. The purpose of this analysis is to gain insight into the effects of the conditions and apparatuses of polymer processing and the type of solvent on the structures and properties of the resulting materials.

The electrospinning of fibers is a dry method of fiber spinning where the nature of the solvent controls the rheological behavior of the spinning solution and the effective viscosity of the solution in turn determines the energy costs of the process and the diameter and morphology of fibers [12].

At present, ultrathin fibers and items formed on their basis enjoy wide application in biomedicine, cellular engineering, separation and filtration processes, the creation of reinforced composites, electronics, analytics, sensor diagnostics, and a number of other innovative areas [12–16]. The choice of the polymers is associated with the use of PU in light industry, the glue-and-sealant industry [6, 10, 17], and medicine [18, 19], while the styrene–acrylonitrile copolymer is used in the manufacture of composite filtering and analytical materials [12]. As was indicated in [10, 16], materials obtained via solution processing of the polymer blends PU–styrene–acrylo-nitrile copolymer, for example, glue compositions and filtering materials, may find a wider application field owing to the improvement of their performance characteristics relative to those of materials based on individual polymers.

In addition, the choice of the blend composites based on PU and the styrene–acrylonitrile copolymer is associated with the interest of the authors in the compatibility of the polymers and the role of the solvent in this problem.

At present, the structure and properties of PU [18–20], as well as blend composites based on PU and a number of polymers of various families (chitosan [21], polyoxybutyrate [22], and the styrene–acrylonitrile copolymer [23–27]), have been studied.

The compatibility of materials based on the blends of thermoplastic PU of various brands and styrene–acrylonitrile copolymers that were obtained via processing of polymer melts was investigated, and a limited range of ratios at which the mentioned polymers were mutually soluble was observed [23–27]. At a mass ratio of 30:70, the thermoplastic PU and the styrene–acrylonitrile copolymer are incompatible [24–26]. As was noted in [25], the styrene–acrylonitrile copolymer shows a higher solubility in the phase enriched with PU than PU shows in the phase enriched with the styrene–acrylonitrile copolymer.

The DSC and DMA studies of the phase states of the blends for films formed from solutions of PU and the styrene–acrylonitrile copolymer with the use of low-molecular-mass liquids of various chemical classes [11] are in conflict with the above-mentioned data. Therefore, the comparative analysis of the effects of solvents on the structures of both films and nonwoven materials obtained via the electrospinning of fibers is of certain interest for fundamental polymer science and use in practice.

4.2 EXPERIMENTAL PART

The objects of research were dilute and concentrated solutions, films, and nonwoven materials based on the Desmocoll 400 thermoplastic PU (Bayer, $Mw = 1.0–10^5$), the SAN 350N styrene–acrylonitrile copolymer (Kumho, $Mw = 1.0–10^5$), and their blends at PU-to-styrene–acrylonitrile mass ratios of 10:90, 30:70, 50:50, and 80:20 for the films and 25:75, 50:50, and 75:25 for the nonwoven materials.

The viscosities of dilute solutions were measured on a VPZh-2 Ostwald viscometer [7, 10] in organic solvents of various chemical classes: ethyl acetate, THF, and acetone. The solution concentrations were in the range 0.1–2.0 g/100 mL.

The above solvents were selected on the basis of their dissolving abilities with respect to PU, the styrene–acrylonitrile copolymer, and their blends in the range of concentrations of the used spinning solutions [10] and on the basis of the setup of the manufacturing process of defect-free fibers via electrospinning.

The films were cast in Petri dishes from concentrated polymer solutions (a concentration of 10 wt %) followed by full removal of the solvent at constant temperature and humidity. The film thicknesses were 300–350 μm. Full removal of the solvent was ensured via evacuation, and the films were checked via measurements of their thicknesses and masses.

The tested polymeric nonwoven materials were prepared via electrospinning from solutions of PU, the styrene–acrylonitrile copolymer, and their blends; the mean diameters of fibers were 1–10 μm. The concentrations of spinning solutions were 8–12 wt%. The samples had unit-area masses of 20–70 g/m^2 and aerodynamic-drag values of 3–30 Pa at an air-flow velocity of 1 cm/s. The nonwoven materials were prepared with the aid of the Nanospider technology. This is a patented technology of capillary-free high-voltage electrospinning of fibers with the free surface of a liquid [13–15]. The electrospinning process was performed under the following conditions: a temperature of 20°C and a relative air humidity of 60%. The distance to the pick-up coil was 0.2–0.3 m.

The calorimetric studies of the materials were performed on a DTAS-1300 thermal analyzer in the temperature range from –90 to +140°C. The heating rate was 20 K/min. The temperature-measurement precision was ±0.5°C.

The molecular mobility was studied via the paramagnetic-probe method. The stable nitroxide radical TEMPO at a concentration of 10^{3}-10^{4} mol/L was used as a probe. The radical was incorporated into the film and nonwoven materials with small amounts of the styrene–acrylonitrile copolymer from vapor at 40°C and into the samples with high contents of the styrene–acrylonitrile copolymer at 70°C. ESR spectra were recorded in the absence of saturation; this condition was verified by the dependence of the signal intensity on the microwave-field power. The correlation times of probe rotation, τ, were calculated from the ESR spectra according to the following formula [28]:

$$\tau = {}_{\Delta}H_{+}\left(\sqrt{I_{+}/I_{-}}-1\right)\times 6.65\times 10^{-10}$$

where $\Delta H+$ is the width of the low-field component of the spectrum and $I+/I-$ is the ratio of the intensities of the low- and high-field components, respectively. The error of the measurement of τ was ±5%.

4.3 RESULTS AND DISCUSSION

Structuring in solutions is determined by the type of used solvent. In this case, the different structures of the resulting materials are primarily related to different interactions of the polymer with the solvent, that is, to different thermodynamic affinities between the solvent and the solute [1, 5, 7, 10].

In this study, the thermodynamic qualities of the solvents with respect to PU, the styrene–acrylonitrile copolymer, and their blends were quantitatively estimated through the determination of intrinsic viscosities [η] of polymer solutions and the Huggins constants. It is believed [7] that the differences in the intrinsic viscosities of solutions of flexible-chain polymers are associated with the fact that, in different solvents, the sizes of molecular coils are different. In solvents that are good from the thermodynamic view-point, coils swell to a higher extent than that in poor solvents; as a consequence, the intrinsic viscosities of dilute solutions in good solvents are higher.

With consideration for the data shown in Figure 4.1 and following the fundamental ideas of the physical chemistry of polymers [7], it was shown that, for the polyurethane thermoplastic and the styrene–acrylonitrile copolymer, THF is a good solvent in terms of thermodynamics. For every polymer, the solvents may be arranged in the order of their decreasing thermodynamic quality as follows: THF, ethyl acetate, acetone. A similar picture is observed throughout the range of compositions of the binary blends of PU and the styrene–acrylonitrile copolymer.

The structures of films and nonwoven materials based on blends of PU and the styrene–acrylonitrile copolymer that were formed in THF and acetone were studied via DSC (Figure 4.2). As was shown in [11], mesomorphic structures are formed in the films. These structures were described in detail, for example, in [29]. The character of change in the fractions and melting temperatures of mesomorphic structures with the compositions of the blends for the films and non-woven materials are different. Here, the data on change in the fraction of mesomorphic structures in a film that

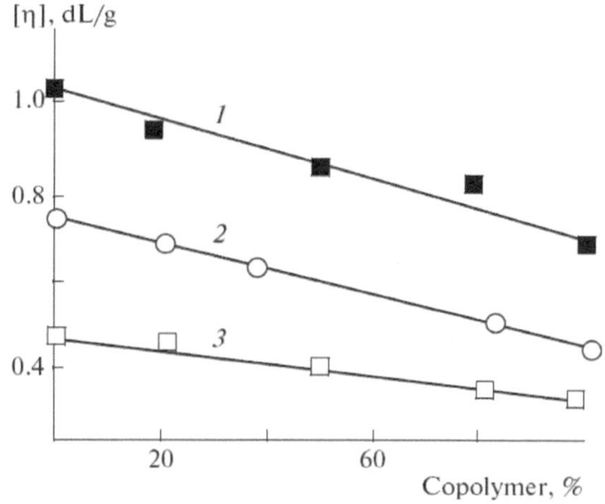

FIGURE 4.1 Effects of the type of solvent and the PU-to-styrene–acrylonitrile polymer mass ratio on [η]. Solvents: (*1*) THF, (*2*) ethyl acetate, and (*3*) acetone.

were reported in [11] and the corresponding data obtained in this study for the nonwoven materials are compared. The fraction of mesomorphic structures was normalized to the content of PU in the blend.

DSC studies (Tables 4.1 and 4.2) showed that the fractions of mesomorphic structures, α, in PU-based films and nonwoven PU materials are similar and amount to 45–47 J/g (the enthalpy of melting). If, in the case of the films, the addition of a small amount of the styrene–acrylonitrile copolymer leads to a marked reduction in the fraction of mesomorphic structures (Figure 4.2a), then, for the nonwoven materials these changes are not so pronounced (Figure 4.2b). At a PU-to-styrene–acrylonitrile mass ratio of 50:50, no mesomorphic structures are observed in the films, while for the fibers, the fraction of these structures is extremely small. At a higher content of the styrene–acrylonitrile copolymer in a blend (above 50%), the fraction of mesomorphic structures in a film increases abruptly, whereas for a nonwoven material, the value of α remains practically the same, regardless of the type of solvent.

Thus, the dependences of α on blend composition for the films and fibers follow different patterns, but all of them have breaks corresponding to the 50% content of PU in the blend. This outcome apparently may be explained by phase inversion. In our opinion, the different patterns of the dependences of α on blend composition for the films and nonwoven

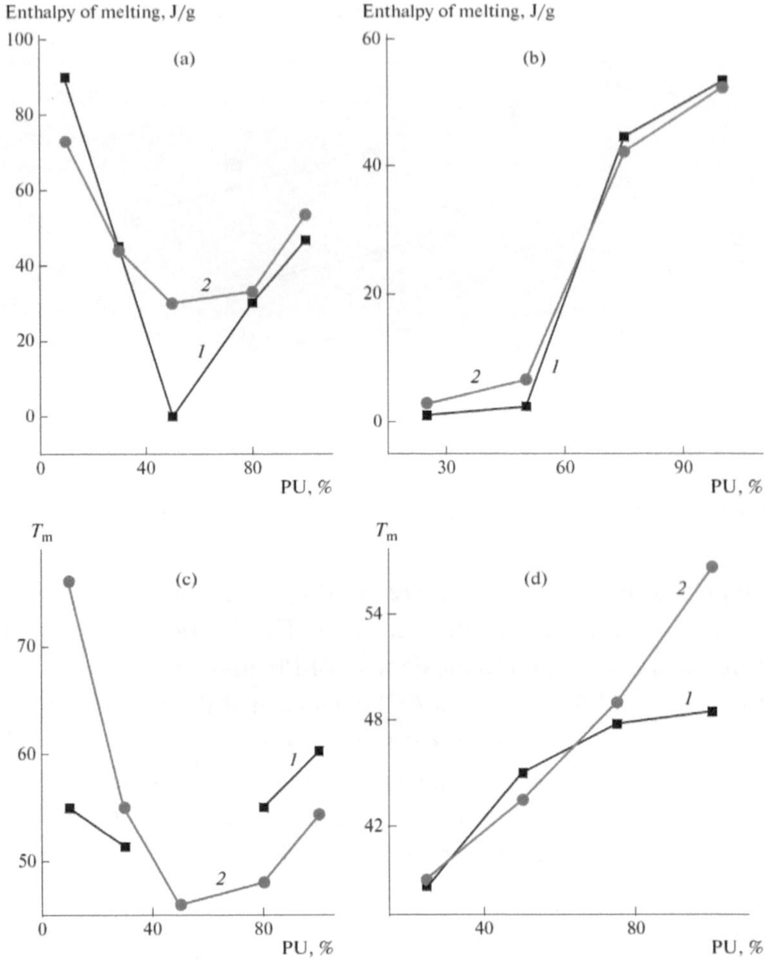

FIGURE 4.2 Plots of (a, b) the enthalpy of melting and (c, d) the temperature of melting of mesomorphic structures vs. blend composition for (a, b) films and (c, d) nonwoven materials: (*1*) THF and (*2*) acetone.

TABLE 4.1 Effects of Polymer (wt%) and Solvent Type on T_g and T_m (°C), as Estimated Via DSC, for Films

Solvent	PU		Mass ratio PU to styrene – acrylonitrile copolymer								Copolymer
			80:20		50:50		30:70		10:90		
	T_m	T_g	T_m	T_g	T_m	T_g	T_m	T_g	T_m	T_g	T_g
THF	60.3	−35	57	−30	–	50	51.4	–	55	−30	80

TABLE 4.1 Continued

Solvent	PU		Mass ratio PU to styrene – acrylonitrile copolymer								Copolymer
			80:20		50:50		30:70		10:90		
	T_m	T_g	T_m	T_g	T_m	T_g	T_m	T_g	T_m	T_g	T_g
Ethyl acetate	54.0	–	53	–20	48	–	65.0	56	80	77	85
Acetone	54.0	–40	48	–	56	–	55.0	–	76	7	115

The data on the firm material were published in [11].

TABLE 4.2 Effect of the Polymer Ratio (wt %) and Solvent Type on T_g and T_m (°C), as Estimated Via DSC, for Fibers

Solvent	PU		Mass ratio PU to styrene – acrylonitrile copolymer						Copolymer
			75:25		50:50		25:25		
	T_m	T_g	T_m	T_g	T_m	T_g	T_m	T_g	T_g
THF	48.5	–40	48	–10	42.0	95 and –35	39	55	110
Acetone	56.6	–38	49	–	43.5	–	41	100	105

materials are associated with different types of structural organization of the blends.

In addition, the characters of change in melting temperature T_m with the composition of the blend in the films and nonwoven materials for the studied polymers are different. If the melting temperatures for the films and the nonwoven materials decrease with an increase in the fraction of the styrene–acrylonitrile copolymer in the blend (up to 50% in the composite), then, at its higher content, the melting temperatures for the films increase; in contrast, for the nonwoven materials, the melting temperatures decrease more abruptly. It is important that, for both films and nonwoven materials, the character of change in the melting temperature does not depend on the type of the solvent. In addition, these relationships provide evidence that phase inversion occurs at a PU-to-styrene–acrylonitrile mass ratio of 50:50 for both films and nonwoven materials. The values of T_m and α change in a symbate manner. For example, after addition of a small amount of the styrene–acrylonitrile copolymer to the blend, defects in the mesomorphic structures of PU appear in the films; as a result, the values of α and *T*m decrease. As to the fibers, α remains practically the same in most cases (the exception being the 50:50 blend). At this composition of the blend, phase inversion occurs and a marked entanglement of chains hampers the formation of mesomorphic structures. Therefore, practically no mesomorphic structures appear in the films formed in THF, while in the case of fibers, the fraction of mesomorphic structures is negligibly small. In blend composites, when the styrene–acrylonitrile copolymer becomes the continuous phase, more and more perfect mesomorphic structures of PU form and, accordingly, the values of α and *T*m increase. For a nonwoven material (probably because of the high rate of spinning), the structure of the polymer has no time to transition to the equilibrium state. It is important that, for the films, the values of *T*m are higher than those for the non-woven material. For example, if for the films based on the 30:70 blend of PU and the styrene–acrylonitrile copolymer, *T*m = 51.4°C, then for the nonwoven material, *T*m = 39°C. For the PU-based films, the melting temperature is 60.3°C, and for the nonwoven material, the melting temperature is 48.5°C. (The data on other blends prepared in THF are similar.) This result additionally suggests that the mesomorphic structures formed in the nonwoven material are farther from perfect than those in the film samples.

Note that, for the films based on the styrene–acrylonitrile copolymer, the glass-transition temperature is 80°C, while for the nonwoven material, this value is 110°C. For the films of the blends, no glass transition of the styrene–acrylonitrile copolymer occurs in the studied temperature range (from –90 to +140°C) except that in the 50:50 blend of PU and the styrene–acrylonitrile copolymer, whereas for the nonwoven material, the glass transition of the copolymer is observed at PU-to-styrene–acrylonitrile mass ratios of 25:75 and 50:50. This outcome testifies that the amorphous phase in this material is more rigid. The absence of the glass transition in the blend composites other than those mentioned above indicates that PU exerts a elasticizing effect on the structures of the blends and that the compatibility of these polymers is high.

Thus, casting of the films from solutions is a slow process; the system occurs in the equilibrium state; and the addition of the styrene–acrylonitrile copolymer (in small amounts) to PU decreases the fraction of mesomorphic structures of PU owing to the interaction between PU and copolymer molecules, as was shown in [5]. At a higher content of the styrene–acrylonitrile copolymer (above 50%), when PU is a discrete phase, disseminations of PU with a dense surface of straightened chains form. The fraction of the mesomorphic structures of PU in a film material increases at a high content of the styrene–acrylonitrile copolymer because of a lower entanglement of chains in the discrete phase than that in the continuous phase. It appears that, in the fibers, the structures of the samples have no time to transition to the equilibrium state, because of the high rate of formation. Therefore, an extremely small fraction of mesomorphic structures of PU form in the blend composites of a nonwoven material that contains a high amount of PU (Figure 4.2a).

The complex character of change in probe mobility was considered in terms of the model of the binary distribution of segments in less and more dense inter-crystalline regions responsible for faster and slower rotations of the TEMPO radical. The study of molecular dynamics in the films and nonwoven materials based on blends of PU and the styrene–acrylonitrile copolymer revealed heterogeneity of the amorphous phase except in the case of PU (Figure 4.3). Heterogeneity of the amorphous phase indicates that it contains structures with low *(I)* and high *(I)* mobileties. In what follows, correlation times were calculated for the fast component solely.

The study of the type of solvent showed that the solvent exerts a strong effect on both the time of rotational mobility of probes, τ, and, hence, the molecular mobility of polymer chains (Figure 4.4). As is seen from Figure 4.4a, the highest mobility of the probe is observed for the films formed in THF, while the lowest mobility of the probe is detected for the films formed in acetone. In the nonwoven material, the solvent has no marked effect on molecular mobility (Figure 4.4b). The addition of the styrene–acrylonitrile copolymer to the polymer matrix entails an increase in τ. In the case of the films, as the amount of the copolymer in the composite is increased to 50%, the correlation time changes insignificantly, and only at a higher content of the polymer, when the styrene–acrylonitrile copolymer is a continuous phase, does τ increase abruptly. This circumstance suggests an increase in the rigidity of amorphous regions. The observed pattern of the dependence of τ on the blend composition in the films may be explained by the structure of interfacial regions (Figure 4.5).

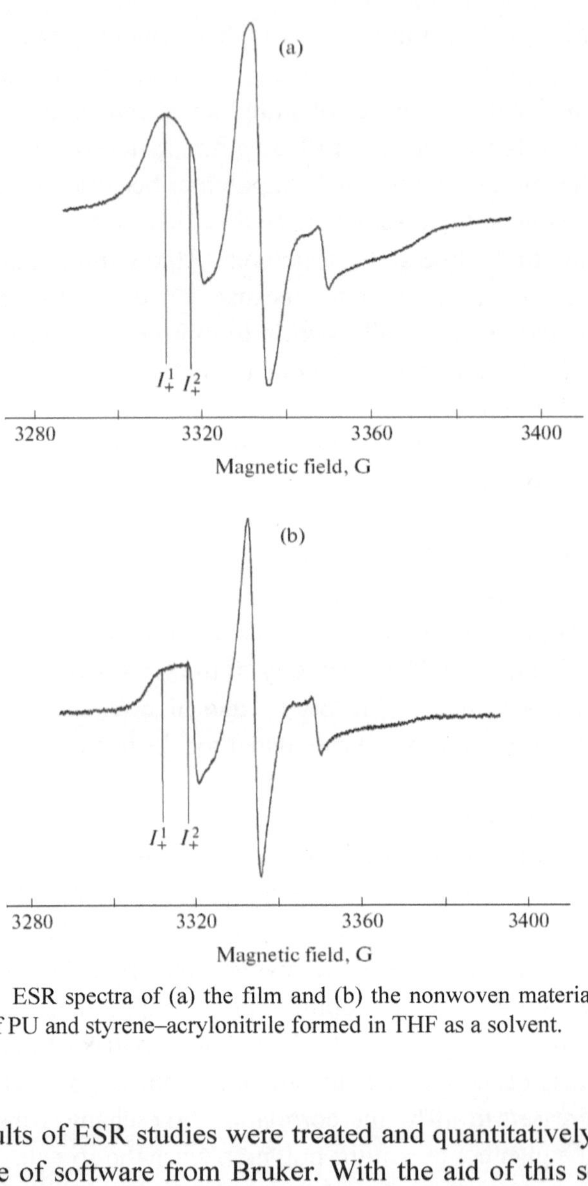

FIGURE 4.3 ESR spectra of (a) the film and (b) the nonwoven material based on the 50:50 blend of PU and styrene–acrylonitrile formed in THF as a solvent.

The results of ESR studies were treated and quantitatively interpreted with the use of software from Bruker. With the aid of this software, the molecular mobilities in the interfacial regions of the composites were calculated. As is clear from Figure 4.5a, in the case of the films, the densest interfacial layers form in the blends containing a small amount of the copolymer. The permeability of the radical across these layers is apparently hindered; therefore, the values of τ change slightly after the addition

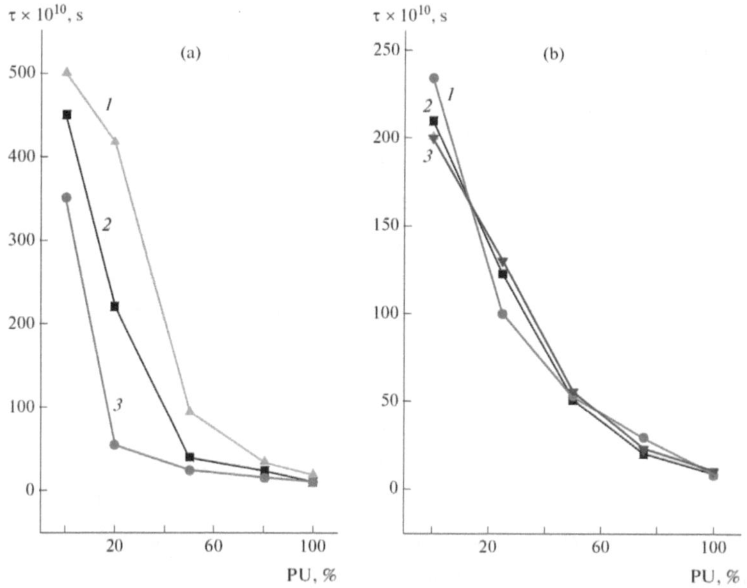

FIGURE 4.4 Changes in τ in (*1*) acetone, (*2*) ethyl acetate, and (*3*) THF in (a) films and (b) nonwoven materials of the blend composition.

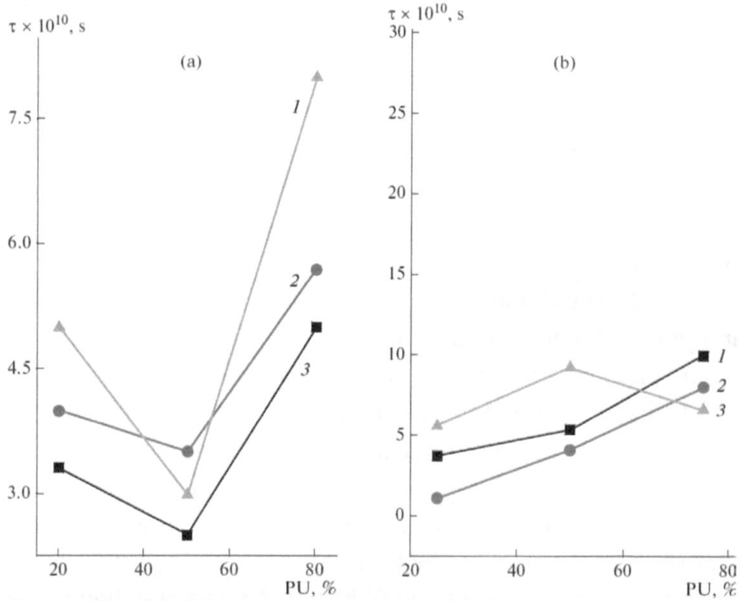

FIGURE 4.5 Changes in τ in interfacial layers for (a) films and (b) nonwoven materials in (*1*) acetone, (*2*) ethyl acetate, and (*3*) THF vs. blend composition.

of the styrene–acrylonitrile copolymer (to 50%). In the blends with high amounts of the copolymer, the interfacial interlayer has a lower τ value, a result that suggests a higher looseness and, as a consequence, a higher permeability to the radical. The interfacial interlayer in the fibers, regardless of composition, has a small value of τ, a circumstance that suggests a high permeability of the interlayer to the radical, and manifests itself on the dependences of τ on composition (Figure 4.4b). For both the films and fibers, the dependences of τ on composition feature a break at a PU-to-copolymer mass ratio of 50:50. This result confirms the above finding that phase inversion occurs at this ratio of the copolymers.

Note that correlation times τ for PU samples formed in different solvents differ insignificantly for both the films and the nonwoven material. At the same time, for the films based on the styrene–acrylonitrile copolymer, these differences are considerable. Thus, the solvent changes the structure of the amorphous regions of the styrene–acrylonitrile copolymer to a much higher extent than that of PU.

As is known [5], for this blend composite, the solvents may be arranged in terms of their decreasing thermodynamic quality as follows: THF, ethyl acetate, acetone. In poor solvents, the degree of interaction between the polymers is much higher than that in good solvents and solvents of different qualities create different polymer structures in solution that are preserved after the removal of solvents. The difference in the interactions of the polymers in a solution leads to a difference in the properties of the films and nonwoven fibrous materials prepared from a solution, because structures existing in a solution are partially preserved in these films and nonwoven fibers.

From the thermodynamic viewpoint, the only difference between good and poor solvents is that they interact with the polymer in a different manner; therefore, interactions between macromolecules are likewise different. A higher convolution of macromolecular coils in a poor solvent facilitates an increase in the number of contacts not only between similar macromolecules but also between dissimilar macromolecules [1, 9]. Hence, the solvent has a strong effect on the structures and molecular mobilities of the polymers. In the best solvent, THF, the structures of the polymer composites are close to equilibrium. As poorer solvents are used, the state of the structures becomes more and more nonequilibrium; the density of chains increases; and, as a result, the rigidity of macromolecules increases in the sequence of the polymers formed in THF, ethyl acetate, and acetone (Figure 4.4a). In the case of fibers, the influence of the solvent on the amorphous structure of the polymer makes itself evident much weaker (Figure 4.4b).

The fraction of macromolecules that form dense amorphous regions in the composite films increases (an estimate was performed on the basis of *I* /*I*, Figure 4.5) on passage from good solvent to poor solvent, as evidenced by the data shown in Figure 4.6a. An increase in the fraction of PU in the blend leads to a reduction in the fraction of regions with hindered motion, and there is only a single spectrum for the blends containing a small amount of the styrene–acrylonitrile copolymer (less than 50%). For the fibers, these differences are less distinct (Figure 4.6b); however, the double spectrum is additionally observed for the composites with high contents of PU. Note also that the break on the above dependences is likewise observed at a PU-to styrene–acrylonitrile mass ratio of 50:50 (Figure 4.6b).

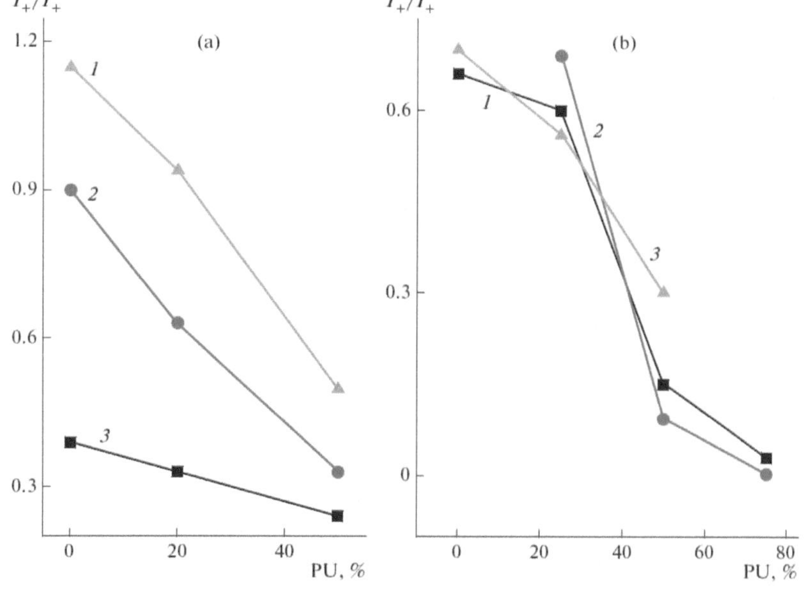

FIGURE 4.6 Plots of *I* /*I* vs. blend composition in (*1*) acetone, (*2*) ethyl acetate, and (*3*) THF for (a) films and (b) nonwoven materials.

The solvent determines the concentrations of the radical in the studied polymers. An increase in the density of the amorphous phase in PU (as well as in the composites) formed in various solvents (Figure 4.7) is attested by a decrease in the concentration of the radical on passage from the good solvent to a poor solvent for both blend films and fibers. However, if the concentration of the radical in composites formed, for example, in THF is analyzed, changes in the films may be explained by both increases in the densities of the films on passage from PU to the styrene–acrylonitrile copolymer and decreases in the fractions of the amorphous phase accessible to the radical; therefore, the concentrations of the radical in the samples decline (Figure 4.7a). Another character of the dependences is observed for the PU–copolymer fibers (Figure 4.7b). The composition of the fiber does not exert such a marked effect on the concentration of the radical as that in the films. Only at a PU-to-copolymer mass ratio of 50:50 is there a jump in the concentration of the radical on its composition dependences. This outcome suggests that the packing of chains in these fibers is the loosest. It is important

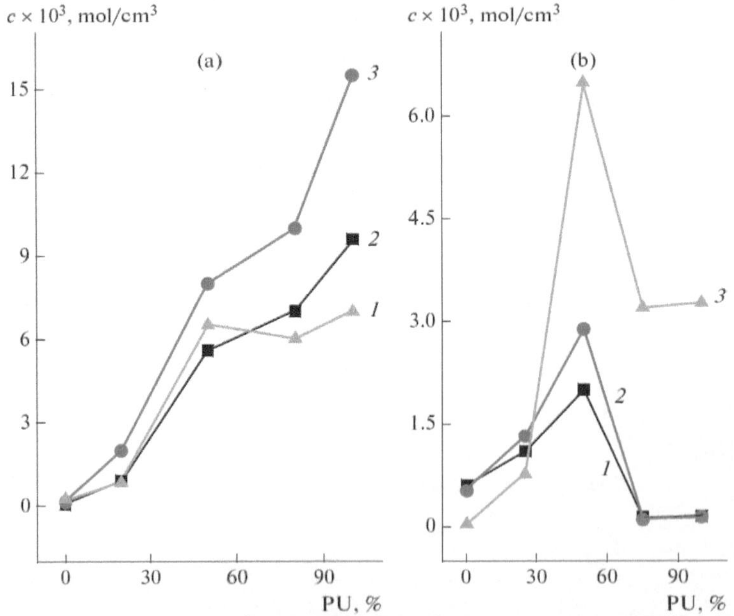

FIGURE 4.7 Variations in the concentration of the radical in (*1*) acetone, (*2*) ethyl acetate, and (*3*) THF in (a) films and (b) nonwoven materials of the blend composition.

that there is an extremum point at this ratio, and this circumstance confirms the finding that phase inversion occurs at this ratio of the polymers.

In addition, the temperature dependences of correlation times for blend films and fibers were investigated. It was found that there is a break at temperatures of 45–55°C that is apparently related to the unfreezing of mesomorphic structures (Figure 4.2). The linear dependence is observed only for films based on the 50:50 PU-to-copolymer blend, and exactly in these composites there are no mesomorphic structures. (In the fibers, the fraction of these structures is negligibly small.)

The variation in activation energy with the composition of the polymer blend depends on the type of solvent and the method of preparing polymeric materials (Figure 4.8). Calculation of the activation energies for the rotational mobility of the probe in the studied polymers showed that, for the PU films, the effect of a solvent on *E*a is small and the addition of the styrene-acrylonitrile copolymer (to 50%) entails a decrease in the activation energy, a result that is probably related to a decrease in the density of the composites (Figure 4.8a). At a higher content of the copolymer in the

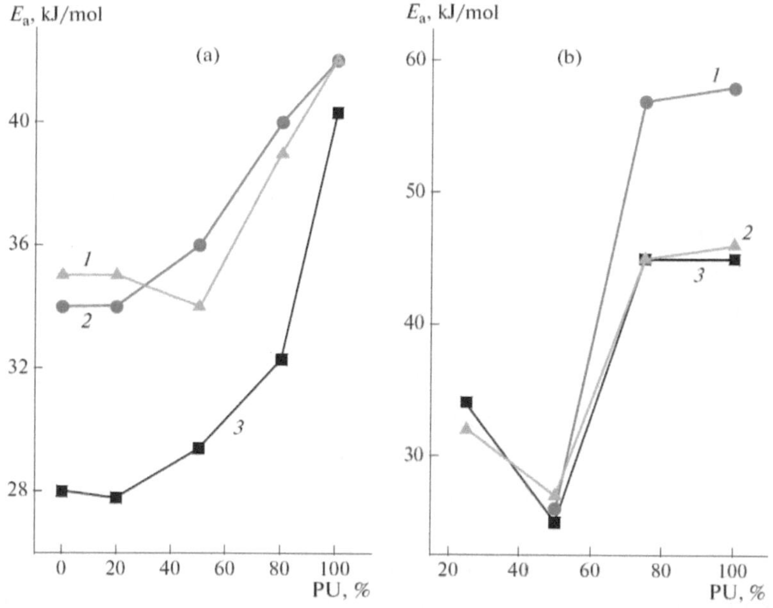

FIGURE 4.8 Plots of *E*a vs. blend composition in (*1*) acetone, (*2*) ethyl acetate, and (*3*) THF in (a) films and (b) nonwoven materials.

blend, changes in *E*a are not so pronounced, regardless of the solvent type. Note that the activation energy increases from the good solvent to a poor solvent.

In the case of fibers, a change in the activation energy follows other regularities (Figure 4.8b). Like parameter α (Figure 4.2b), the values of *E*a for the blend (up to 50% PU) are practically constant. At a 50% content of PU, the value of *E*a decreases abruptly, and in the composites with higher contents of the styrene–acrylonitrile copolymer, the activation energies increase slightly. It should be emphasized that the values of *E*a for fibrous polymers formed in various solvents other than acetone are similar, whereas for the films, these values differ appreciably (Figure 4.8a). It is vital to note that, for all blends except the 50:50 blend, there is a break on the temperature dependences of correlation time in the range 45–55°C. Because, at the same temperatures, there is the peak of melting of the mesomorphic structures, the break on the temperature dependences of τ may be explained by the unfreezing of such structures. A gain in *E*a at temperatures above 50°C may be associated with a higher activation energy precisely in mesomorphic structures.

The effect of the oxidizer ozone on the amorphous phase of these polymers was investigated with the use of the microprobe method. For example, regardless of the method of preparing composites, polymers containing high amounts of the styrene–acrylonitrile copolymer are subject to oxidation. PU and the blends with small amounts of the copolymer are more resistant to the action of the aggressive medium of ozone (Figure 4.9).

The correlation times change insignificantly for PU and the 80:20 blend of PU and the styrene–acrylonitrile copolymer during ozonation of the films and nonwoven materials formed in THF. For the composites containing high amounts of the styrene–acrylonitrile copolymer, the values of τ increase at low ozonation times (to 2 h) for both the films and the nonwoven materials. Further ozone oxidation leads to decreases in τ, and these changes are the most pronounced for the 25:75 PU–copolymer blend and the 30:70 PU–copolymer film. It is known that chemical and physical crosslinking of macromolecules and degradation of chains occur during oxidation. Chemical crosslinking implies the formation of new entanglements as a result of branches that appear during ozone oxidation. At the initial step of oxidation, crosslinking processes prevail; therefore, the molecular mobility decelerates and, as a consequence, τ increases.

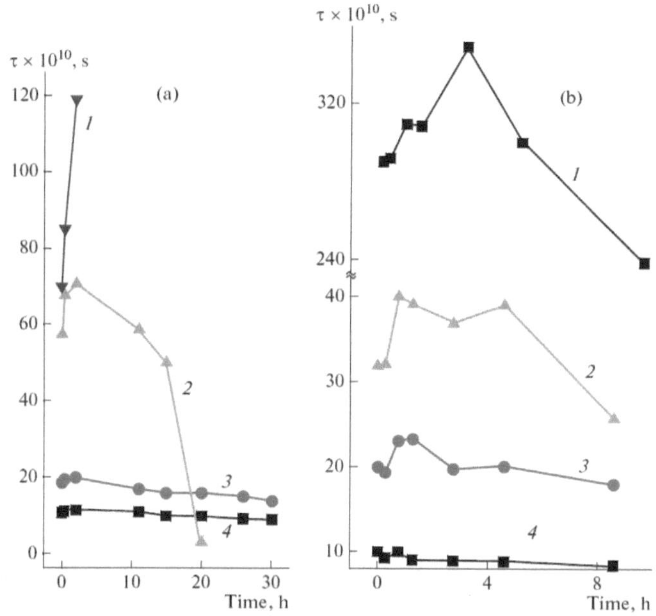

FIGURE 4.9 Variation in τ with the time of ozonation for (a) films and (b) nonwoven materials. (a) PU-to-copolymer mass ratios of (*1*).

For a longer effect of ozone, the processes of chain degradation begin to prevail and this tendency manifests itself as a decrease in τ. Hence, it may be inferred that the styrene–acrylonitrile co-polymer is the most prone to ozone oxidation. Similar dependences were obtained for the samples formed in acetone and ethyl acetate.

A comparison of the properties of the films and nonwoven materials made it possible to analyze change in molecular mobility of polymer molecules at the early stages of their interaction with oxidative aggressive media.

In addition, a change in ratio *I*/*I* with an increase in the time of ozonation for the fibers formed in THF was studied. As is seen from Figure 4.10, at the initial step of ozonation (before 2 h), this ratio first increases and then decreases. An increase in *I* /*I* is evidence that the fraction of dense amorphous regions increases apparently owing to the physical crosslinking of macromolecules during ozonation; a decrease in this parameter suggests that the destruction of these regions occurs. Correlation time and

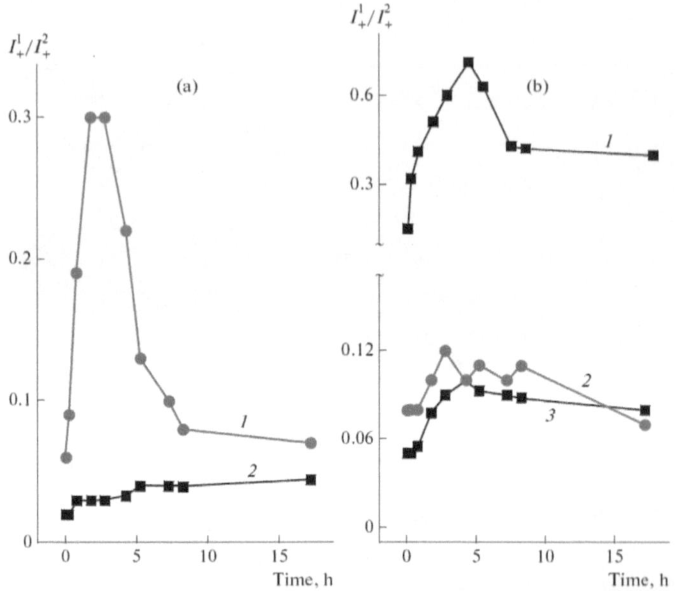

FIGURE 4.10 Plot of I^{1+}/I^{2-} vs. time of ozonation for (a) films and (b) nonwoven materials. (a) PU-to-copolymer mass ratios of (*1*) 30:70 and (*2*) 50:50; (b) PU-to-copolymer mass ratios of (*1*) 25:75, (*2*) 50:50, and (*3*) 75:25.

I/*I* change in a symbate manner. This fact confirms the finding that cross-linking between chains prevails at the initial stage of ozone oxidation.

Thus, it has been shown that differences in the structures of the resulting materials are primarily related to differences in the thermodynamic qualities of the solvents. The type of solvent has the most pronounced effect on the structures and molecular dynamics in the case of the styrene–acrylonitrile copolymer and the blend composites containing high amounts of the copolymer. In this case, mesomorphic structures form in both films and nonwoven materials. At a 50:50 mass ratio of PU and the styrene–acrylonitrile copolymer, there are breaks on the dependences of τ, *Ea*, the concentration of the radical, and the fraction of mesomorphic structures on the blend composition for both films and nonwoven materials as a result of phase inversion. With the use of the spin-probe method, it has been found that the oxidizer ozone affects amorphous phases in the films and nonwoven materials. It has been shown that, regardless of the method of blend formation, the polymers with high amounts of the styrene–acrylonitrile copolymer are subject to oxidation. PU and the blends with low amounts of the copolymer are more resistant to the aggressive medium of ozone. If, for a film material, the type of solvent shows a strong effect on the molecular mobility in samples, then, in a nonwoven material, this effect is much less distinct. In addition, it is shown that, if the concentrations of the radical in the films decrease by almost an order of magnitude on passage from PU to the styrene–acrylonitrile copolymer, then, in a nonwoven material, this change is insignificant.

KEYWORDS

- **molecular dynamics**
- **nonwoven materials**
- **polymers**
- **polyurethane**
- **styrene–acrylonitrile**
- **thermo-physical**

REFERENCES

1. Kuleznev, V. N., *Blends and Melts of Polymers* (Scientific bases and Technologies, St. Petersburg, 2013) [in Russian].
2. Miroshnikov, Yu. P., Letuchyii, M. A., Lemstra, P. J., Govaert-Spoelstra, A., Engelen, Y. M. T., Polym. Sci., Ser. A 42 (7), 795 (2000).
3. Papkov, S. P., *Physicochemical Foundations of Processing of Polymer Solutions* (Chemistry, Moscow, 1971) [in Russian].
4. Pocius, A., *Adhesion and Adhesives Technology* (Hanser, Munchen, 2012).
5. Drinberg, S. A., Itsko, E. F., *Solvents for Paint Technology: Handbook* (Chemistry, St. Petersburg, 2003) [in Russian].
6. Stoyer, D., Feitag, W., *Paints, Coating, and Solvents* (Wiley-VCH Verlag CmbH&Co, Wienheim, 1998).
7. Tager, A. A., *Physics and Chemistry of Polymers* (Scientific world, Moscow, 2007) [in Russian].
8. Krokhina, L. S., Kuleznev, V. N., Lyusova, L. R., and Glagolev, V. A., Macromolecular compounds, Ser. A 18 (3), 663 (1976).
9. Kuleznev, V. N., Wolf, B., and Pozharnova, N. A., Polym. Sci., Ser. B 44 (3–4), 67 (2002).

10. Boksha, M. Yu. PhD thesis of Technical Sciences (Lomonosov Moscow State University of Fine Chemical Technologies, Moscow, 2010).
11. Karpova, S. G., Naumova, Yu. A., Lukanina, Yu. K., Lyusova, L. P., Popov, A. A., Polym. Sci., Ser. A 56 (4), 472 (2014).
12. Filatov, Y., Budyka, A., and Kirichenko, V., *Electrospinning of Micro- and Nanofibers: Fundamentals in Separation and Filtration Processes* (Begell House Inc., New York, 2007).
13. Kapustin, I. A., Filatov, I. Y., and Filatov, Y. N., Fiber Chem. 44 (5), 299 (2013).
14. Sokolov, V. V., Kildeeva, N. R., I. Yu. Filatov, and Yu. Filatov, Fiber Chem. 44, 1 (2013).
15. Ol'khov, A. A., Staroverova, O. V., Filatov, Yu. N., G. M. Kuz'micheva, Iordanskii, A. L., Stoyanov, O. V., Zaikov, G. E., Zenitova, L. A., Vestn. Kazanskogo Tekhnol. Un-Ta 16 (8), 157 (2013).
16. Filatov, Yu. N., I. Yu. Filatov, and M. Yu. Nebratenko, RF Patent No. 2357785, Byull. Izobret., No. 16 (2009).
17. Lyusov, Yu. N., Rubber, No. 19, 37 (1985).
18. Shtil'man, M. I. *Polymers for Medical and Biological Purposes* (Akademkniga, Moscow, 2006) [in Russian].
19. Prisacariu, C., TPE Magazine 4, 244 (2012).
20. Kanapitsas, A., Pissis, P., and J. L. Gomez Ribelles, J. Appl. Polym. Sci. 71 (8), 1209 (1999).
21. Karpova, S. G., Iordanskii, A. L., and Klenina, N. S., Russ. J. Phys. Chem. B 7 (3), 225 (2013).
22. Karpova, S. G., Iordanskii, A. L., and Chvalun, S. N., Dokl. Phys. Chem. 446 (2), 176 (2012).
23. Ratzsch, M., Haudel, G., Pompe, G., and Meyer, E., J. Macromol. Sci., Part A: Pure Appl. Chem. 27 (13), 1631 (1990).
24. Zerjal, B., Musil, V., Smit, I., and Jeli, Z., Malavasic, T., J. Appl. Polym. Sci. 50 (4), 719 (1993).
25. Erjal, B., Jeli, Z., and Malavaši, T., Eur. Polym. J. 32 (11), 1351.
26. Kanapitsas, A., Pissis, P., and A. Garcia Estrella, Eur. Polym. J. 35 (5), 923 (1999).
27. Ulcnik, M., Zerjal, B., and Malavasic, T., Thermochim. Acta, No. 276, 175 (1996).
28. Vasserman, A. M., Kovarskii, A. L., *Spin Labels and Probes in Physical Chemistry of Polymers* (Science, Moscow, 1986) [in Russian].
29. Marikhin, V. A., Myasnikova, L. P., *Supramolecular Structure of Polymers* (Chemistry, Leningrad, 1977) [in Russian].

CHAPTER 5

INVESTIGATION OF POLYPROPYLENE/LOW-DENSITY POLYETHYLENE BLENDS

E. E. MASTALYGINA,[1] N. N. KOLESNIKOVA,[2] S. G. KARPOVA,[2] and A. A. POPOV[1]

[1]*Laboratory of Advanced Composite Materials and Technologies, Department of Chemistry and Physics, Plekhanov Russian University of Economics, Russia; E-mail: elena.mastalygina@gmail.com, popov@sky.chph.ras.ru*

[2]*Laboratory of Physical Chemistry of Synthetical and Natural Polymers, Emanuel Institute of Biochemical Physics of Russian Academy of Sciences, Russia; E-mail: kolesnikova@sky.chph.ras.ru, karpova@sky.chph.ras.ru*

CONTENTS

ABSTRACT

Thermal and morphological study of blends based on isotactic polypropylene (iPP) and low-density polyethylene (LDPE) in a wide range of compositions were investigated by differential scanning calorimetry and electron paramagnetic resonance spectroscopy (paramagnetic probe method). The partial compatibility in the amorphous regions of iPP and LDPE providing the interface layer formation was observed for the blends containing 30–95 wt% of iPP. There was a plasticizing effect of LDPE on iPP, increasing the segmental mobility of the macromolecules chains in its amorphous phase. If the content of iPP in the blend was less than 30 wt%, the non-equilibrium molecular structure of the iPP/LDPE composition with a more rigid interface layer was formed.

5.1 INTRODUCTION

Polyolefins are the most important and most widely used synthetic polymers; their annual production exceeds 130 million metric tons [1]. An enormous scale of industrial production and a wide variety of application of polyolefins, particularly polyethylene and polypropylene, cause the importance of developing new materials based on polyolefins. The combination of polyolefins by blending allows varying their properties and producing materials having the appropriated characteristics, avoiding an expensive stage of synthesis [2].

The range of polyolefins can be significantly extended by compounding polyethylene (PE) and polypropylene (PP) for a variety of ratios. The addition of small amounts of PE to PP increases its frost resistance [3], impact strength [4] and oxidation resistance [5]. The addition of PP to PE, in its turn, allows increasing mechanical strength and rigidity of the composition [6, 7], as well as resistance to high temperatures [8].

Polyethylene and polypropylene are commonly used in industries such as agriculture, construction, and domestic uses [9, 10], so both polymers could bring possible environmental problem after their usages. They are very stable and hence remain inert to degradation and deterioration leading to their accumulation in the environment.

The recycling of plastic wastes could partly solve environmental problems. Plastic recovery starts with the separating of the different types of plastics [11]. The structural similarity and close density values of PP and PE cause the technical difficulties and economic irrationality of separating these polymers in the recycling process [12]. So the resulting product of plastics separation contains both PE and PP and can be used only for low quality recycled materials [13]. The PP/PE blend developing expands the opportunities for improving properties of recycled polyolefin mix. Thus, the recycling is a stimulating factor for creation of PP/PE blends.

The second direction in solving the environmental problems of polyolefin wastes is the production of composite materials with rapid degradation in natural conditions [14–16]. It is considered that biodegradation is more suitable route for waste disposal rather than recycling [17]. There are works devoted to creation of the biodegradable compositions based on PP/PE blends with natural biodegradable fillers such as wood flour [18], flax fibers [19], rape straw [20] and others. So, creating of biodegradable composites based on polypropylene/polyethylene blend matrix is also a currently important research trend.

Creation of materials with definite properties based on PP/PE blends, including biocomposites with natural filler, requires studying morphology PP/PE blend matrix. Morphology of polymer blends depends on many factors, including the molecular structure, the melt flow index, the ratio of components in the blend, and the presence of additives improving compatibility [21]. Generally, the creation of compositions of these polymers implies the formation of the dispersed system, in which a smaller phase is dispersed in the matrix of a bigger phase. The size of dispersed domains of the smaller phase increases with addition of the dispersed polymer, as well as with deterioration of blending, more continuous annealing or slower cooling of the material [21]. Phase inversion occurs in a wide range of polymer concentrations, approximately 30–70 wt%, due to the high viscosity of PP/PE blends and a practical impossibility of achieving the phase equilibrium [22]. Both polymer phases are continuous in this concentration range.

A large number of studies were devoted to the research of PP/PE blends. Nevertheless, there are still discussions on the formation of the structure and compatibility of these polymers in the blend. First of all, it is connected with the fact that the morphology of the blend strongly depends on the type of PP and PE which are compounded. According to market studies, three types of PE that conquer the plastic market in sold volume are high density polyethylene (HDPE) and low density polyethylene (LDPE) [9]. Among polypropylenes the isotactic polypropylene (iPP) is the most widespread type in domestic uses. But most studies available in literature are devoted to the research of the blends of iPP with high-density polyethylene (HDPE) or linear low-density polyethylene (LLDPE).

Some researchers believe that, despite the incompatibility of PP and PE, both of these polymers can influence the processes of crystallization and the structure formation of the other component. The study [23] has shown the decrease in crystallinity and heat of fusion upon the incorporation of PP into HDPE is due to the fact that the formation of crystallites in the blend was affected by the presence of PP. Some researchers have proven not only the mutual influence of PP and PE on the structure and properties of the blend, but also the partial compatibility of these polymers at particular ratios of the components. Galeski et al. reported in their study [24] about the partial compatibility of iPP and HDPE in the melt for the blends with the content of iPP less than 10 wt% and more than 60 wt%. Moreover, they argued that there was a mutual influence of iPP and HDPE on the composition structure at these ratios of the components. Investigating the nucleation process in iPP/HDPE blends the authors determined that the heterogeneous nucleation was characteristic for the blend compositions with a low content of iPP. Jose et al. [7] stated that the crystallization temperatures of polymers do not depend on the composition of the blend, while the degree of crystallinity of both polymers decreases with the increase of the amount of the other component, which also indicates the mutual influence of the blend components on each other.

In the studies, devoted to the blends of LLDPE and iPP, the researchers also pointed to the partial compatibility of LLDPE and iPP. It was attributed to the structural similarity between the blend components, particularly the low degree of branching. Jun Li and Shanks [25] found the reduction in the growth rate of spherulites during the process of iPP crystallization for the blend of iPP/LLDPE = 20/80. This result was explained by the decrease in the supercooling temperature of iPP indicating the partial solubility of iPP in LLDPE.

LDPE is used as widely as HDPE and LLDPE; however, iPP/LDPE blend compositions have given less attention by researchers. In addition, most studies investigate only three or four iPP/LDPE blends, and the results of some studies differ with each other, which makes it difficult to obtain complete information. For example, Ujhelyiová et al. [26] reported the partial iPP and LDPE compatibility at the temperature above the melting point of the components and at low LDPE content (less 5–10 wt%). Moreover, the authors of the study [26] judged about it by reduction of the total melting enthalpy of the blend. Dong et al. [27] showed the different range of compositions characterized by iPP solubility in LDPE – less 15 wt% of iPP in the blend. Besides, the structural changes in the iPP/LDPE blends at low content of iPP were explained not only by iPP solubility in LDPE, but also by the fact that the concentration of iPP was too small for its proper crystallization. Murín et al. [28] and Gorrasi G. et al. [29] by contrast, argued in favor of the complete incompatibility of iPP and LDPE; however, they pointed to the possibility of the mutual influence of these polymers on the composition properties.

To date, PP/PE blends are already widely used in industry for manufacturing ropes, nets, packaging materials, as well as engineering plastics [8, 30, 31]. Determination of the effects of the iPP/LDPE blend composition on the structure and properties will allow creating composite materials with required characteristics. This makes it possible to predict the behavior of these materials during the operation and to define possible directions of their use, which, in turn, will expand fields of application of material based on polyolefin blends.

5.2 EXPERIMENTAL PART

5.2.1 MATERIALS

The blends under investigation consisted of isotactic polypropylene (iPP, TM 01030 Caplen from Gazpromneft-MNPZ, OJSC, Russia) and low-density polyethylene (LDPE, TM 15803–020 from Neftekhim-sevilen, OJSC, Russia) with melt flow index of 1.2 ± 0.1 g/10 min and 1.65 ± 0.1 g/10 min (at 190°C and 2.16 kg load), respectively, were used in this study. Close values of melt flow indexes provided the possibility of qualitative compounding of the polymers with the formation of the homogeneous mixture [2]. The average molecular weight and molecular-weight distribution for iPP were $M_w = 2.1 \times 10^5$, $M_n = 4.6 \times 10^4$, $M_w/M_n = 4.6$ and for LDPE were $M_w = 1.0 \times 10^5$, $M_n = 1.5 \times 10^4$, $M_w/M_n = 6.7$ (1,2,4-trichlorobenzene, 140°C, water 150°C GPC). The iPP content in the blends was 0, 5, 10, 20, 30, 40, 50, 60, 70, 80, 90, 95, 100 wt%. A wide range of compositions allowed to determine accurately the changes in the properties in the case of varying the blend composition.

5.2.2 PROCESSING

All the blends used for this study were prepared by the Plasticorder PLD-651 (Brabender, GmbH & Co, Germany) in an argon atmosphere (State Standard GOST 10157–79) at a temperature of 190°C and a rotor rotational speed of 30 rev min^{-1}. A weighted amount of iPP was placed in the mixing chamber, 2 minutes later LDPE sample weight was added. Mixing of polymers was performed for 5 min. The inert argon atmosphere was used to reduce the oxidation of polymers. After cooling, the obtained material was ground in the knife mill RM-120 (Vibrotechnik, LLC, Russia). The isotropic films were obtained by pressing the ground material in the hydraulic hand press PRG-10 (VNIR, LLC, Russia) at a temperature of 190°C and a pressure of 7.8 MPa (80 kgf cm^{-2}) on a cellophane substrate, followed by quenching in water at 20°C. The thickness of the films was 130 ± 10 μm.

5.2.3 METHODS

5.2.3.1 Differential Scanning Calorimetry

The behavior of polymers during melting and crystallization was investigated by differential scanning calorimetry (DSC) using a differential scanning microcalorimeter DSM-10M (Institute for Biological Instrumentation,

Russia). The scanning speed was 8°C min^{-1}, sample weight was 10 ± 0.1 mg. The temperature scale was scaled by indium (melting temperature T_m = 156.6°C, specific heat of fusion ΔH = 28.44 J g^{-1}).

The temperatures of melting and crystallization, T_m and T_{cr}, were determined by the endothermic maximum of the melting peak and the exothermic maximum of the crystallization peak on DSC thermograms, respectively. To obtain cooling curves, the samples of iPP/LDPE blends were heated up to 200°C, stored at this temperature for 5 min, then cooled at a speed of 8°C min^{-1} to 40°C. The enthalpy of fusion of the samples (ΔH_i) was calculated from the area of the melting peak limited by the base-line. To calculate the degree of crystallinity χ(PP) and χ(PE), the Eq. (1) was used [32]:

$$\chi = \frac{\Delta H_i}{\Delta H_0} \times 100 \qquad (1)$$

where ΔH_i is the specific heat of melting calculated per the content of polymer *i* (iPP or LDPE) in the blend; ΔH_o(PP) = 147 J g^{-1} – specific heat of melting of a completely crystalline PP [33], ΔH_o(PE) = 295 J g^{-1} – specific heat of melting of a completely crystalline PE [34]. Each value of the parameters ΔH_i, T_m, T_c was obtained by averaging of five measurements.

5.2.3.2 Hydrostatic Weighing Method

The density of the samples was measured by the method of hydrostatic weighing using an analytical balance KERN ALT 220–4M (KERN & SOHN GmbH, Germany). The test temperature was 25°C. The sample density of the iPP/LDPE blends was determined based on experimental weighting data by the Eq. (2):

$$\rho_{sam} = \frac{m_a}{m_f} + \rho_f \qquad (2)$$

where ρ_{sam} is a density of the sample; ρ_f = 0.8070 g cm^{-3} is a density of the working fluid (ethyl alcohol of 95 wt%); m_a and m_f are sample masses in the air and in the working fluid. The density of the amorphous regions of iPP and LDPE was calculated on the basis of the reference values of the density of completely crystalline PP and PE – 0.936 g cm^{-3} and 0.999 g cm^{-3} [35], relatively.

5.2.3.3 Electron Paramagnetic Resonance Spectroscopy

The structure of amorphous regions of the samples was investigated using electron paramagnetic resonance spectroscopy (paramagnetic probe method) [36] by the ESP spectrometer (Institute of Chemical Physics, Russia). The stable nitroxide radical 2,2,6,6-tetramethylpiperidine-1-oxyl (TEMPO-1) was used as a paramagnetic probe. The radical was introduced into the films from its vapor at 30°C. The molecular mobility in the initial polymers, blend compositions and in the interface layer, as well as the radical concentration in these materials were defined. The rotational mobility of the probe was determined by the correlation time τ_c. To calculate the values of τ_c the Eq. (3) was used [37]:

$$\tau_c = 6,65 \times 10^{-10} \left(\sqrt{\frac{I_+}{I_-}} \right) \times \Delta H_+ \qquad (3)$$

where I_+ and I_- are intensities of the first and the third peaks in the EPR spectrum; ΔH_+ is a half-width of the EPR spectrum component located in the weak field.

The amount of the radical absorbed by the amorphous phase of the sample was evaluated by the area of the EPR spectrum normalized to the fraction of the amorphous phase in this sample [36]. The proportion of the amorphous phase was calculated proceeding from the degree of crystallinity calculated based on the DSC data.

5.3 RESULTS AND DISCUSSION

According to the data available in literature, PP and PE in the blend are crystallized separately and form crystalline lattices typical of these polymers [3]. Moreover, both polymers can have a mutual effect on the process of crystallization and the morphology of the blend.

The processes of melting and crystallization of the iPP/LDPE blend samples of different compositions were investigated by the DSC method. DSC cooling curves of the samples are shown in Figure 5.1. The two endothermic peaks in the temperature range corresponding to the melting of the individual polymers are observed for all the thermograms at the heating of iPP/LDPE blend samples.

The melting temperatures T_m of iPP and LDPE in the blends and their degrees of crystallinity were determined from the obtained experimental data. The results are shown in Table 5.1. The melting temperatures of iPP and LDPE do not significantly depend on the composition of the blend. The decrease in T_m of the minor component by 1–2°C in relation to the homopolymer is observed at the low content of iPP or LDPE in the blend, which indicates a reduction in the perfection of the formed crystallites.

The analysis of the DSC curves and the dependence of crystalline phase on the cooling temperature of the iPP/LDPE blend samples allow to observe the abnormalities in the process of crystallization for the blends containing 5–30 wt% of iPP. There are two crystallization peaks

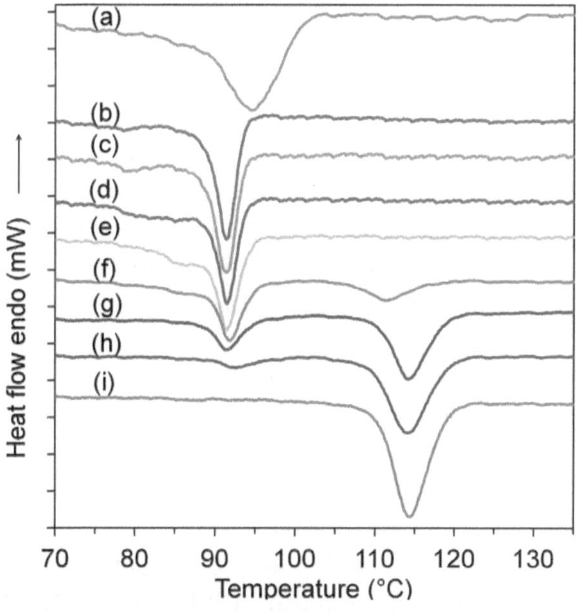

FIGURE 5.1 DSC cooling curves of LDPE (a), iPP (i) and iPP/LDPE blends with the iPP content of 5 (b), 10 (c), 20 (d), 30 (e), 40 (f), 60 (g), 80 (h) wt%.

TABLE 5.1 Melting Temperatures (T_m), Crystallization Temperatures (T_{cr}) and Degree of Crystallinity (χ) of LDPE and iPP in the Blends of Different Compositions

PP, wt%	Melting temperature, °C (±0.3°C)		Crystallization temperature, °C (±0.3°C)		Degree of crystallinity, % (±2%)	
	iPP	LDPE	iPP	LDPE	iPP	LDPE
0	–	106.6	–	94.0	–	23.8
5	161.0	106.6		91.4	56.5	23.7
10	162.4	107.0		91.4	55.9	23.1
20	162.8	106.8		91.5	59.9	21.5
30	163.2	107.0		91.5	58.2	22.8
40	162.9	106.9	111.3	91.8	54.0	22.2
50	162.8	106.7	113.0	91.6	44.0	18.6
60	163.3	106.9	113.9	91.5	51.3	21.6
70	163.0	106.5	113.8	91.9	53.6	21.5
80	163.2	106.0	113.9	92.5	52.6	21.1
90	163,1	105.0	113.3	92.5	53.6	24.7
95	163.5	–	113.5	92.6	53.6	24.5
100	163.1	–	114.3	–	63.3	–

corresponding to the temperature areas of crystallization of individual polymers on the cooling thermograms for the blends when the iPP content is equal to 40 wt% or more, while only one crystallization peak is observed in the thermograms for the blends with the PP content of 5–30 wt%. Upon the second heating of the studied samples with the iPP content of 5, 10, 20 and 30 wt%, two peaks corresponding to the melting peaks for homopolymers LDPE and iPP are observed on the melting thermograms. The crystallization abnormalities of iPP described above are also at other cooling rates of the samples, namely, 2, 4, and 16°C min^{-1}. In other words, despite the absence of the peak in the temperature range of crystallization of the individual iPP, both polymers in this mixture form crystalline phases.

It should be noted, that the crystallization peak of iPP shifts to lower temperatures area gradually, starting from the blend containing 50 wt% of iPP. For example, T_{cr}(PP) in the blend of iPP/LDPE = 60/40 is 113.9°C, iPP/LDPE = 50/50–113.0°C, iPP/LDPE = 40/60–111.3°C. Further, the iPP crystallization peak apparently moves to the area of lower temperatures and overlaps the crystallization peak of LDPE.

The iPP concentration in the blend melts of these compositions is low, so the compactness of iPP segments is not sufficient for the formation of crystallization nuclei. Only immediately after the onset of crystallization of LDPE, the latter forces iPP out of its phase, and the local concentration of iPP macro chains in the melt increases, allowing iPP to crystallize. Moreover, LDPE crystallites apparently can act as additional crystallization nuclei for iPP, as evidenced by the appearance of a low-temperature diffusion region on the cooling curves for the blends of these compositions. The possibility of heterogeneous iPP nucleation at its low content in the iPP/HDPE blend was described in the study [24]. Furthermore, the mentioned above abnormalities may be associated with a partial compatibility of iPP and LDPE in a melt at a low iPP content in the blend (less than 30 wt%), as shown in the studies [23, 24].

It is noteworthy that the crystallization peak of LDPE in all the blend compositions shifts by 1.5–2.5°C to the area of lower temperatures relatively to the crystallization peak of homopolymer LDPE (Table 5.1). The regularities of the crystallization processes for both components in the blends described above indicate a mutual influence of polymers on the composition.

The crystallinity of LDPE χ(PE) for different blend compositions slightly varies with the maximum for the individual LDPE – 23.8%. At the same time, the changes in the crystallinity of iPP χ(PP) are extreme (Table 5.1). Thus, LDPE has a greater impact on the crystallinity of iPP.

The individual iPP is characterized by the maximum degree of crystallinity of 63.3%. Even the addition of a small amount of LDPE (5–10 wt%) to iPP reduces the degree of crystallinity of iPP by 10%, which is associated with the difficulties of iPP crystallization in the presence of LDPE.

When the content of iPP in the blend is from 30 to 70 wt%, the phase inversion takes place, which affects the crystallization process and the properties of the crystalline phase. The composition with the equal content of components is characterized by the minimum degree of crystallinity of both polymers. When the content of iPP in the blend is 50 wt%, the decrease in χ (PP) by ~ 20% and χ (PE) by ~ 5% relatively to homopolymers occurs. In this composition the interpenetrating polymer networks form. The resulting steric hindrances make the crystallization of both polymers complicated.

When the content of iPP in the blend is 40 wt% and less, iPP is a dispersed phase in the LDPE matrix. The crystallization of iPP for these compositions occurs in the same temperature range as for LDPE, the number of defective crystallites of iPP, including those crystallized on the LDPE nuclei, increases, which leads to an increase in χ (PP) (55.9–59.9%).

The study of the amorphous phase of iPP and LDPE, as well as of the blended compositions of these polymers, was conducted using paramagnetic probe. The correlation time τ_c of the rotational motion of a stable nitroxyl radical in the iPP/LDPE blends, as well as the radical concentration in the amorphous phase of the samples and the interface region were defined.

The concentrations of the radical are high in individual LDPE and in the blend with the iPP content of 10 wt% after 7-hour saturation with the radical, while in individual iPP and in the iPP/LDPE blends of the other compositions the concentrations of the radical are low. With the increase in the saturation time in LDPE and in the blend of iPP/LDPE = 10/90, the radical concentration varies slightly (about 1.5 times), reaching the limit value on the 3rd or 4th day. In contrast, in the other blend compositions and in iPP the concentration increases approximately by 15–20 times, reaching saturation after 25 days. Moreover, the saturation time of the samples increases with the increase in the content of iPP. The observed retardation of the diffusion process in iPP is associated with its lower molecular mobility (25×10^{-10} s), as compared with LDPE (4×10^{-10} s).

However, after saturation is reached, the radical concentration is higher with the increase of the iPP content in the blend (Figure 5.2, curve f). Obviously, this can be attributed to the lower density of the iPP amorphous regions, as compared with LDPE, and, hence, to the larger free fluctuation volume, where the radical can be adsorbed.

The observed regularities can be explained from the viewpoint of nonhomogeneity of the iPP amorphous phase, while the amorphous phase of LDPE has a homogeneous structure. At the initial stage of the storage of the samples in radical vapors, it is absorbed in the loosest regions of the iPP amorphous phase, as evidenced by the low correlation time (Figure 5.3, curve a). In time the growing amount of radical penetrates

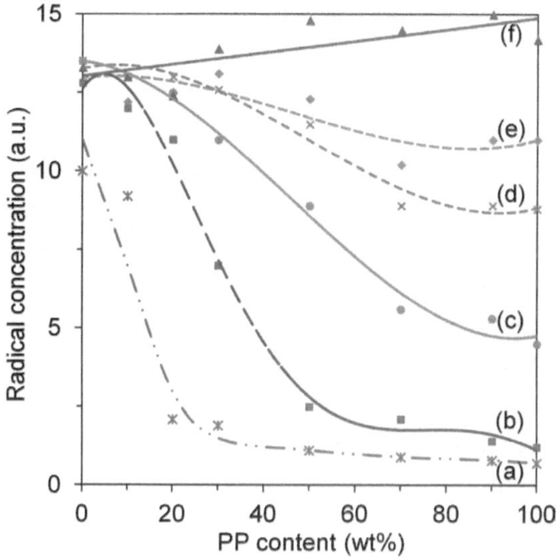

FIGURE 5.2 The dependence of the radical concentration in the amorphous phase of the samples on the iPP content in iPP/LDPE blends at 0.3 (a), 1.7 (b), 5.5 (c), 7.1 (d), 11.9 (e), 25 (f) days of saturation.

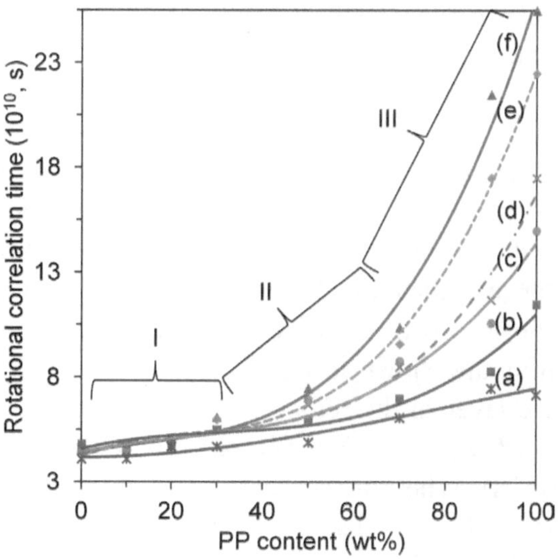

FIGURE 5.3 The dependence of the correlation time of the radical rotation on the iPP content in iPP/LDPE blends at 0.3 (a), 1.7 (b), 5.5 (c), 7.1 (d), 11.9 (e), 25 (f) days of saturation.

into denser structures of the iPP amorphous phase, and, as a result, the correlation time gradually increases (from 6.9×10^{-10} to 25×10^{-10} s). The saturation is reached on the 25th day of radical diffusion in the samples, after that the correlation time does not change.

Curve f on the Figure 5.3 shows the dependence of the correlation time of the radical rotation on the blend composition after the radical saturation. As mentioned above, the rotational correlation time of the radical probe in iPP is higher than in LDPE. First of all, it is associated with the peculiarities of the spatial organization of polymers. The optimum chain conformation of iPP is a 3/1 helix, while the LDPE chain has a planar zigzag conformation, which causes a more free rotation of the macrochain segments [3]. Therefore, despite a higher density, LDPE has a lower rigidity of segments than iPP.

Blends of iPP/LDPE show the total deviation of the mentioned above dependence on the additivity towards LDPE. Curve f (Figure 5.3) can be divided into 3 regions depending on the amount of the deviation from additivity: I – from 0 to 30 wt%; II – 30 to 70 wt% and III – from 70 to 100 wt% of iPP in the iPP/LDPE blend.

The non-additive dependence τ_c on the composition (Figure 5.3, curve f) indirectly indicates a partial compatibility of iPP and LDPE with formation of the interface layer. The studies of Jose S. et al. [7], Zhou et al. [37] pointed out the possibility of forming the interface layer when polyolefins are compounded in a certain ratio, which indicated a partial compatibility of polymers. Different values of deviation from additivity in this case can be explained by changing in the density and the segmental mobility in the interface region due to the varying of the blend composition.

The dependence of the radical rotational correlation time (curve a) and the amount of the absorbed radical (curve b) in the interface region on the blend composition are shown in Figure 5.4. Curve a (Figure 5.4) shows that the interface layer in the region I are characterized by a lower segmental mobility ($7–9 \times 10^{-10}$ s) as compared with those in regions II and III ($3.5–4 \times 10^{-10}$ s). When the content of iPP in the blend is lower than 30 wt% (region I), it forms discrete domains in the LDPE matrix. After saturation the radical easily penetrates into the amorphous phase of LDPE, but its penetration into iPP is likely to be hindered by a sufficiently rigid interface layer between the iPP and LDPE amorphous regions. This is also

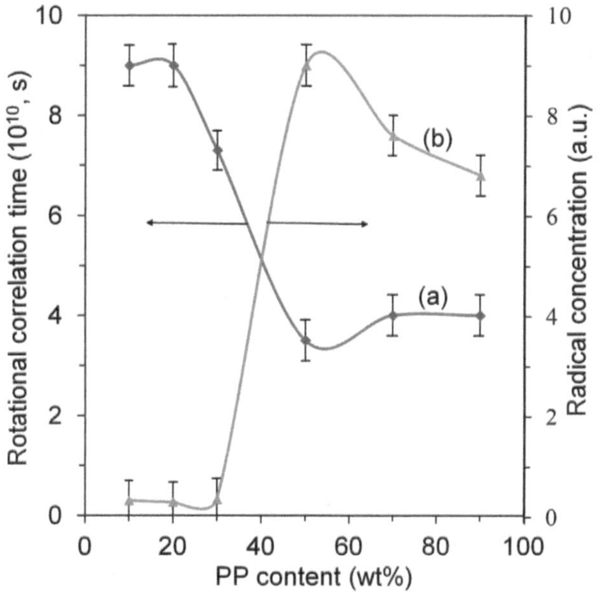

FIGURE 5.4 The dependence of the correlation time of the radical rotation (a) and the radical concentration (b) in the interphase layer on the iPP content in iPP/LDPE blends.

evidenced by a low concentration of the radical in the interface regions of the blends of these compositions (Figure 5.4, curve b) and the same value of the τ_c of the radical as in the individual LDPE (~4.5×10^{-10} s) (Figure 5.3).

This is most likely to be associated with the abnormalities in the crystallization process of the blends of these compositions, which results in formation of the nonequilibrium structure of the material with a large number of rigid strained tie chains. It should be noted that the study [5] shows a significant increase in the rate of ozone oxidation of iPP/LDPE blends with the iPP content of 10–30 wt%, which also indicates the formation of non-equilibrium structures.

When the content of iPP in the blend is 30–70 wt% (region II), the samples are characterized by the interpenetrating networks structure, the rigidity of interface layer is reduced, and more favorable conditions for the penetration of the radical in iPP exist. Therefore, an increase in τ_c with the increase of the iPP content in the mixture (from 10 to 5.5×10^{-10} s) is observed in this region (Figure 5.3, curve f).

When the content of iPP is 50 wt%, the minimum value of τ_c and the maximum concentration of the radical in the interface regions are observed in the blend as compared with other blend compositions (Figure 5.4). This indicates the formation of a less dense border layer, which is apparently associated with interpenetration of the segments of iPP and LDPE macrochains and the increase in the share of the interface region. It could also be the reason for the observed reduction in iPP and LDPE crystallinity.

On the region III (Figure 5.3, curve f) the material is the iPP matrix with distributed LDPE domains. However, the interface layer is characterized by a high segmental mobility (Figure 5.4, curve a), so the radical is able to easily penetrate into LDPE. The value of τ_c on this region (Figure 5.3, curve f) increases from 10×10^{-10} s (iPP/LDPE = 70/30) to 25×10^{-10} s (iPP). The negative deviation from the additivity law is, obviously, associated with the possibility of LDPE to modify iPP, leading to the increase of the segmental mobility in its amorphous regions and the reduction of the degree of crystallinity as compared with homopolymer of iPP. In other words, small additions of LDPE act as a plasticizer for iPP, as pointed out in the papers [4, 37].

5.4 CONCLUSION

The analysis of the results obtained by DSC and EPR spectroscopy methods has made it possible to follow the changes in the structure and properties of the crystalline and amorphous phases of iPP and LDPE blends in the wide range of compositions. The remarkable mutual influence of both components on the formation of the morphology of the composition was observed.

Despite the fact that iPP and LDPE are incompatible polymers, their partial compatibility can occur in the least ordered amorphous regions with the formation of an interface layer.

Three ranges of compositions differing in structure and properties have been identified. The blends with a low content of iPP (5–30 wt%) in the form of a discrete phase have a non-equilibrium structure with a more rigid interface layer. The abnormal iPP crystallization process in the temperature range of LDPE crystallization causes it.

The blends containing from 30 to 70 wt% of iPP have structure of interpenetrating polymer networks, in which both polymers form a continuous phase. In this range of compositions, LDPE and iPP are partially compatible in the amorphous phase with formation of a developed interface layer.

In the blends with a high content of iPP (70–95 wt%), iPP forms a continuous phase with dispersed LDPE domains. Moreover, LDPE has a plasticizing effect on iPP, increasing the segmental mobility of the macromolecules chains in its amorphous phase. A similar plasticizing effect of LDPE is observed for the range of compositions where interpenetrating polymer networks (30–70 wt% of iPP) are formed.

ACKNOWLEDGMENTS

The authors are grateful to the Center of shared scientific equipment of Emanuel Institute of Biochemical Physics of Russian Academy of Sciences for the equipment provision.

KEYWORDS

- **blends**
- **differential scanning calorimetry**
- **electron paramagnetic resonance spectroscopy**
- **polyethylene**
- **polypropylene**
- **structure**

REFERENCES

1. Pasch, H., Malik, M. I., (2014) Advanced Separation Techniques for Polyolefins. Springer International Publishing, Switzerland.
2. Boudenne, A., Ibos, L., Candau, Y., Thomas, S., (2011) Handbook of Multiphase polymer systems. John Wiley and Sons Ltd, Chichester, UK.

3. Harutun, G. K., (2003) Handbook of Polypropylene and Polypropylene Composites. Revised and Expanded, CRC Press.
4. Na, B., Wang, K., Zhang, Q., Du, R., Fu, Q., (2005) Tensile properties in the oriented blends of high-density polyethylene and isotactic polypropylene obtained by dynamic packing injection molding. Polymer 46:3190–3198.
5. Shibryaeva, L. S., Popov, A. A., Zaikov, G. E., (2006) Thermal Oxidation of Polymer Blends. VSP International Science Publishers, Leiden-Boston.
6. Yousef, B. F., Mourad, A. I., Hilal-Alnaqb, I. A., (2011) Prediction of the mechanical properties of polyethylene/polypropylene blends using artificial neural networks, Procedia Engineering, 10: 2713–2718.
7. Jose, S., Aprem, A. S., Francis, B., Chandy, M. C., Werner, P., Alstaedt, V., Thomas, S., (2004) Phase morphology, crystallization behavior and mechanical properties of isotactic polypropylene/high density polyethylene blends. Eur Polym, J., 40:2105–2115.
8. Zahran, R. R., El-Latif, M. M., Konsowa, A., Awwad, M., (2010) Development of isotropic compatible HDPE/PP blends for structural applications. J Appl Polym Sci 115 (3):1407–1418.
9. Yusak, N. M., Mohamed, R., Ramli, M. A., (2014) Mechanical Analyses of Polyethylene/Polypropylene Blend with Photodegradant. J Appl Sci Agric 9(11):300–305.
10. Camisa, W., Mantell, S. C., Davidson, J. H., Singh, G., (2010) Prediction of Degradation of Polyolefins Used in Solar Domestic Hot Water Components. ASME 2010–4th International Conference on Energy Sustainability, Phoenix, Arizona, USA Vol 2:301–308.
11. Al-Salem, S. M., Lettieri, P., Baeyens, J., (2009) Recycling and recovery routes of plastics solid waste (PSW): a review. Waste Manage 29:2625–2643.
12. Maio FDi, Rem, P., Hu, B., Serranti, S., Bonifazi, G., (2010) The W2Plastics Project: Exploring the Limits of Polymer Separation. Open Waste Manage, J., 3:90–98.
13. Serranti, S., Luciani, V., Bonifazi, G., Hu, B., Rem, P. C., (2015) An innovative recycling process to obtain pure polyethylene and polypropylene from household waste. Waste Manage 35:12–20.
14. Pantyukhov, P., Kolesnikova, N., Popov, A., (2014) Preparation, structure, and properties of biocomposites based on low density polyethylene and lignocellulosic fillers. Polym Compos. Doi: 10.1002/pc.23315.
15. Faruka, O., Bledzkia, A. K., Finkb, H. P., Saind, M., (2012) Biocomposites reinforced with natural fibers: 2000–2010. Prog Polym Sci 37 (11): 1552–1596.
16. Chaudhuri, S., Chakraborty, R., Bhattacharya, P., (2013) Optimization of biodegradation of natural fiber (Chorchorus capsularis): HDPE composite using response surface methodology. Iran Polym, J., 22:865–875.
17. Jose, J., Nag, A., Nando, G. B., (2014) Environmental aging studies of impact modified waste polypropylene. Iran Polym, J., 23:619–636.
18. Dikobe, D. G., Luyt, A. S., (2010) Comparative study of the morphology and properties of PP/LLDPE/wood powder and MAPP/LLDPE/wood powder polymer blend composites. Express Polym Lett 4 (11):729–741.
19. Toupe, J. L., Trokourey, A., Rodrigue, D., (2014) Simultaneous Optimization of the Mechanical Properties of Postconsumer Natural Fiber/Plastic Composites: Phase Compatibilization and Quality/Cost Ratio. Polym Compos 35 (4):730–746.

20. Kijenska, M., Kowalska, E., Palys, B., Ryczkowski, J., (2010) Degradability of composites of low density polyethylene/polypropylene blends filled with rape straw. Polym Degrad Stabil 95:536–542.
21. Sharifzadeh, E., Ghasemi, I., Karrabi, M., Azizi, H., (2014) A new approach in modeling of mechanical properties of binary phase polymeric blends. Iran Polym, J., 23:525–530.
22. Harrats, C., Mekhilef, N., Harrats, C., Thomas, S., Groeninckx, G., (2006) Micro- and Nanostructured Multiphase Polymer Blend Systems, Taylor & Francis, Ch. 3: 98–115.
23. Madi, N. K., (2013) Thermal and mechanical properties of injection molded recycled high density polyethylene blends with virgin isotactic polypropylene. Mater Design 46: 435–441.
24. Bartczak, Z., Galeski, A., Pracella, M., (1986) Spherulite nucleation in blends of isotactic polypropylene with high-density polyethylene. Polymer 27:537–543.
25. Li, J., Shanks, R. A., Long, Y., (2001) Miscibility and crystallization of polypropylene–linear low density polyethylene blends. Polymer 42 (5):1941–1951.
26. Ujhelyiova, A., Marcincin, A., Legen, J., (2005) DSC analysis of polypropylene-low density polyethylene blend fibers. Fibers Text East Eur 13 (5):129–132.
27. Dong, L. S., Olley, R. H., Bassett, D. C., (1998) On morphology and the competition between crystallization and phase separation in polypropylene-polyethylene blends. *J Mater Sci* 33(16):4043–4048.
28. Murin, J., Uhrin, J., Sevcovic, L., Horvath, L., Chodak, I., Nogellova, Z., (2005) Mechanical and nuclear magnetic resonance study of low density polyethylene, polypropylene and their blends. Acta Phys Slovaca 55 (6):577–587.
29. Gorrasi, G., Pucciariello, R., Villani, V., Vittoria, V., Belviso, S., (2003) Miscibility in Crystalline Polymer Blends: Isotactic Polypropylene and Linear Low-Density Polyethylene. J Appl Polym Sci 90:3338–3346.
30. Nwabunma, D., Kyu, T., (2007) Polyolefin blends. John Wiley & Sons, Inc., Canada.
31. Mathew, M. T., Novo, J., Rocha, L. A., Covas, J. A., Gomes, J. R., (2010) Tribological, rheological and mechanical characterization of polymer blends for ropes and nets. Tribology Int 43 (8):1400–1409.
32. Lobo, H., Bonilla V. Jose (2003) Handbook of Plastics Analysis. Marcel Dekker Inc., New York.
33. Wunderlich, B., (2005) Thermal analysis of Polymeric Materials. Springer, Berlin.
34. Abdel-Bary, M., (2003) Handbook of Plastic Films. Rappa Technology Lmt., Shrewsbury, UK.
35. Akay, M., (2012) Introduction to Polymer Science and Technology. Mustafa Akay & Ventus Publishing ApS.
36. Drescher, Malte, Jeschke, Gunnar (2012) EPR Spectroscopy. Applications in Chemistry and Biology, Springer.

CHAPTER 6

ELASTIC MODULUS OF POLY(ETHYLENE TEREPHTHALATE)/ POLY(BUTYLENE TEREPHTHALATE) BLENDS

M. A. MIKITAEV,[1] G. V. KOZLOV,[1] A. K. MIKITAEV,[1] and G. E. ZAIKOV[2]

[1]*Kh.M. Berbekov Kabardino-Balkarian State University, Nal'chik – 360004, Chernyshevsky St., 173, Russian Federation, E-mail: i_dolbin@mail.ru*

[2]*N.M. Emanuel Institute of Biochemical Physics of Russian Academy of Sciences, Moscow 119334, Kosygin St., 4, Russian Federation, E-mail: chembio@sky.chph.ras.ru*

CONTENTS

ABSTRACT

The quantitative interpretation of the extreme dependence of elastic modulus on composition for blends poly(ethylene terephthalate)/poly(butylene terephthalate) has been offered, which uses the percolation theory and fractal analysis. It has been shown that elastic modulus extreme increasing is due to the corresponding growth of shear strength of blends components autohesional bonding. The micromechanical models do not give the indicated effect adequate description.

6.1 INTRODUCTION

The maximum of elastic modulus at equal contents of components in blends is one from outstanding features of the blends poly(ethylene terephthalate)/poly(butylene terephthalate) (PET/PBT) [1, 2]. In addition it is important to note, the elastic moduli of initial PET and PBT are practically equal – the discrepancy between them makes up ~ 1% by absolute value, that is smaller than their determination experimental error. The authors [1, 2] supposed that variation of elastic modulus of blends PET/PBT at composition change was due to blends components miscibility variation. It is significant that PET and PBT are miscible partly, namely, amorphous phase miscibility (single glass transition temperature) can be realized, but crystalline phases nonmiscibility (two crystallization temperatures) is observed [2]. In work [3] it has been proposed to consider semicrystalline polymers as composites, in which amorphous phase is played by matrix role and filler role – by crystallites. However, in such treatment the extreme change of crystalline phase characteristics is necessary, whereas regardless of blends PET/PBT production mode these characteristics are changed monotonously and not very significantly [4]. Nevertheless, the blends PET/PBT can be considered as polymer/polymeric composites [5], particularly at the condition, that one polymeric phase is dispersed in another as disperse particles with the size of 0.2–1.5 mcm [6]. With appreciation of the stated above considerations the purpose of the present work is the treatment of blends PET/PBT as polymer/polymeric composites within the frameworks of micromechanical [7] and percolation [8] models for quantitative description of the extreme dependence of their elastic modulus on composition.

6.2 EXPERIMENTAL PART

The industrial production polymers were used: PET PELPET, grade G5801 (intrinsic viscosity [η]=0.8 dL/g), procured from firm Reliance Industries Ltd (India), and PBT LUPOX, grade GP-1000 ([η]=1.0 dL/g), supplied by firm LG Polymers India Pvt Ltd (India). PET and PBT pellets were manually mixed and dried at temperature 393 K for 8 hours in a hot air circulating oven [2].

The blends with PET:PBT ratio of 80:20, 70:30, 60:40, 50:50, 40:60 and 20:80 by weight were prepared by components mixing in melt using twin-screw extruder Haake Rheocord 9000 of mark TW100 at the screw rate rotation of 40 rpm in the range of temperatures of 423–533 K. Then the extrudate was water cooled and granulated. The extruded pellets were molded into standard mechanical tests specimens by injection molding mode on molding machine Boolani Industries Ltd., production of India, within the range of temperatures 493–553 K [2].

The mechanical tests of blends PET/PBT on three-pointed bending were carried out on universal testing machine LR-50K, Lloyds Instrument according to ASTM 790M-90 at temperature 293 K and cross-head speed of 2.8 mm/min [2].

6.3 RESULTS AND DISCUSSION

Let us consider the micromechanical models application of the description of blends PET/PBT elastic modulus. In the simplest from the possible cases two models were proposed [7]. For the case of parallel arrangement the uniform strain in both phases is assumed and upper boundary of elastic modulus of blends E_{bl}^{up} is given as follows [7]:

$$E_{bl}^{up} = E_n\varphi_n + E_m\varphi_m, \tag{1}$$

where E_n and E_m are elastic moduli of filler and matrix, respectively, φ_n and φ_m are volume contents of filler and matrix, accordingly (φ_m=1–φ_n).

In case of series arrangement the stress is assumed to be uniform in them and the lower boundary of elastic modulus of blends E_{bl}^{l} is determined according to the equation [7]:

$$E_{bl}^{l} = \frac{E_n E_m}{E_n \varphi_m + E_m \varphi_n}. \tag{2}$$

At the condition $E_n = E_m$, which is true for the considered blends, the equations (1) and (2) give the trivial result: $E_{bl}^{up} = E_{bl}^{l} = E_n = E_m$, for example, the indicated equations are not capable of describing the experimentally observed maximum on the dependence of elastic modulus on composition for blends PET/PBT [1, 2]. This is explained by the fact, that all micromechanical models require fulfillment of the condition $E_n > E_m$ for their correct application. One more obvious deficiency of micromechanical models is *a priori* used in them assumption of perfect adhesion between phases of composite, that is far from always being fulfilled for real composites [9].

The percolation model gives the following relationship for E_{bl} determination [8]:

$$\frac{E_{bl}}{E_m} = 1 + 11(\varphi_n)^{1.7} \tag{3}$$

The relationship (3) does not also take into consideration interfacial adhesion level between composite phases and therefore the authors [5] modified it as follows:

$$\frac{E_{bl}}{E_m} = 1 + 11(b_\alpha \varphi_n)^{1.7} \tag{4}$$

where b_α is a dimensionless parameter, characterizing the interfacial adhesion level (in case of perfect adhesion $b_\alpha = 1.0$ [5]).

By its physical essence interfacial adhesion between PET and PBT represents the formation of autohesional bonding, shear strength of which τ_c can be determined as follows [10]:

$$\tau_c = 6.28 \times 10^{-5} N_c^{3.83}\text{, MPa} \tag{5}$$

where N_c is the number of intersections (contacts) of macromolecular coils in boundary layer of two polymers (in the considered case – PET and PBT).

Within the frameworks of fractal analysis the value N_c can be determined according to the relationship [11]:

$$N_c \sim R_g^{2D_f - d} \tag{6}$$

where R_g is gyration radius of macromolecular coil, D_f is its fractal dimension, d is the dimension of Euclidean space, in which a fractal is considered (it is obvious, that in our case d=3).

The value D_f is calculated according to the following technique. First Poisson's ratio value ν was estimated with the aid of the formula [12]:

$$\frac{\sigma_Y}{E_{bl}} = \frac{1-2\nu}{6(1+\nu)} \tag{7}$$

where σ_Y is blend yield stress.

Then the structure fractal dimension d_f for blends PET/PBT was determined according to the equation [13]:

$$d_f = (d-1)(1+\nu) \tag{8}$$

And at last for linear polymers the value D_f is determined as follows [14]:

$$D_f = \frac{2d_f}{3} \tag{9}$$

In Figure 6.1, the relation between parameter b_α, determined with the aid of the equation (4), and shear strength τ_c of autohesional contact PET-PBT is adduced. As it was to be expected from the most general considerations, between parameters b_α and τ_c the linear correlation is observed, which passing through coordinates origin, described by the following empirical equation:

$$b_\alpha = 4.6\tau_c \tag{10}$$

The Eqs. (4) and (10) combination allows to obtain the following relationship for determination of blends PET/PBT elastic modulus:

$$E_{bl} = E_m\left[1+11\left(4.6\tau_c\varphi_n\right)^{1.7}\right] \qquad (11)$$

It is obvious, that PBT relative fraction at its content smaller than 50 mass % and PET relative fraction at its very same content is accepted as φ_n. In Figure 6.2, the comparison of theoretically calculated according to the Eq. (11) E_{bl}^T and experimentally obtained E_{bl} elastic modulus values for blends PET/PBT is adduced, which has show their good correspondence (the average discrepancy between E_{bl}^T and E_{bl} makes up smaller than 5%).

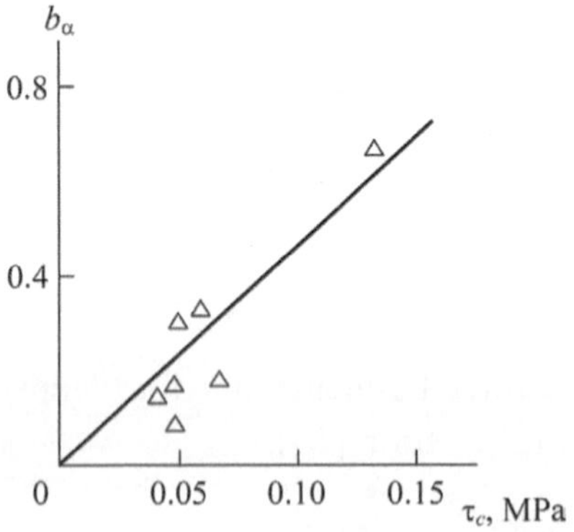

FIGURE 6.1. The relation between parameter b_α, characterizing interfacial adhesion level, and shear strength τ_c of autohesional bonding for blends PET/PBT.

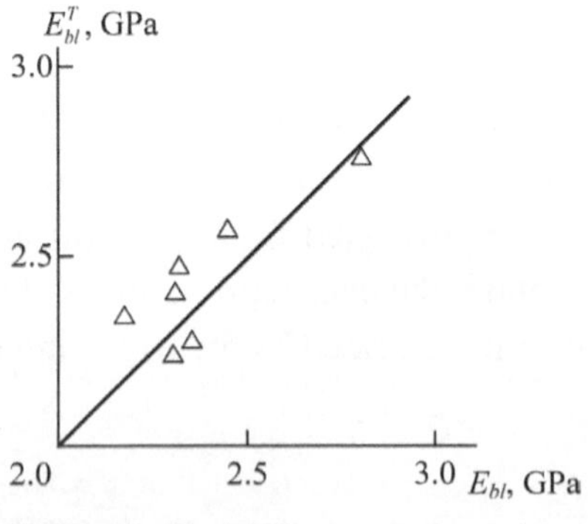

FIGURE 6.2. The comparison of theoretically calculated according to the equation (11) E_{bl}^T and experimentally measured E_{bl} values of elastic modulus for blends PET/PBT.

6.4 CONCLUSIONS

Thus, the present work results have demonstrated, that micromechanical models cannot describe correctly the dependence of elastic modulus on composition for blends PET/PBT. This is due to non-fulfillment of the conditions, obligatory for the indicated models: essentially larger value of filler elastic modulus in comparison with corresponding characteristic for polymer matrix and assumption of perfect interfacial adhesion (as far as we know, the Sato-Furukawa model [15] is sole micromechanical model, taking into consideration real level of interfacial adhesion in composites). The considered blends elastic modulus enhancement is due to the growth of contacts number of macromolecular coils of PET and PBT in boundary layer and corresponding enhancement of interfacial adhesion level. The offered model, using percolation theory and fractal analysis notions, allows enough precise quantitative description of the experimental data.

ACKNOWLEDGEMENTS

Work is performed within the complex project on creation of hi-tech production with the participation of the Russian higher educational institution, the Contract of JSC "Tanneta" with the Ministry of Education and Science of the Russian Federation of February 12, 2013 No. 02.G25.31.0008 (Resolution of the Government of the Russian Federation No. 218).

KEYWORDS

- **blend**
- **elastic modulus**
- **fractal analysis**
- **interfacial adhesion**
- **micromechanical model**
- **percolation**

REFERENCES

1. Avramova, N. (1995). Polymer, 36, 801–808.
2. Aravinthan, G., Kale, D. D. (2005). J. Appl. Polymer Sci., 98, 75–82.
3. Kardos, J. L., Raisoni, J. (1975). Polymer Engng. Sci., 15, 183–190.
4. Szostak, M. (2004). Mol Cryst. Liq. Cryst., 416, 209–215.
5. Mikitaev, A. K., Kozlov, G. V., Zaikov, G. E. (2008). Polymer Nanocomposites: Variety of Structural Forms and Applications (p. 319). New York, Nova Science Publishers, Inc.
6. Maruhashi, Y., Tida, S. (2001). Polymer Engng. Sci., 41, 1987–1995.
7. Ahmed, S., Jones, F. R. (1990). J. Mater. Sci., 25, 4933–4932.
8. Bobryshev, A. N., Kozomazov, V. N., Babin, L. O., Solomatov, V. I. (1994). Synergetic of Composite Materials (p. 154). Lipetsk, NPO ORIUS.
9. Kozlov, G. V., Yanovskii Yu. G., Zaikov, G. E. (2010). Structure and Properties of Particulate-Filled Polymer Composites: The Fractal Analysis (p. 282). New York, Nova Science Publishers, Inc.
10. Yakh'yaeva Kh.Sh., Kozlov, G. V., Magomedov, G. M. (2014). Fundamental Problems Modern Engineering Science, 11, 206–209.
11. Vilgis, T. A. (1988). Physica A, 153, 341–354.
12. Kozlov, G. V., Sanditov, D. S. (1994). Anharmonic Effects and Physical-Mechanical Properties of Polymers (p. 261). Novosibirsk, Nauka.
13. Balankin, A. S. (1991). Synergetics of Deformable Body (p. 404). Moscow, Publishing Ministry of Defense SSSR.
14. Kozlov, G. V., Mikitaev, A. K., Zaikov, G. E. (2014). The Fractal Physics of Polymer Synthesis (p. 359). Toronto, New Jersey, Apple Academic Press.
15. Sato, Y., Furukawa, J. (1963). Rubber Chem. Techn., 36, 1081–1089.

CHAPTER 7

LOW-TOXIC NITROGEN-CONTAINING ANTIOXIDANT FOR POLYVINYL CHLORIDE

R. M. AKHMETKHANOV, I. T. GABITOV, A. G. MUSTAFIN, V. P. ZAKHAROV, and G. E. ZAIKOV

Bashkir State University, 32 Zaki Validi St., 450076 Ufa, Russia, E-mail: rimasufa@rambler.ru

CONTENTS

ABSTRACT

Kinetic regularities of thermooxidative dehydrochlorination of rigid and plasticized PVC in the presence of 5-hydroxy-6-methyluracil have been studied. The high antioxidant efficacy of 5-hydroxy-6-methyluracil in the process of polymer degradation has been revealed. It is shown that the studied uracil significantly slows down the process of accumulation of hydroperoxides in oxidation of the plasticizer of PVC, dioctyl phthalate, which is the cause of a significant slowdown in the rate of decomposition of the plasticized polymer.

7.1 INTRODUCTION

A number of derivatives of uracil is used as drugs in the practical medicine [1]. For some uracils, in particular, for 5-hydroxy-6-methyluracil and 5-amino-6-methyluracil, antioxidant activity in reactions of radical-chain oxidation of isopropanol and 1,4-dioxane [2, 3] has been revealed. In this regard, it is of great scientific and practical interest to study the inhibitory efficiency of 5-hydroxy-6-methyluracil in thermooxidative degradation of rigid and plasticized polyvinyl chloride.

7.2 EXPERIMENTAL PART

The object of the study is uracil derivative 5-hydroxy-6-metiluratsil:

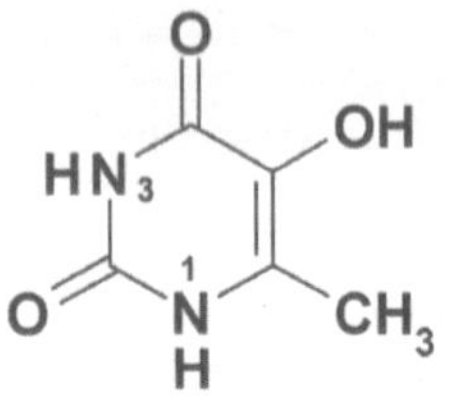

Thermooxidative dehydrochlorination of rigid and plasticized PVC was performed at 175°C in a bubbling type reactor with a stream of oxygen (3.5 liters per hour). Duration of the thermal stability of PVC (τ) was determined by the color change of the indicator "Congo red" in the allocation of HCl during the degradation of the polymer (175°C) in accordance with GOST 14041–91. The rate of dehydrochlorination of PVC was determined by the same means as in [4]. The kinetics of accumulation of hydroperoxides was estimated by iodometric method [5].

Polyvinyl chloride PVC S-7059M was purified by washing with ethanol in a Soxhlet apparatus. Ester plasticizer dioctyl phthalate (DOP) was purified by filtration through a column filled with alumina. 5-hydroxy-6-methyluracil (99.0% of main substance) was not subjected to further purification.

7.3 RESULTS AND DISCUSSION

The input of 5-hydroxy-6-methyluracil in rigid PVC in conditions of oxidative degradation leads to a noticeable decrease in the rate of dehydrochlorination of polymer (Figure 7.1). The maximum decrease in the rate of elimination of HCl was observed at concentration of uracil equal to 2 mmol per mol of PVC (critical concentration of antioxidant). Beyond the critical concentration of the antioxidant, an increase in the rate of degradation of the polymer is observed.

The decrease in the rate of thermooxidative dehydrochlorination of polymer in the presence of uracil is observed up to values corresponding to the value of the speed of the thermal elimination of HCl from PVC in an inert atmosphere, which is typical for stabilizers-antioxidants. The stabilizing efficacy of 5-hydroxy-6-methyluracil by the level of reduction rate of thermooxidative dehydrochlorination of PVC is comparable to the industrial efficiency of antioxidant diphenylolpropane (Figure 7.1).

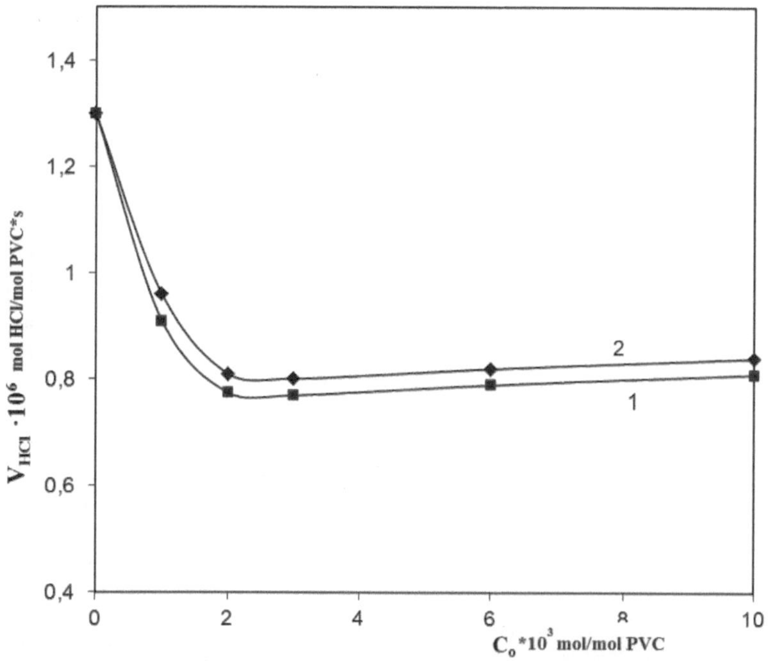

FIGURE 7.1 The dependence of the rate of thermooxidative dehydrochlorination of PVC on the content of diphenylolpropane (1) and 5-hydroxy-6-methyluracil (2) (175° C, O_2, 3.5 L/h).

The problem of stabilization of plasticized PVC is largely associated with the prevention of oxidative decomposition of the plasticizer, since plasticizers, in particular ester plasticizers, in the presence of oxygen become easily engaging in free-radical oxidation reactions, activating the process of elimination of HCl from the polymer.

The process of thermooxidative dehydrochlorination of PVC plasticized with dioctyl phthalate is accompanied by autocatalysis. The input of 5-hydroxy-6-methyluracil in the plasticized polymer leads to a sharp decrease in the rate of thermooxidative dehydrochlorination of polymer and to translation oh the process from autocatalytic to stationary mode (Figure 7.2).

The maximum decrease in elimination rate of HCl from the polymer containing 40 pbw/100 pbw of PVC dioctyl phthalate, as in the case

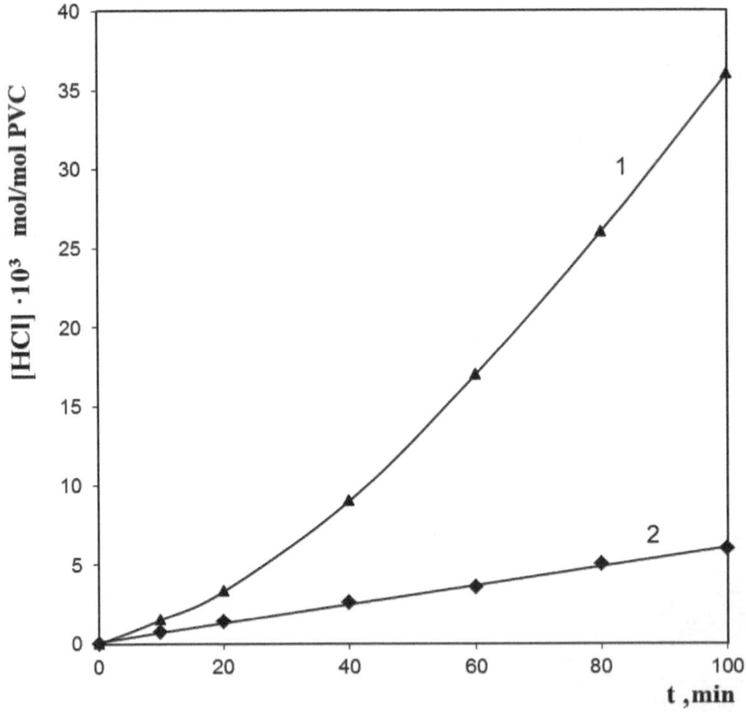

FIGURE 7.2 Kinetic curves of the process of dehydrochlorination of PVC plasticized with dioctyl phthalate (40 pbw/100 pbw PVC) in the presence of 5-hydroxy-6-methyluracil (content 5-hydroxy-6-methyluracil 1–0 mmol per mol of PVC, 2–2 mmol per mol of PVC), (175°C, O_2, 3.5 L/h).

of non-plasticized PVC, is observed at concentration of 5-hydroxy-6-methyluracil equal to 2 mmol per mol of PVC. Higher concentration of uracil leads to accelerated degradation of the polymer (Figure 7.3).

The decrease in the rate of thermooxidative decomposition of plasticized PVC in the presence of 5-hydroxy-6-methyluracil is observed up to values corresponding to the rate of the thermooxidative degradation of non-plasticized polymer. Evidently, uracil protects the plasticizer from oxidation, which, in its turn, due to solvation stabilization increases the thermal stability of polyvinyl chloride (a known effect of "echo"-stabilization) [6].

Evidently, as in the case of radical-chain reactions of oxidation of model organic substrate in the presence of uracils [7], an active center in position N_1 (link N_1-H) in 5-hydroxy-6-methyluracil is responsible for the elementary act of inhibiting oxidation of the ester.

The stabilizing efficacy of uracil is also confirmed by duration of the thermal stability of PVC-compositions comprising metal-containing

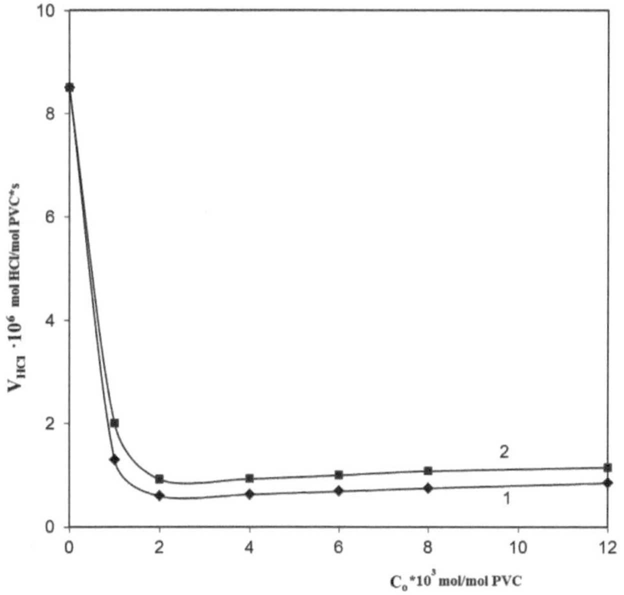

FIGURE 7.3 The dependence of the rate of oxidative dehydrochlorination of PVC plasticized by 40 pbw/100 pbw PVC dioctyl phthalate on the content of diphenylolpropane (1) and 5-hydroxy-6-methyluracil (2) (175°C, O_2, 3.5 L/h).

stabilizer. Additional insertion of 5-hydroxy-6-methyluracil to the plasticized polymer compositions increases the value of "time of thermal stability" by 1.4–1.6 times (Table 7.1).

Thermooxidative stability of PVC plasticized by ester is mainly determined by the plasticizer resistance to oxidation. In the process of thermooxidation of ester, the resulting hydroperoxides decompose into radicals that during the degradation of PVC-plasticat have an accelerating influence on the process of dehydrochlorinating of the polymer. In this regard, it is important to evaluate the effect of uracil on the process of accumulation of hydroperoxides during thermooxidation of ester.

The regularities of the process of accumulation of hydroperoxides in thermooxidative decomposition of dioctyl phthalate, the ester plasticizer of polyvinyl chloride, in the presence of 5-hydroxy-6-methyluracil have been revealed. The high efficiency of antioxidative action of uracil comparable with the efficiency of diphenylolpropane is expressed in a significant reduction of rate of accumulation of hydroperoxides in the oxidation of ester plasticizer (Figure 7.4).

7.4 CONCLUSIONS

In conditions of thermooxidative dissolution of rigid and plasticized PVC, 5-hydroxy-6-methyluracil is an effective antioxidant-stabilizer and by

TABLE 7.1 Time of Thermal Stability of PVC Compounds

Components, indicator	**Composition, pbw**				
	1	**2**	**3**	**4**	**5**
PVC	100	100	100	100	100
Dioctyl phthalate	40	40	40	40	40
Calcium stearate	2	2	2	2	2
Diphenylolpropane	—	0.05	0.1	—	—
5-hydroxy-6-methyluracil	—	—	—	0.05	0.1
τ, min, 175°C	27	42	49	38	44

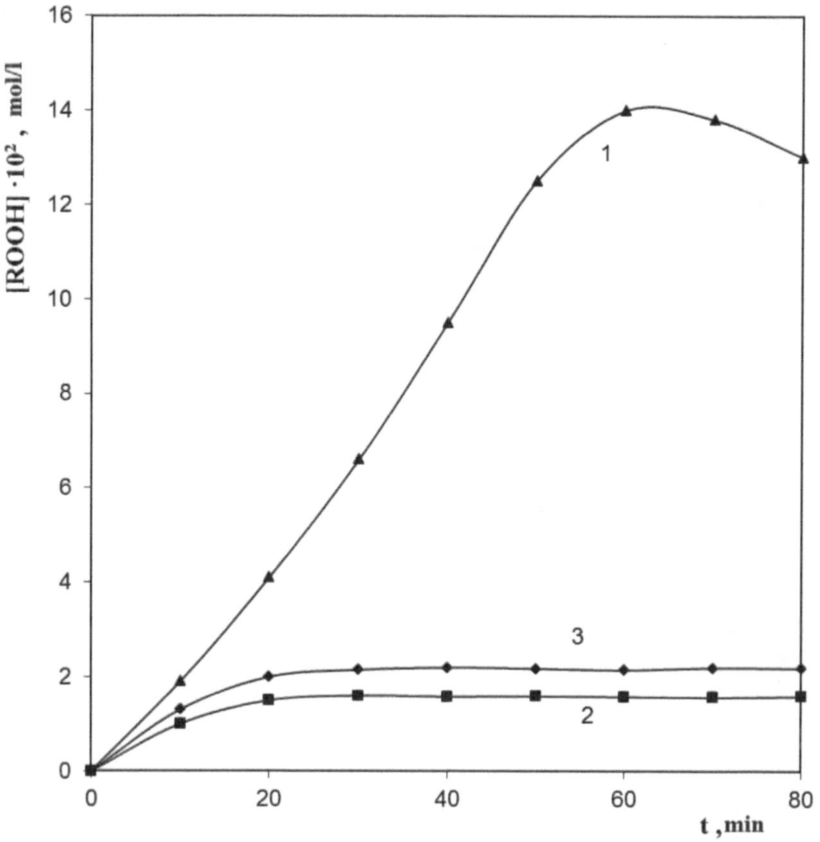

FIGURE 7.4 Kinetic curves of accumulation of hydroperoxides in the process of oxidation of dioctyl phthalate in the presence of diphenylolpropane (2), 5-hydroxy-6-methyluracil (3) (concentration of additives is 2×10^{-3} mol per mol of DOF, the original DOF (1), 165°C, O_2, 3.0 L/h).

stabilizing efficiency it is almost equal to industrial phenolic antioxidant diphenylolpropane.

High antioxidant activity of 5-hydroxy-6-methyluracil in the process of destruction of plasticized PVC is explained by the ability of uracil to effectively slow down the process of accumulation of hydroperoxides in the oxidation of ester plasticizer.

High antioxidant activity and low toxicity of 5-hydroxy-6-methyluracil (hazard class 4 [8]) allows using it as an antioxidant-stabilizer in formulations of plasticized PVC materials for medical and food purposes.

KEYWORDS

- **5-hydroxy-6-methyluracil**
- **antioxidant**
- **polyvinyl chloride**
- **thermooxidative dehydrochlorination**

REFERENCES

1. Lazareva, D. N., Alekhin, E. K. Immunity Stimulants. Moscow: Meditsina, 1985. 256 p.
2. Gerchikov, A. Ya., Garifullina, G. G., Sultanaeva, I. V. et al. Chemical and Pharmaceutical Journal. 2000. Vol. 34. №10. pp. 28–30.
3. Yakupova, L. R., Calakhutdinova, R. A., Pankrat'ev, E. Yu., Safiullin, R. L. Kinetics and Catalysis. 2012. Vol. 53. №6. Pp. 708–715.
4. Gabitov, I. T., Akhmetkhanov, R. M., Kolesov, S. V. et al. Bulletin of Bashkir University 2012. Vol. 17. №1. Pp 48–51.
5. Antonovsky, V. L., Buzlanova, M. M. Analytical Chemistry of Organic Peroxide Compounds. Moscow: Khimiya, 1978. 309 p.
6. Minsker, K. S., Abdullin, M. I. Reports of the USSR Academy of Sciences. 1982. Vol. 263. №1. Pp. 140–143.
7. Dautova, I. F., Akhatova, I. V., Safarova, I. V. et al. Reports of Academy of Sciences. 2010. Vol. 431. №4. Pp. 1–3.
8. Berkhin, E. B., Lampatov, V. V., Ul'yanov, G. P. Effect of 5-Hydroxy-6-Methyluracil and Dimetophosphon on Tubular Secretion of Xenobiotics in Kidney. Pharmacology and Toxicology. 1987. № 5. Pp. 37–38.

CHAPTER 8

IMPACT OF ORGANOSILICONE MODIFIERS ON THE PROPERTIES OF ETHYLENE COPOLYMERS

S. N. RUSANOVA,[1] N. E. TEMNIKOVA,[1] S. YU. SOFINA,[1] O. V. STOYANOV,[1] and G. E. ZAIKOV[2]

[1]*Kazan National Research Technological University, 68 K. Marksa Str., Kazan 420015, Russia*

[2]*Institute of Biochemical Physics N.M. Emanuel Russian Academy of Sciences, 4 Kosygina Str., Moscow 119334, Russia*

CONTENTS

ABSTRACT

The possibility of using of alkoxysilanes, containing various functional groups, as structuring agents of ethylene copolymers was studied and their influence on the properties of binary and ternary ethylene copolymers was revealed.

8.1 INTRODUCTION

Ethylene copolymers are widely used for the preparation of materials and products for various applications, including coatings and adhesives. To expand the scope of the industrially-produced ethylene copolymers is possible by their modification.

A large number of studies suggest the successful modification of polyolefins by unsaturated alkoxysilanes in the presence of radical forming compounds and obtaining materials based on them, which can be cross-linked in the presence of water. Traditionally, amino- and glycidoxyalkoxysilanes are used as modifiers of epoxy resins, polyurethane sealants, polysulfide oligomers, as well as coupling agents to improve adhesion of dispersed and fibrous fillers to the polymer matrix. We carried out a chemical modification of copolymers of ethylene with vinyl acetate by limiting alkoxysilanes, including those containing amino- and glycidoxy groups [1, 2].

Let's analyze the impact of various alkoxysilanes on properties of ethylene copolymers.

8.2 SUBJECTS AND METHODS

As the objects of study were used copolymers of ethylene with vinyl acetate (EVA) brands Evatane 20–20, Evatane 28–05 and copolymers of ethylene with vinyl acetate and maleic anhydride (EVAMA) brands Orevac 9305 and Orevac 9307 by "Arkema." Main characteristics of the copolymers are given in Table 8.1.

As the modifiers were used:

Tetraethoxysilane (ETS-32) (TC 6–02–895–78) – low-viscosity waxy transparent liquid of light yellow color. The content of silicon in terms of silicon dioxide is 30–34%, tetraethoxysilane – 50–65%. Density is 1062 kg/m^3. Viscosity is 1.6 cP.

[3-(2-aminoethylamino)propyl]trimethoxysilane (DAS) – transparent yellowish liquid with a molecular mass of 222. Density is 1028 kg/m^3. The refractive

TABLE 8.1 Characteristics of the Copolymers of Ethylene

Brand	**Evatane 20–20**	**Evatane 28–05**	**Orevac 9305**	**Orevac 9307**
Symbol	EVA 20	EVA 27	EVAMA 26	EVAMA 13
VA content, %	19–21	27–29	26–30	12–14
MA content, %	—	—	1.5	1.5
Melt flow index, g/10 min t=190°C load 5 kg	59.29	18.16	92.77	26.76
Melt flow index, t=125°C load 2.16 kg	2.67	0.76	14.72	1.02
Density, kg/m³	938	950	951	939
Tensile stress at failure, MPa	14	23.88	7	19.5
Elongation at break, %	740	800	760	760
Elastic modulus, MPa	31	17	7	61

index n^D_{20} = 1.028, the content of amine groups is 13%. Melting point is −36°C, flashpoint is 104°C and boiling point is 262°C.

Aminopropyltriethoxysilane (AGM-9) (TC 6–02–724–77) – transparent colorless liquid with a molecular mass of 221. The density is 962 kg/m³, the refractive index n^D_{20} = 1.4223, the content of amine groups is 7–7.5%. Melting point is 70°C, decomposition temperature is 217°C and boiling point is 194°C.

(3-glycidoxypropyl)trimethoxysilane (GS) – transparent colorless liquid with a molecular mass of 236. The density is 1070 kg/m³, the refractive index n^D_{20} = 1.4367, the content of glycidoxy groups is 31%. Melting point is 70°C, flashpoint is 135°C and boiling point is 264°C.

The reaction mixture of polymers with the modifier was performed on laboratory rollers at a rotational speed of the rolls of 12.5 m/min and a friction of 1:1.2. The conditions of obtaining the compositions are shown in Table 8.2. The modifier content was varied in the range of 0–10 wt.%. The rolled mass was kept overnight at room temperature for stress relief and then was pressed in accordance with GOST 12019–66 to obtain test samples.

TABLE 8.2 Technological Parameters of Rolling and Pressing

Polymer	$t_{roll.}$, °C	$t_{press.}$, °C	Specific pressure, MPa	$\tau_{heating.}$, min/mm	$\tau_{holding.}$, min/mm	$\tau_{cooling.}$, min/mm
EVA20	100	160	15	5	5	1
EVA27	100	160	15	5	5	1
EVAMA26	120	160	15	5	5	1
EVAMA13	120	160	15	5	5	1

Physical and mechanical tests of samples were carried out according to GOST 11262–80. Viscosity was determined by a standard technique [3]. Melt flow index (MFI) was determined on a capillary viscometer IIRT in accordance with GOST 11645–73 at temperatures of 125°C, 190°C and at a load of 2.16 kg and 5 kg. To determine the gel fraction the samples were extracted by chloroform in a Soxhlet apparatus for 8 hours. Sorption and diffusion properties were determined in accordance with GOST 12020–72. Adhesion tests were performed according to GOST 411–77 (Method A).

8.3 RESULTS AND DISCUSSION

In the Refs. [4–8] was described the modification of copolymers of ethylene with vinyl acetate by alkoxysilanes containing different functional groups under intensive thermomechanical impact. Analysis of the IR spectra of the modified copolymers revealed that the ester groups of the copolymers chemically react with the alkoxy groups of the modifiers. Bounor-Legaré et al. [9] have established that one molecule of tetraalkoxysilane reacts with two groups of vinyl acetate that allows to obtain cross-linked and branched polymers.

Modification of polyolefins by the compounds capable to graft to macromolecules naturally leads to increasing of the molecular weight and, consequently, affects the processes of the polymer flow. Molecular weight may change due to the formation of cross-linked or graft structures. The presence of the cross-linked structures formed as a result of chemical interaction between the polymer and the modifier can be assessed from the change of solubility and fluidity of their melts and solutions. Should take into account the fact that the sol fraction is largely affects the processes of the solutions flow, and the gel fraction of the polymer affects the melt flow.

Analysis of the results showed (Figure 8.1) that the introduction of small amounts of ETS in the investigated copolymers leads to increasing the intrinsic viscosity of the polymer and reducing the MFI, because of increasing the molecular weight due to the formation of silanol bridges between macromolecules. Further increasing of ethyl silicate concentration leads to the formation of the grafted siloxane and the long and the short chain branching. This reduces the viscosity of the modified copolymer and increases the melt index.

When introducing the amine organosilicon modifiers into double ethylene copolymers an increase of the intrinsic viscosity of polymers (more than in 2 times) is observed. This is due to increase of the molecular weight of the polymer during modification. Parallel to this, there is a decrease in the melt flow index for the systems EVA – aminosilane (Figure 8.2).

Introduction of ETS-32 into the terpolymers of ethylene has no effect on the rheological characteristics of the material, whereas the introduction of alkoxysilanes, containing aminogroups, leads to a significant increase in the proportion of the insoluble fraction (G_F> 50%) and to a sharp decrease in the melt flow index (Figures 8.3 and 8.4).

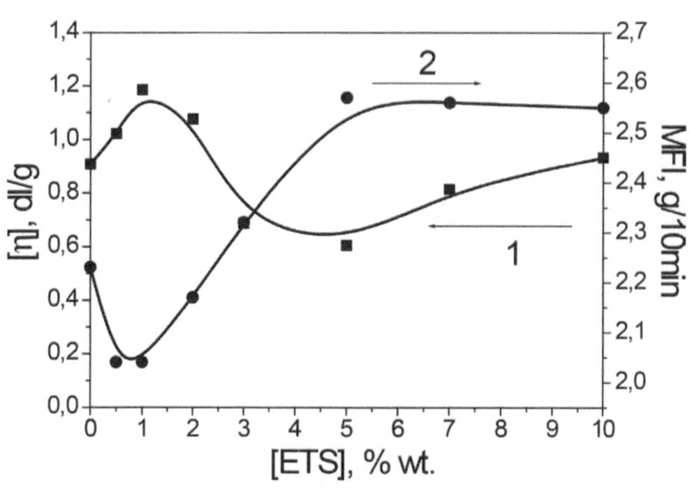

FIGURE 8.1 The dependence of MFI and the viscosity of the modified EVA20 on the content of ETS.

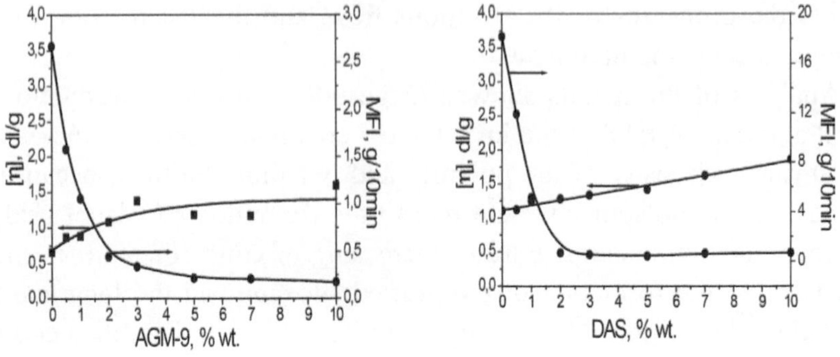

a б

FIGURE 8.2 The dependence of the intrinsic viscosity and MFI on the modifier content: a – EVA20 – AGM-9 (MFI 125°C – 2.16 kg); b – EVA27 – DAS (MFI 190°C – 5 kg).

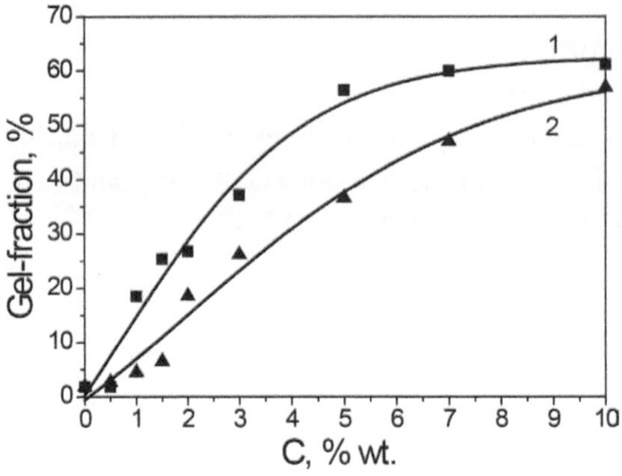

FIGURE 8.3 The dependence of EVAMA26 solubility in chloroform on the modifier content: 1 – AGM-9; 2 – DAS.

Extraction with boiling chloroform of the copolymers, modified by glycidoxysilane, showed that all the samples are completely soluble, and G_F content does not exceed 6%. This indicates the preferential formation of branched structures and the absence of cross-linked structures in the polymer. The physical linking, formed in the modified polymers, unlike the cross-linked structures, do not prevent the flow of the polymer. As well as at the introduction of aminosilanes, there is a decrease of the melt flow (Figure 8.5) of the modified polymers (by 30%) with simultaneous growth

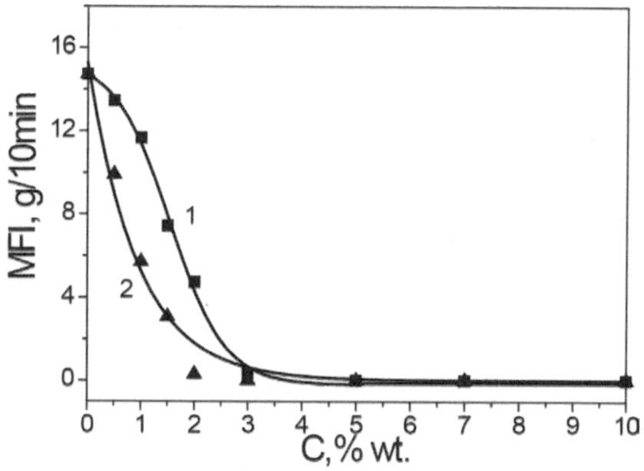

FIGURE 8.4 The dependence of EVAMA26 MFI (125°C – 2.16 kg) on the modifier content: 1 – AGM-9; 2 – DAS.

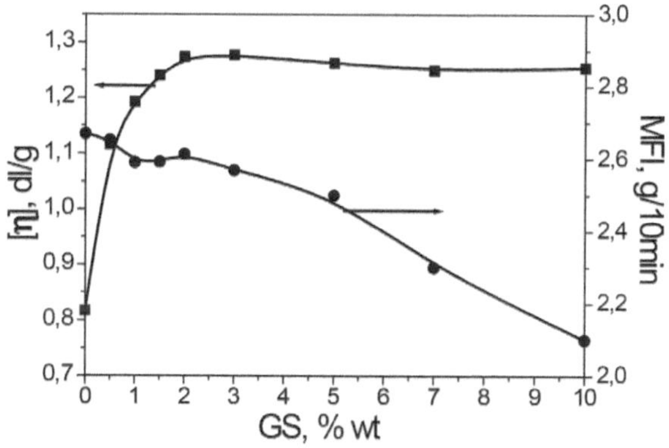

FIGURE 8.5 The dependence of the intrinsic viscosity and MFI of the system EVA20 – GS (MFI 125°C – 2.16 kg) on the modifier content.

of the intrinsic viscosity (in 1.3–1.5 times), which confirms the preferential formation of branches and intermolecular bridges, which increase the length of the macromolecule.

Introduction of organosilicon modifier affects the change of operational properties of the compositions. Physical and mechanical tests have shown that the introduction of ethyl silicate into the copolymers of ethylene with vinyl acetate improves their deformation and strength characteristics (Figure 8.6): for EVA20 by 10–15%, EVA27–15–20%, EVAMA13 and EVAMA26 by 15–20% and 10–15%. The magnitude of the modifying effect depends on the concentration of ester groups in the copolymer: the less the number of the acetate groups, the greater the amount of ethyl silicate spending on the growth of the siloxane chains, elasticizing the polymer.

Stress at break and elongation of EVAMA26 copolymers are practically unchanged (Figure 8.7). Since the modifier reacts both with acetate and with anhydride group, as a result of competing reactions appear a large number of short-chain branching which do not significantly affect he physical and mechanical properties.

In this case, the introduction of a modifier in the investigated copolymers more than 2–3% by weight is inexpedient because a further increase of its content has no significant effect on the properties of the copolymers.

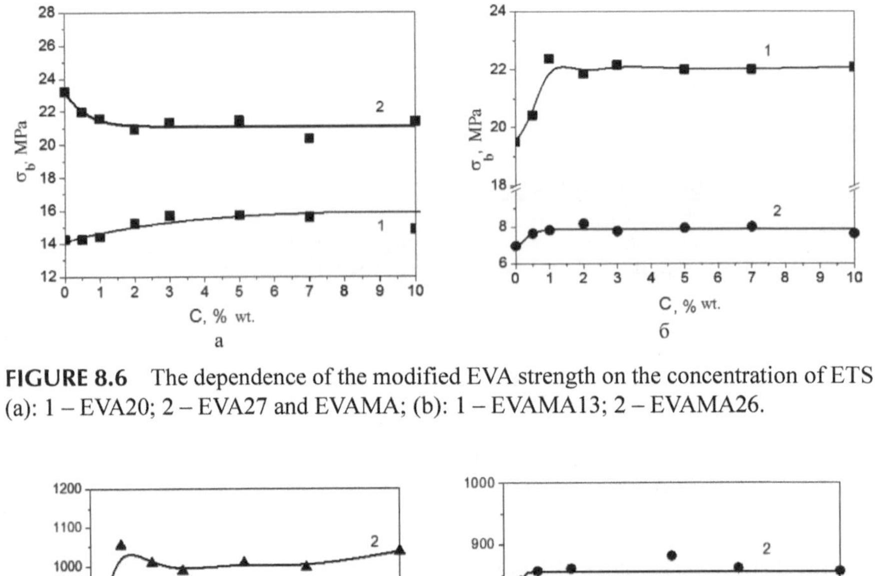

FIGURE 8.6 The dependence of the modified EVA strength on the concentration of ETS (a): 1 – EVA20; 2 – EVA27 and EVAMA; (b): 1 – EVAMA13; 2 – EVAMA26.

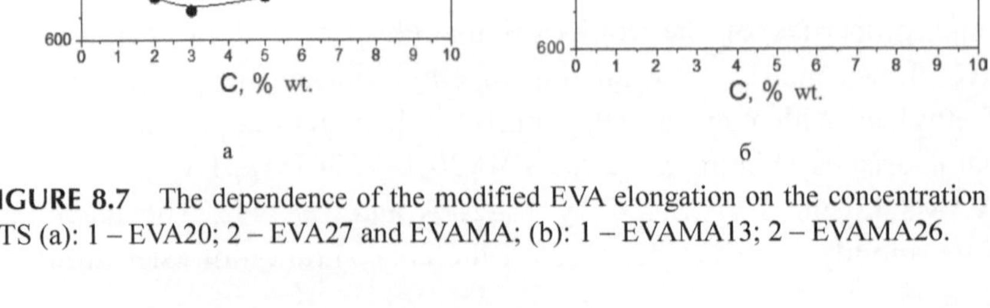

FIGURE 8.7 The dependence of the modified EVA elongation on the concentration of ETS (a): 1 – EVA20; 2 – EVA27 and EVAMA; (b): 1 – EVAMA13; 2 – EVAMA26.

With the addition of AGM-9 to EVA20 an increase in strength properties is observed in the entire range of concentrations, with the addition of DAS a maximum is observed at the modifier content of 7%. Consequently, the introduction of the modifier DAS in the amount of more than 7% is not effective, because it does not lead to an increase in strength properties (Figure 8.8).

The introduction of AGM-9 into EVA20 in the amount of 7% by weight leads to an increase of breaking stress by 47%, and in the case of DAS by 68% (Figures 8.9 and 8.10). In the case of GS strength variation has an extreme character, and the maximum value is observed at low (1% by weight) content of the modifier. A similar dependence is observed in the modification of EVA27 by aminosilane AGM-9, the strength of the

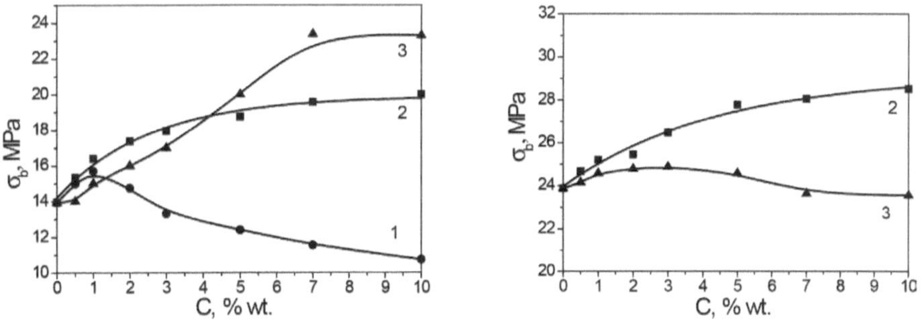

a б

FIGURE 8.8 The dependence of the modified EVA20 (a) and EVA27 (b) strength on the modifier content: 1 – GS; 2 – AGM-9; 3 – DAS.

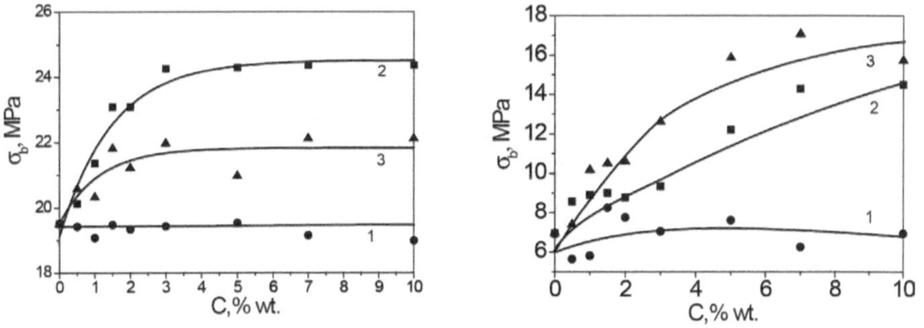

a б

FIGURE 8.9 The dependence of the modified EVAMA13 (a) and EVAMA26 (b) strength on the modifier content: 1 – GS; 2 – AGM-9; 3 – DAS.

polymer is increased by 25%. Effect of DAS on the strength of EVA27 is less expressed than in the modification of EVA20.

In all cases there is a decrease in elongation of the modified sevilen (Figure 8.10). Decrease in elongation at simultaneous growth of strength is typical for systems containing long- and short-chain branching or forming weakly-bonded chemical grid or grid of physical links.

Modification of terpolymers of ethylene with vinyl acetate and maleic anhydride by amine-containing alkoxysilanes is accompanied by a significant growth of the strength and deterioration of the deformation properties (Figures 8.10 and 8.11). Introduction of DAS into EVAMA13 increases its strength by 13%, whereas its elasticity reduces by 78%. In the case of

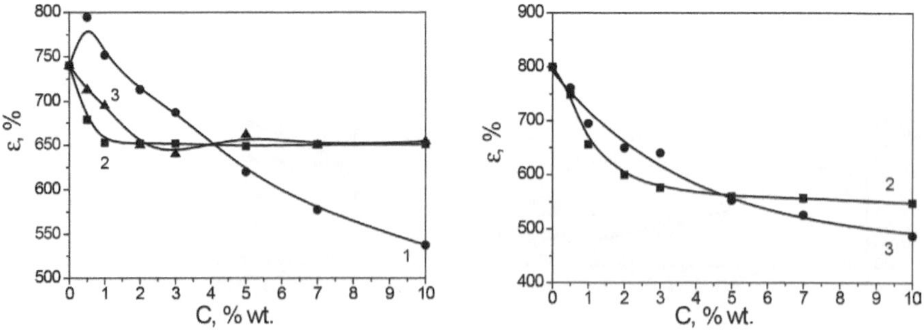

0

a б

FIGURE 8.10 The dependence of EVA20 (a) and EVA27 (b) elongation on the concentration of the modifier: 1 – GS; 2 – AGM-9; 3 – DAS.

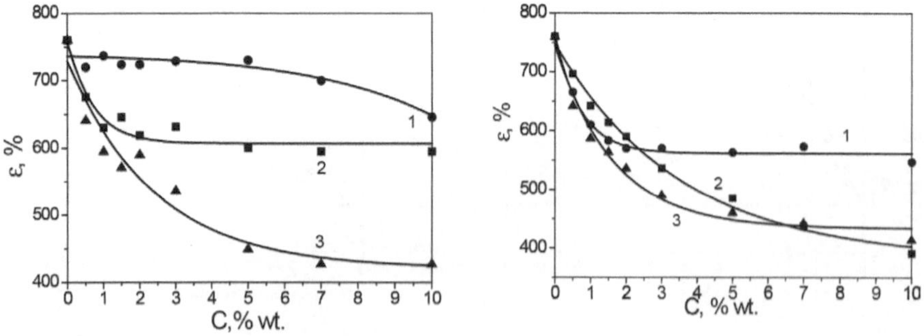

a б

FIGURE 8.11 The dependence of EVAMA13 (a) and EVAMA26 (b) elongation on the concentration of the modifier: 1 – GS; 2 – AGM-9; 3 – DAS.

EVAMA26 the strength of the composition increases ~ by 2.5 times, elongation reduces by 83%. Influence of AGM-9 is similar but is less pronounced. In the case of EVAMA13 its strength increases by 30% and elongation decreases by 13%. In the case of EVAMA26 its strength increases ~ by 2 times, elongation decreases ~ by 2 times.

Such relationships are typical for materials subjected to cross-linking, which confirms the assumption of chemical network formation, formed by the interaction of the functional groups of the polyolefin and the modifier. The use of glycidoxyalkoxysilane as a modifier practically does not change the strength characteristics of the ethylene copolymers.

8.4 CONCLUSION

Interest in the problem of silanol functionalization of polyolefins by limit organosilicon compounds is due to applied value of the materials on the base of the modified polymers and general fundamental problems of studying the chemical interactions in the systems silanes – polyolefins. The studies of influence of alkoxysilanes containing different functional groups on the complex of deformation and strength properties of binary and ternary copolymers of ethylene with vinyl acetate showed that aminoalkoxysilanes are more efficient modifiers than tetraalkoxy- and glycidoxysilanes.

ACKNOWLEDGEMENT

This work was financially supported by the Ministry of Education and Science of Russia in the framework of the theme №693 "Structured composite materials based on polar polymer matrices and reactive nanostructured components."

KEYWORDS

- **adhesion**
- **alkoxysilane**

- **copolymers of ethylene**
- **diffusion-sorption properties**
- **gel fraction**
- **melt index**
- **modification**
- **physico-mechanical properties**
- **viscosity**

REFERENCES

1. Temnikova, N. E., Rusanova, S. N., Yu. S. Tafeeva, S.Yu. Sofina, Stoyanov, O. V., The Effect of an Amino-Containing Modifier on Properties of Ethylene Copolymers. Polymer Science, Series D, 6(1), 19–25 (2013).
2. Rusanova, S. N., Yanaeva, A. O., Stoyanov, O. V., Shcherbina, A. A., Chalykh, A. E., Gerasimov, V. K., The effect of ethyl silicate on the adhesion strength of ethylene copolymers. International Polymer Science and Technology, 40(1), 47–49 (2013).
3. Temnikova, N. E., Thesis Cand. Chem. Sciences, KNRTU, Kazan, 2013. 154 p.
4. Stoyanov, O. V., Rusanova, S. N., Petukhova, O. G., Remizov, A. B., The chemical structure of ethylene-vinyl acetate copolymers, modified by limit alkoxysilane, according to IR. Journal of Applied Chemistry, 7, 1774–1177 (2001).
5. Rusanova, S. N., Thesis Cand. Tech. Sciences, KNRTU, Kazan, 2000. 120 p.
6. Petukhova, O. G., Thesis Cand. Chem. Sciences, KNRTU, Kazan, 2005. 113 p.
7. Rusanova, S. N., Stoyanov, O. V., Remizov, A. B., Yanaeva, A. O., Gerasimov, V. K., Chalykh, A. E., IR spectroscopic study of ethylene copolymers silanol modification. Herald of Kazan Technological University, 9, 346–352 (2010).
8. Temnikova, N. E., Rusanova, S. N., Yu. S. Tafeeva, Stoyanov, O. V., Research of ethylene copolymers modification by aminosilanes by IR spectroscopy FTIR method. Herald of Kazan Technological University, 19, 112–124 (2011).
9. Bounor-Legaré, V., Ferreira, A. Verbois, Ph. Cassagnau: A. Michel: New transesterification between ester and alkoxysilane groups: application to ethylene-co-vinyl acetate copolymer crosslinking. Polymer, 43, 6085–6092 (2002).

CHAPTER 9

UV SPECTROSCOPY STUDY OF 1,2-DIHYDRO-C60-FULLERENES IN POLAR SOLVENT

YULIYA N. BIGLOVA,[1] VADIM V. ZAGITOV,[1] MANSUR S. MIFTAKHOV,[2] RAISA Z. BIGLOVA,[1] and VLADIMIR A. KRAIKIN[2]

[1]*Chemistry Department, Bashkir State University, Ufa, Russia, E-mail: bn.yulya@mail.ru*

[2]*Institute of Organic Chemistry, URC RAS, Ufa, Russia*

CONTENTS

ABSTRACT

We have carried out the spectroscopic study of C_{60} and its mono-substituted derivatives (methanofullerenes) of process dissolution in concentrated (98%) sulfuric acid. There has been found a number of absorption maxima, whose position does not depend on the type of substituent. Several maxima coincide with the absorption maxima of anion- and cation-radicals C_{60} whereas the distance between them (~80 nm) coincides with $\Delta\lambda$ for ionized polyenes.

9.1 INTRODUCTION

Fullerenes, for example, specific carbon clusters with a closed system of double bonds with extremely low LUMO-energy are majorly of interest as n-type components of heterojunction phototransformators. Among functionalized fullerenes there were detected promising compounds for the practical use of renewable energy sources, molecular electronics and materials for medicine [1–3]. However the low solubility of C_{60} and its derivatives in polar and non-polar environments hinders their practical implementation. Water-soluble substances are required for medical use and film-forming properties, solubility in organic solvents, and stability in air are required for photovoltaic devices. Necessary and sometimes unique properties of C_{60} derivatives are built, first of all, by the functionalization of the fullerene molecule. Extraordinary properties of fullerene are often manifested only in the ionized form.

It is of interest that fullerene and fullerene cations radical are present as reasonably common components in a series of planetary nebulae in the interstellar medium and around certain astrophysical-objects [4, 5]. With the use of IR-spectroscopy and the follow-up calculations C_{60} (ionized form) was discovered in the cosmic space in the amount of 0.35% by weight of the total carbon.

There are several ways of generating C_{60} cations. For example, the cation and cation-radical of C_{60} were observed by spectroscope after irradiation of C_{60} in frozen matrix of argon, freon, carbon tetrachloride [6–9]; and generated by the matrix photoionization method [10]. C_{60} cation-radical were registered to be formed at dissolution of fullerene in oleum by way of EPR-spectroscopy [11] and UV spectroscopy [12]. The authors of the work [13] demonstrated that C_{60} cation-radicals are easily formed at room temperature in solutions of superacids (trifluoromethanesulfonic acid with potassium persulfate or oleum). The results obtained are confirmed by using other synthetic and analytical methods [14, 15]. There is spectroscopic evidence of such cations formation, based on the use of the model compounds spectra, as well as electronic spectra existing in scientific literature and theoretical calculations of C_{60} ion-radical [6, 10, 16, 17]. Cations are registered as well in organic solvents with the addition of Lewis acids ($SbCl_5$ and SbF_5) [18, 19]. Thus to generate superacid $HSbF_6$ weak Bronsted acid HF is mixed with SbF_5 [14].

Most carbocations are not sufficiently stable, which maintains their existence in the solution in such concentrations that are necessary to determine the electrical conductivity of the solution. UV-spectroscopy is most commonly used to register carbocations.

In this work, we carried out a spectroscopic study of a number of mono-substituted methanofullerenes and their initial C_{60} dissolved in 98% sulfuric acid. We discovered characteristic absorption bands whose maxima coincide with the absorption maxima of ionized fullerene C_{60} and polyenes.

9.2 EXPERIMENTAL PART

The objects of study were selected to be fullerene C_{60} (CJSC "Innovations of Leningrad Institutes and Enterprises," 99.5% of basic substance) and mono-substituted 1,2-dihydro-C_{60}-fullerenes (known as [60]PCBM) which we previously produced **(I)** [20] and synthesized by modified Bingel-Hirsch method **(II – IV)** (ESI†):

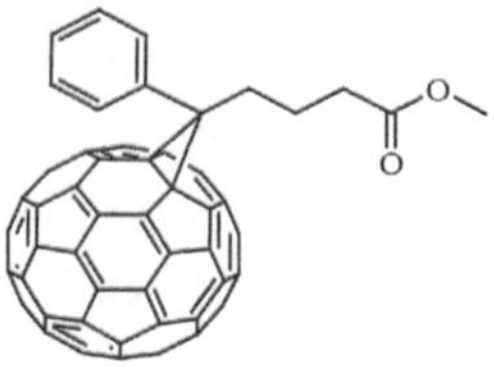

methyl ether of [6,6]-phenyl-C_{61}-butyl acid **(I)**

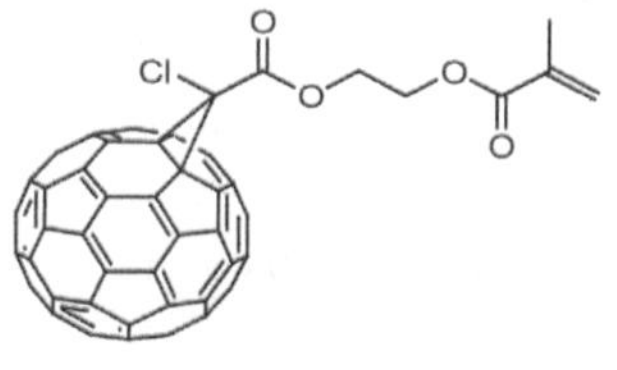

{1-chloro-1-[2-(methacryloyloxy)ethoxycarbonyl]-1,2-methane}-1,2-dihydro-C_{60}-fullerene **(II)**

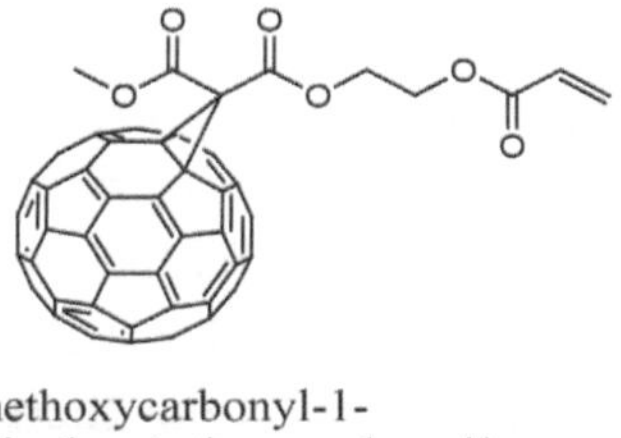

{(1-methoxycarbonyl-1-[(acryloyloxy)ethoxycarbonyl]-1,2-methane]}-1,2-dihydro-C_{60}-fullerene **(III)**

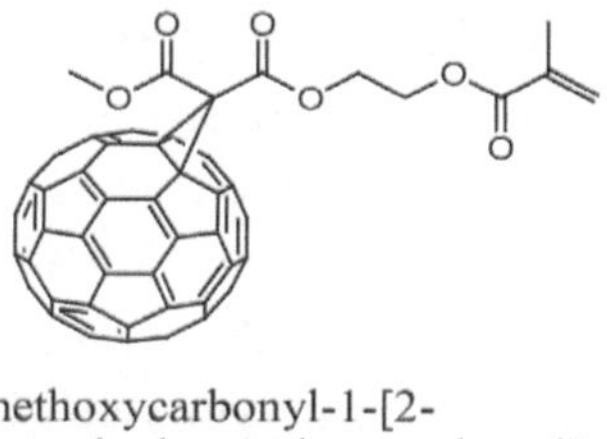

{(1-methoxycarbonyl-1-[2-(methacryloyloxy)ethoxycarbonyl]-1,2-methane]}-1,2-dihydro-C_{60}-fullerene **(IV)**

A test portion of fullerene (methanofullerenes) was dissolved at room temperature in chloroform. The aliquot of yellowish-brown solution with a concentration of the compound 10^{-4} M was placed in a volumetric flask, the solvent was evaporated and the dry residue was filled with 98% sulfuric acid. The concentration of completely dissolved methanofullerenes amounted to 1.70×10^{-5} M.

The electronic spectra of fullerene and methanofullerenes sulfuric acid solutions absorption were registered within the grid coordinates: optical density (A) – wavelength (λ) on UV-spectrophotometer UV-mini 1240 produced by "Shimadzu" with UV-Probe software. The spectra were recorded at room temperature in the wavelength range from 190 to 1100 nm (2.0 nm slit width, fast scanning speed) using 1 cm thick quartz cuvette. The start of the spectra capture was 5 min after the acid was added and it was repeated every 10 min.

C_{60} fullerene and its derivatives were weighed on Sartiorius M2P scales (sensitivity 10^{-6} g). The micropipettes Biohit Proline (Finland) with sample ranges of 10–100 and 100–1000 μL were used for aliquots intake. 98% sulfuric acid was distilled out of 94% H_2SO_4 in air at atmospheric pressure. Chloroform was purified by common methods [21].

UV spectra were researched by Zindo and TDDFT methods for fullerene and methanofullerenes cation structures. Geometrical parameters of claimed compounds studies were optimized by B3LYP/6–31 g (d). In the case of Zindo research, 500 transitions were taken into account for the analysis of the spectrum and in the case of TD research there were 50 transitions. All calculations were done using the software package Gaussian-09.

9.3 RESULTS AND DISCUSSION

Due to its high hydrophobicity crystalline fullerene C_{60} is poorly soluble in 98% sulfuric acid. Apparently, only a small portion of the surface of the fullerene nuclei in the crystal contacts with the acid. This is insufficient for a comprehensive protonation and solvation, and, therefore, for the transition of fullerene molecules in the solution. There are 4 characteristic absorption bands with maxima at 317, 260, 218 and 196 nm in the short-wave range in the electronic spectra of solutions derived by dissolving suspended fullerene in sulfuric acid. In the long-wave range, the characteristic bands are unidentifiable due to insufficient background indices (Figure 9.1). As the fullerene dissolves, the optical density in the absorption maxima grows linearly with different and very low speeds (Figure 9.2).

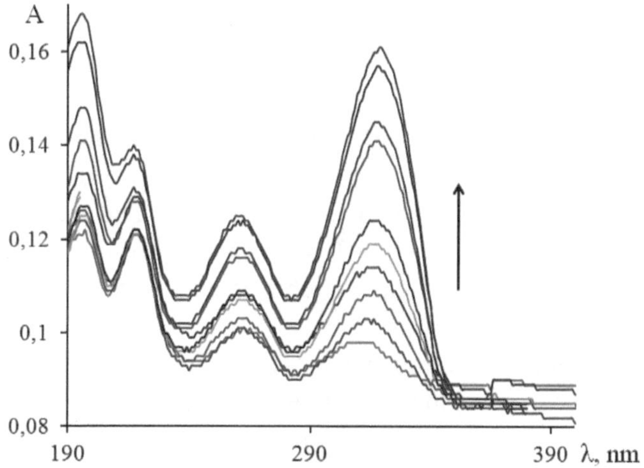

FIGURE 9.1 The evolution of fullerene C_{60} absorption electron spectrum dissolved in 98% sulfuric acid.

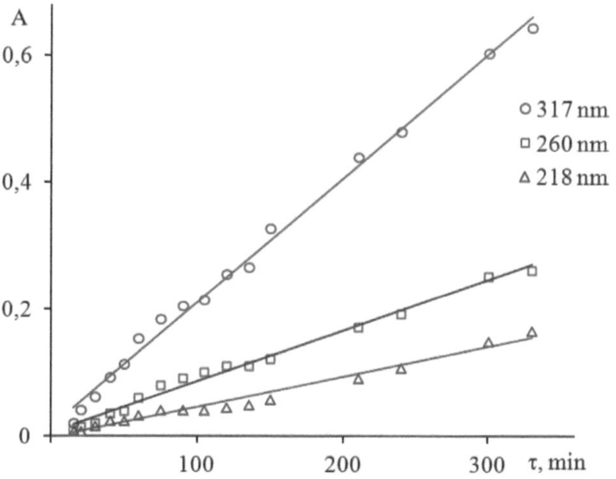

FIGURE 9.2 Change of optical density in maxima of absorption bands in the process of C_{60} fullerene dissolution in 98% sulfuric acid.

Methanofullerenes dissolve in sulfuric acid much better than the original C_{60}. For the functionalized fullerenes under analysis, the solubility increases sequentially as follows: {(1-methoxycarbonyl-1-[2-(methacryloyloxy)ethoxycarbonyl]-1,2-methane]}-1,2-dihydro-C_{60}-fullerene (**IV**) > {(1-methoxycarbonyl-1-[(acryloyloxy) ethoxycarbonyl]-1,2-methane]}-1,2-dihydro-C_{60}-fullerene (**III**) > methyl ether of [6,6]-phenyl-C_{61}-butyl acid (**I**) > {1-chloro-1-[2-(methacryloyloxy)ethoxycarbonyl]-1,2-methane}-1,2-dihydro-C_{60}-fullerene (**II**) (Figure 9.3). The best solubility is demonstrated by methanofullerenes with the largest number of easily protonatable oxygen atoms in the substituent (simple etheric and carbonyl ones), the second best is the compound **I** with phenyl group α-positioned in cyclopropane ring, and the least soluble is adduct **II** containing chlorine atom in the same position.

Regardless of the nature and length of the substituent in the C_{60} nucleus, the spectral curves of mono-substituted 1,2-dihydro-C_{60} fullerenes are identical: there are three absorption maxima (at 218, 251 and 317 nm) recorded in the near ultraviolet range, their intensity increases over time and after some time it reaches a constant value (Figure 9.4a). Moreover, the higher the solubility, the earlier the curve of the optical density dependent on time plateaus. After that, the optical density does not change, which may indicate the completion of the dissolution process of 1,2-dihydro-C_{60} fullerenes (the maximum is missing in the 196 nm range, presumably

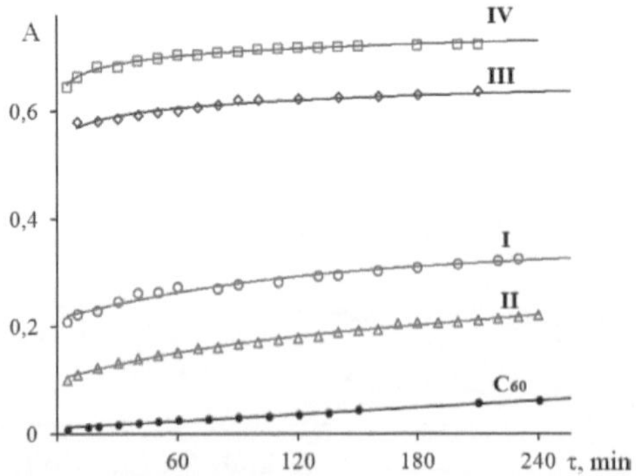

FIGURE 9.3 Optical density dependent on the duration of dissolution at a fixed wavelength of 317 nm and unchanging final concentration of compounds 1.7×10^{-5} M.

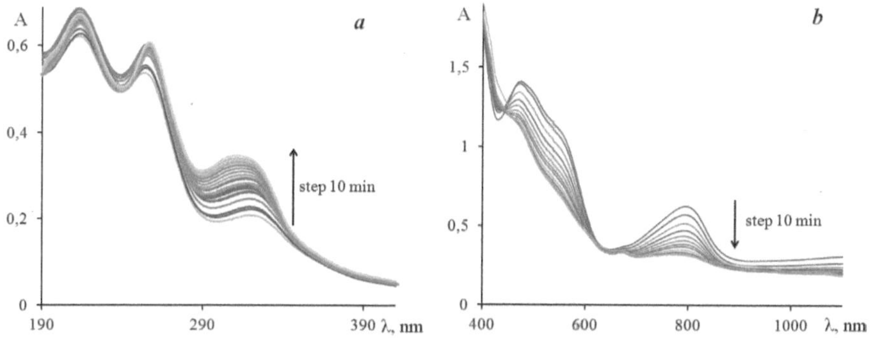

FIGURE 9.4 The evolution of the electron absorption spectrum in the short-wave (*a*) and long wave (*b*) ranges at the dissolution of mono-substituted adduct **I** in 98% sulfuric acid (similar character of spectrum changes was observed for all of the compounds under study).

due to the shift to the unregistable short-wave range). In the long-wave range of the spectrum, the opposite pattern is observed: the intensity of appeared absorption bands decreases over time. The changes in the range of 450–850 nm appear to be the most significant (Figure 9.4b).

If the dissolution of the original C_{60} in sulfuric acid is accompanied only by the increase in optical density of the solution, then, in case of methanofullerenes, there is as well a hypsochromic shift of the maxima for all of the compounds under study (Figure 9.5). The authors registered similar process in the long-wavelength range of the electron when fullerene is dissolved in oleum [12, 13]. In our case, the amount of the peaks displacement in the short wavelength range of the spectrum is a fraction of that in the long-wavelength region. The phenomena observed can be caused by processes similar to the processes occurring during the dissolution of polyenes in high-ionizing environments: protonation and cationic polymerization.

Indeed, the ionizations of fullerene and of polyene have a lot in common. This is a close coincidence of the series of absorption maxima and the distances at which they occur (Figure 9.6, Table 9.1). And in homologous series of polyenes, and in the series of the studied methanofullerenes, lengthening the chain of conjugation by 1 component leads to the shift of the absorption maximum by approximately 80 nm (Figure 9.7). For all methanofullerenes the value of $\Delta\lambda$ calculated by regression equations (Table 9.2) match well with calculated values for polyenes [23].

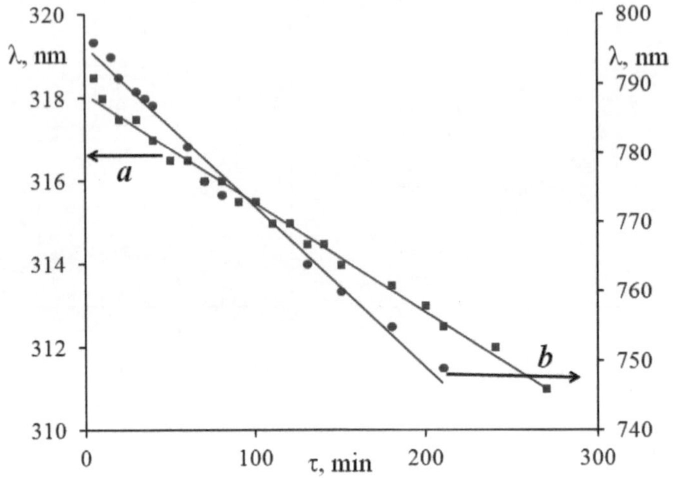

FIGURE 9.5 Shifts in time of short-wave (314 nm) (*a*) and long-wave (797 nm) (*b*) maxima of the absorption bands for methanofullerene **I**.

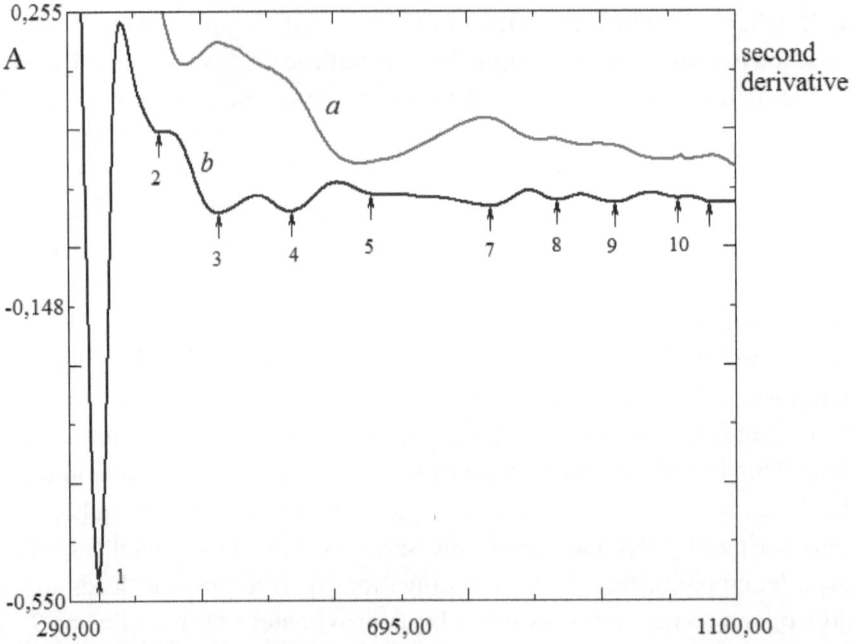

FIGURE 9.6 Electronic absorption spectrum (*a*) and its second derivative (*b*) of methanofullerene **IV** in 98% sulfuric acid.

TABLE 9.1 Wavelengths in the Absorption Band Maxima (Resulted from Experimental and Calculation Methods) and Number of Conjugated Bonds for Cations Formed by Dissolving Adducts and Polyenes in 98% Sulfuric Acid

Scheme	Number of conjugated bonds, n	Polyenes[22]	Experimental method, λ_{max}^{*} (nm)				Calculation method by Zindo, λ_{max} (nm)	
			I	II	III	IV	C_{60}^{+}	IV
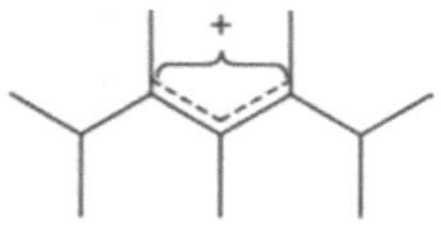	1	305.0	325.5	324.0	326.0	326.0	317.0	328.2
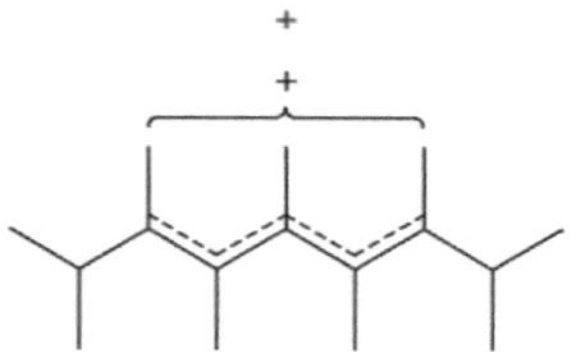	2	397.0	400.5	398.0	402.0	401.5	388.7	400.3
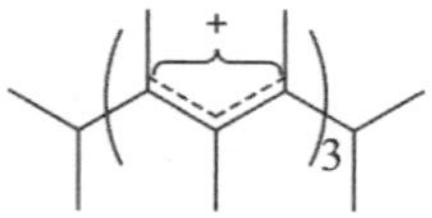	3	473.0	480.5	474.0	475.5	474.5	470.0	485.5
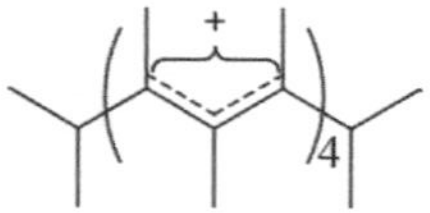	4	550.0	559.5	559.5	563.0	562.0	552.2	553.6
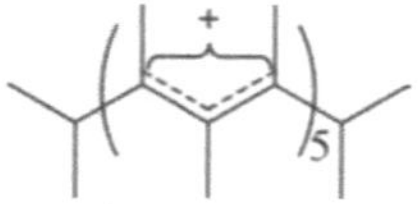	5	630.0***	628.0	628.0	626.5**	628.0**	620.9	636.1

Scheme	Number of conjugated bonds, n	Polyenes[22]	Experimental method, λ_{max}^{*} (nm)				Calculation method by Zindo, λ_{max} (nm)	
			I	II	III	IV	C_{60}^{+}	IV
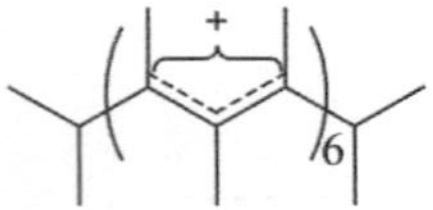	6	710.0***	692.5**	694.0**	—	—	681.5	694.2
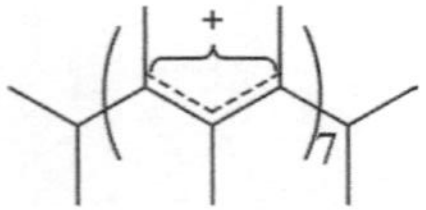	7	790.0***	806.0	804.5	803.5	805.0	—	825.8
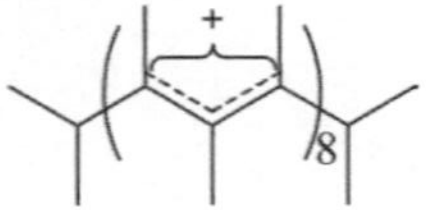	8	870.0***	886.5	885.5	886.5	886.0	853.6	—
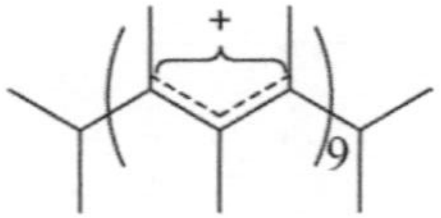	9	950.0***	956.5	953.0	954.5	955.0	964.1	965.2
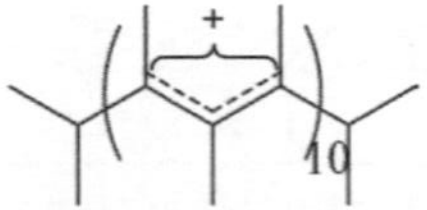	10	1030.0***	1029.5	1033.0	1031.0	1030.5	—	—

* for methanofullerenes defined by the second derivative 5 min after sulfuric acid was introduced.

** arm on the absorption spectra.

***calculated continuable series [22].

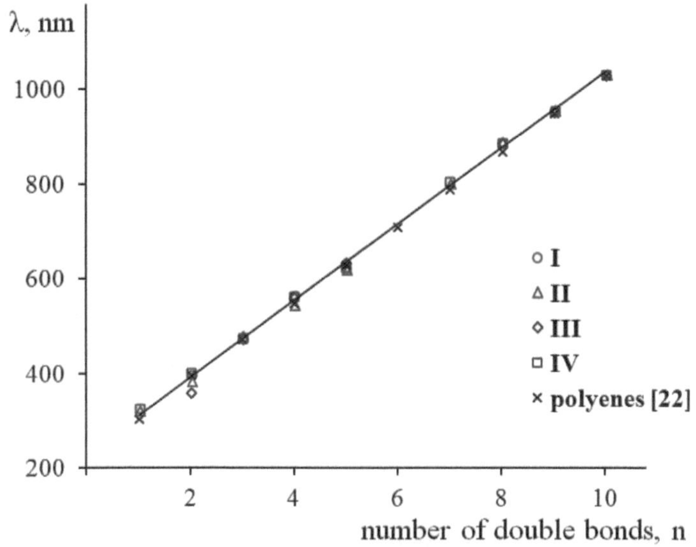

FIGURE 9.7 Dependence of wavelength in the absorption maxima on the number of conjugated bonds in polyenes and methanofullerenes.

TABLE 9.2 Regression Equation and Correlation Coefficients for Methanofullerenes and Polyenes

Compound	Equation	R^2
I	y=79.1x+236.6	0.9918
II	y=79.0x+240.7	0.9960
III	y=78.8x+232.6	0.9954
IV	y=78.5x+228.0	0.9916
Ref. [22]	y=81.1x+228.5	0.9979

This match allows us to conclude that in the case of the fullerene core of methanofullerenes non-localized electron cloud splits into several oscillating linear vibrators (similar to polyene ones) of different lengths, each of which is characterized by its absorption band in the electronic spectrum. It can mean that, by analogy with polyenes, non-localized electron cloud does not split into separate vibrators, but oscillates as a whole. Thus, apparently, there is a set of fullerene nuclei with various lengths of adjacent sequences, whose specific content differs. Besides, the more atoms are harnessed by π-molecular orbital, e.g., the longer the system of conjugated sequences is, the lower the energy difference between the ground and upper states is and, therefore, the more visible is the shift of the absorption maximum toward longer wavelengths, and the higher the intensity of absorption is.

A significant part of the absorption bands in the electronic spectra registered in methanofullerenes sulfuric acid solutions electron spectra is present in the spectra of fullerene C_{60}^+ core of these compounds, derived by calculation (Figure 9.8, Table 9.1).

As for the comparative analysis of the data obtained in the present work and in the scientific literature per se, namely in the overview [15], the spectroscopically measured values of the maximum absorption of C_{60}^+ in various publications are very different. The authors as a rule confine themselves to haphazard statement of the absorption maxima of emerging C_{60}^+. The research [13] determined the peaks of absorption maxima at 477.4, 633.3 nm, and research [12] identified the peaks corresponding to wavelengths of 320.9 and 478.2 nm, which agrees with the values (Table 9.1) determined in the present paper. Bearing in mind that the fullerene is a linear structure, we identified the patterns for C_{60}^+ correlating with those for carbocation structures with long system of conjugated double bonds.

Some of the absorption bands that we have identified before were attributed to both anion-radicals and cation-radicals [23].

Thus, the modification of fullerene by functionalization of its core by easily protonating double bonds and oxygen atoms can significantly increase the local concentration of fullerene cores in concentrated (98%) sulfuric acid. The latter, in its turn, allows to reliably identify a series of

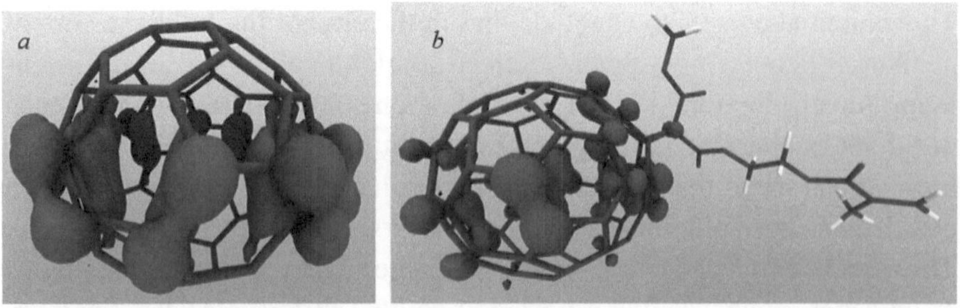

FIGURE 9.8 Visualizations of the C_{60}^+ (*a*) and carbocation of methanofullerene **IV** (*b*).

the absorption bands in the visible spectrum, many of the bands are identical to the absorption bands of fullerene cation-radicals derived by dissolving it in oleum and in superacids.

ACKNOWLEDGMENTS

The study was carried out under a financial support of the Russian Foundation for Basic Research (grants № 14-03-31610 mol_a, 14-02-97008).

KEYWORDS

- **fullerene**
- **ionization**
- **mono-substituted 1,2-dihydro-C60-fullerenes**
- **[60]PCBM**
- **sulfuric acid**
- **UV spectroscopy**

REFERENCES

1. Chen, Y. C., Hsu, C. Y., Lin, R., Y. Y., K-Ho, C., Lin, J. T., *Chem. Sus. Chem.*, 2013, *6*, 20.
2. Jariwala, D., Sangwan, V. K., Lauhon, L. J., Marks, T., J., Hersam, M. C., ***Chem. Soc. Rev.***, 2013, ***42***, 2824.
3. Bianco, A., Ros, T. D., *Fullerenes: Principles and Applications*, 2012, *2*, 507.
4. Foing, B., H., Ehrenfreund, P., *Nature*, 1994, 369, 296.
5. Cami, J., Bernard-Salas, J., Peeters, E., Malek, S. E., *Science*, 2010, 329, 1180.
6. Kato, T., Kodama, T., Shida, T., Nakagawa, T., Matsui, Y., Suzuki, S., Shi-romaru, H., Yamauchi, K., Achiba, Y., *Chem. Phys. Lett.*, 1991, 180, 446.
7. Gasyna, Z., Andrews, L., Schatz, P. N., *J. Phys. Chem.*, 1992, 95, 1525.
8. Fulara, J., Jakobi, M., Maier, J. P., *Chem. Phys. Lett.*, 1993, 211, 227.
9. Fulara, J., Jakobi, M., Maier, J. P., *Chem. Phys. Lett.*, 1993, 206, 208.
10. Gasyna, Z., Andrews, L., Schatz, P. N., *J. Phys. Chem.*, 1992, 96, 1525.
11. Olah, G. A., Bucsi, I., Aniszfeld, R., Prakash, G., K. S., *Carbon*, 1992, 30, 1203.

12. Cataldo, F., Iglesias-Groth, S., Hafez, Y., *Eur. Chem. Bull.*, 2013, 2, 1013.
13. Cataldo, F., *Spectrochimica Acta*, 1995, 51A, 405.
14. Reed, C. A., Kim, K. C., Bolskar, R., D., Mueller, L. J., *Science*, 2000, 289, 101.
15. Reed, C. A., Bolskar, R. D., *Chem. Rev.*, 2000, 100, 1075.
16. Sahnoun, R., Nakai, K., Sato, Y., Kono, H., Fujimura, Y., Tanaka, M., *Chem. Phys. Lett.*, 2006, 430, 167.
17. Mattesini, M., Ahuja, R., L.Sa, H. Hugosson, W., Johansson, B., Eriksson, O., *Physica B*, 2009, 404, 1776.
18. Baumgarten, M., Gherghel, L., *Appl. Magn. Reson.* 1996, 11, 171.
19. Bolskar, R. D., Ph.D. Dissertation, University of Southern California, Los Angeles, CA, 1997.
20. Hummelen, J. C., Knight, B. W., LePeq, F., Wudl, F., Yao, J., Wilkins, C. L., *J. Org. Chem.*, 1995, 60, 532.
21. Rabinovich, V. A., Khavin, Y. Z., Kratkii khimicheskii spravochnik [Concise Handbook of Chemistry] 3rd ed., rev. *Leningrad: Chemistry*, 1991. 432 p.
22. Dneprovsky, A., S., Temnikova, T. I., Teoreticheskie osnovy organicheskoi khimii [*Theoretical Bases of Organic Chemistry*]. Chemistry. 1991. 560 p.
23. Kato, T., Kodama, T., Shida, T., *Chem. Phys. Lett.*, 1991, 180, 446.

CHAPTER 10

HEXAGONAL STRUCTURES IN PHYSICAL CHEMISTRY AND PHYSIOLOGY

G. A. KORABLEV,[1] YU. G. VASILIEV,[1] and G. E. ZAIKOV[2]

[1]*Izhevsk State Agricultural Academy, Russia,*
E-mail: korablevga@mail.ru, devugen@mail.ru

[2]*N.M. Emmanuel Institute of Biochemical Physics, Russian Academy of Sciences, Moscow, Russia, E-mail: chembio@sky.chph.ras.ru*

CONTENTS

ABSTRACT

Some principles of forming carbon cluster nanosystems are analyzed based on spatial-energy ideas. The dependence nomogram of the degree of structural interactions on coefficient α is given, the latter is considered as an analog of entropic characteristic. The attempt is made to explain the specifics of forming hexagonal cell clusters in biosystems.

10.1 INTRODUCTION

Main components of organic compounds constituting 98% of cell elemental composition are: carbon, oxygen, hydrogen and nitrogen. The polypeptide bond formed by COOH and NH_2 groups of amino acid CONH acts as the binding base of cell protein biopolymers.

Thus, carbon is the main conformation center of different structural ensembles, including the formation of cluster compounds. In the Nobel lecture in physiology Edvard Moser [1] pointed out such analogy and presented some trial data, which, probably need to have additional theoretical confirmation. For further discussion of these problems the idea of spatial-energy parameter (P-parameter) is introduced in this paper.

10.2 INITIAL CRITERIA

The idea of spatial-energy parameter (P-parameter) which is the complex characteristic of the most important atomic values responsible for interatomic interactions and having the direct bond with the atom electron density is introduced based on the modified Lagrangian equation for the relative motion of two interacting material points [2].

The value of the relative difference of P-parameters of interacting atoms-components – the structural interaction coefficient α is used as the main numerical characteristic of structural interactions in condensed media:

$$\alpha = \frac{P_1 - P_2}{(P_1 + P_2)/2} 100\% \tag{1}$$

Applying the reliable experimental data we obtain the nomogram of structural interaction degree dependence (ρ) on coefficient α, the same for a wide range of structures (Figure 10.1). This approach gives the possibility to evaluate the degree and direction of the structural interactions of phase formation, isomorphism and solubility processes in multiple systems, including molecular ones.

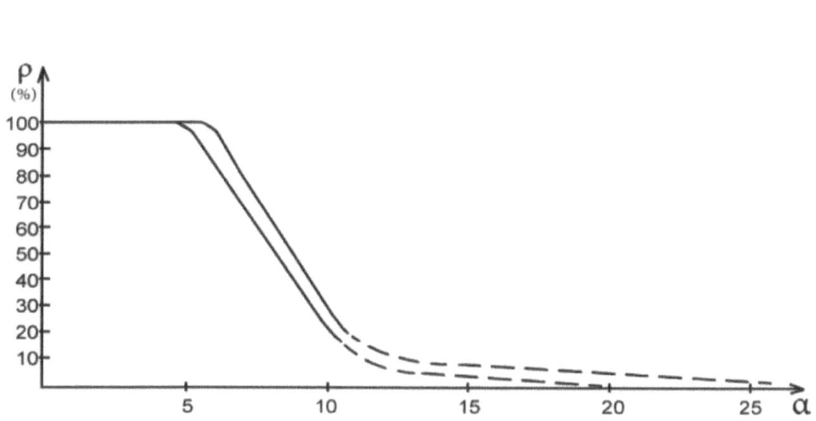

FIGURE 10.1 Nomogram of structural interaction degree dependence (ρ) on coefficient α.

Such nomogram can be demonstrated [2] as a logarithmic dependence:

$$\alpha = \beta\left(\ln \rho\right)^{-1} \tag{2}$$

where coefficient β – the constant value for the given class of structures. β can structurally change mainly within ± 5% from the average value. Thus coefficient α is reversely proportional to the logarithm of the degree of structural interactions and therefore, by analogy with Boltzmann equation, can be characterized as the entropy of spatial-energy interactions of atomic-molecular structures [3].

Actually the more is ρ, the more probable is the formation of stable ordered structures (e.g., the formation of solid solutions), e.g., the less is the process entropy. But also the less is coefficient α.

The Eq. (2) does not have the complete analogy with entropic Boltzmann's equation as in this case not absolute but only relative values of the corresponding characteristics of the interacting structures are compared, which can be expressed in percent. This refers not only to coefficient α but also to the comparative evaluation of structural interaction degree (ρ), for example – the percent of atom content of the given element in the solid solution relatively to the total number of atoms.

In conclusion, the relative difference of spatial-energy parameters of the interacting structures can be a quantitative characteristic of the interaction entropy: $\alpha \equiv S$.

10.3 FORMATION OF CARBON NANOSTRUCTURES

After different allotropic modifications of carbon nanostructures (fullerenes, tubules) have been discovered, a lot of papers dedicated to the investigations of such materials, for instance were published, determined by the perspectives of their vast application in different fields of material science.

The main conditions of stability of these structures formulated based on modeling the compositions of over thirty carbon clusters are given [4]:

1. Stable carbon clusters look like polyhedrons where each carbon atom is three-coordinated.
2. More stable carbopolyhedrons containing only 5- and 6-term cycles.
3. 5-term cycles in polyhedrons – isolated.
4. Carbopolyhedron shape is similar to spherical.

Let us demonstrate some possible explanations of such experimental data based on the application of spatial-energy concepts. The approximate equality of effective energies of interacting subsystems is the main condition for the formation of stable structure in this model based on the following equation:

$$\left(\frac{P_0}{KR}\right)_1 \approx \left(\frac{P_0}{KR}\right)_2; \quad P_1 \approx P_2 \tag{3}$$

where K – coordination number, R – bond dimensional characteristic.

At the same time, the phase-formation stability criterion (coefficient α) is the relative difference of parameters P_1 and P_2 that is calculated following the Eq. (1) and is αST<(20–25)% (according to the nomogram).

During the interactions of similar orbitals of homogeneous atoms we have

$$K_1 R_1 \approx K_2 R_2 \tag{3a}$$

Let us consider these initial notions as applicable to certain allotropic carbon modifications:

1. Diamond. Modification of structure where $K_1=4$, $K_2=4$; $R_1=R_2$, $P_1=P_2$ and $\alpha=0$. This is absolute bond stability.

2. Non-diamond carbon modification for which, $K_1=1$; $R_1=0.77$ Å; $K_2=4$; $R_2^{4+}=0,2$Å, $\alpha=3,82\%$. Absolute stability due to ionic-covalent bond.
3. Graphite, $K_1=K_2=3$, $R_1=R_2$, $\alpha=0$ – absolute bond stability.
4. Chains of hydrocarbon atoms consisting of the series of homogeneous fragments with similar values of P-parameters.
5. Cyclic organic compounds as a basic variant of carbon nanostructures. Apparently, not only inner-atom hybridization of valence orbitals of carbon atom takes place in cyclic structures, but also total hybridization of all cycle atoms.

But not only the distance between the nearest similar atoms by bond length (d) is the basic dimensional characteristic, but also the distance to geometric center of cycle interacting atoms (D) as the geometric center of total electron density of all hybridized cycle atoms.

Then the basic stabilization equation for each cycle atom will take into account the average energy of hybridized cycle atoms:

$$\left(\frac{\Sigma P_0}{Kd}\right)_i^{'} \approx \left(\frac{\Sigma P_0}{KД}\right)_i^{''} \tag{4}$$

$$P' \approx P'' \tag{4a}$$

where $\Sigma P_0=P_0N$; N – number of homogeneous atoms, P_0 – parameter of one cycle atom, K – coordination number relatively to geometric center of cycle atoms. Since in these cases $K'=K''$ and $N'=N$, $K=N$, the following simple correlation for paired bond appears:

$$\frac{P_0^{'}}{d} \approx \frac{P_0^{''}}{Д}; \quad P_E^{'} \approx P_E^{''} \tag{5}$$

During the interactions of similar orbitals of homogeneous atoms $P_0^{'} \approx P_0^{''}$, and then:

$$d \approx D \tag{5a}$$

Equation (5) reflects a simple regularity of stabilization of cyclic structures:

In cyclic structures the main condition of their stability is an approximate equality of effective interaction energies of atoms along all bond directions.

The corresponding geometric comparison of cyclic structures consisting of 3, 4, 5 and 6 atoms results in the conclusion that only in 6-term cycle (hexagon) the bond length (d) equals the length to geometric center of atoms (D): d=D.

Such calculation of α following the equation analogous to (1), gives for hexagon $\alpha=0$ and absolute bond stability. And for pentagon $d\approx1.17D$ and the value of $\alpha=16\%$, e.g., this is the relative stability of the structure being formed. For the other cases $\alpha>25\%$ – structures are not stable. Therefore hexagons play the main role in nanostructure formation and pentagons are additional substructures, spatially limited with hexagons. Based on stabilization equation hexagons can be arranged into symmetrically located conglomerates consisting of several hexahedrons.

It is assumed that defectless carbon nanotubes (NT) are formed as a result of rolling the bands of flat atomic graphite net. The graphite has a lamellar structure, each layer of which is composed of hexagonal cells. Under the center of hexagon of one layer there is an apex of hexagon of the next layer.

The process of rolling flat carbon systems into NT is, apparently, determined by polarizing effects of cation-anion interactions resulting in statistic polarization of bonds in a molecule and shifting of electron density of orbitals in the direction of more electronegative atoms.

Thus, the aforesaid spatial-energy notions allow characterizing in general the directedness of the process of carbon nanosystem formation [5].

10.4 HEXAGONAL STRUCTURES IN BIOSYSTEMS

In the full-on report by Edvard Moser [1] the following problem results can be pointed out:

1. Cluster structures of cells form geometrically symmetrical hexagonal systems.
2. Cells themselves statistically concentrate along coordinate axes of symmetry with deviations not exceeding 7.5% (Figure 10.2).
3. For independent cluster systems in different excitation activity phases four modules which differ scale-wisely on coefficients can be pointed out: 1.4–1.421.

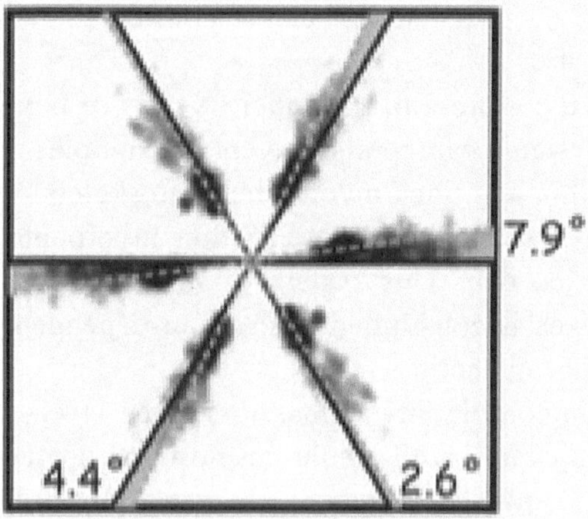

FIGURE 10.2 Statistic distribution of cells along coordinate axes [1].

Cluster C_{60} containing 60 three-coordinated atoms and 180 effective bonds is the smallest stable carbon cluster. The similar structure is most probable in biosystems even with the availability of three-coordinated bonds of nitrogen atoms due to its $2P^3$-orbitals. This is cluster K_{60}. The second module of clusters with the coefficient 1.4 has 252 bonds, which means 72 additional bonds and corresponds to 12 new hexagons or 24 new atoms of the system forming cluster K_{84}. By the way, such cluster as C_{84} is among stable ones in carbon systems. If the rectified coefficient 1.421 is used in calculations, four more additional bonds are formed, which, apparently, play a binding role between the cluster subsystems.

The carbon cluster C_{60} contains 12 pentagons separated with 20 hexagons. The pentagons can be considered as a defect of graphite plane but structurally stabilizing the whole system. Still it is unknown if there are similar formations in biosystems.

It can be assumed that entropic statistics of distribution of activity degree of structural interactions given in section 2 based on the nomogram (Figure 10.1) is also fulfilled in biosystems. Thus, according to the nomogram if $\alpha < 7\%$, the maximum of structural interactions is observed, and their sharp decrease – if $>7\%$.

Therefore the maximal deviation angle of cell statistic distribution from coordinate axes equal to 7.5° can be considered the demonstration of entropic regularity.

Apart from the aforesaid, a number of facts of hexagonal formation of biological systems can be also given as examples. For instance, the collocation of thin and thick myofilaments in skeletal muscle fibers and cardiomyocytes. At the same time, 6 thin myofilaments are revealed around each thick one. This system of functionally linked macromolecular complexes is constituted of calcium-dependent transient bonds between myosins and actins.

Also mechanotoropic interactions in surface layers of multilayer flat epitheliums conjugated with ample desmosomal contacts creating force fields on the background of the available hydrostatic pressure in epithelial cells are naturally followed by the formation of ordered epidermal columns with flat cells in the surface mainly of hexagonal shape and more rarely – of pentagonal one. There are also other options of revealing the aforesaid regularities.

10.5 CONCLUSION

The comparisons and calculations carried out based on spatial-energy ideas allow explaining some features of forming hexagonal structures in biosystems.

KEYWORDS

- **cell systems**
- **entropy**
- **hexagonal clusters**
- **spatial-energy parameter**

REFERENCES

1. Moser, E. Nobel lecture in physiology: 11.03.2015, TV channel "Science."
2. Korablev, G. A. Spatial-Energy Principles of Complex Structures Formation. Brill Academic Publishers and VSP, Netherlands, 2005, 426 pp.
3. Korablev, G. A., Petrova, N. G., Korablev, R. G., Osipov, A. K., Zaikov, G. E. On Diversified Demonstration of Entropy. Polymers Research Journal. 2014. Vol. 8, № 3. 145–153.
4. Sokolov, V. I., Stankevich I. F. Fullerenes – new allotropic forms of carbon: structure, electron composition and chemical properties. Successes in chemistry, 1993, v. 62, №5, 455–473.
5. Korablev G. A., Zaikov, G. E. Formation of carbon nanostructures and spatial-energy criterion of stabilization. Mechanics of composite materials and structures: RAS – IPM. 2009. V. 15, № 1. 106–118.-

CHAPTER 11

COMPLEX FORMATION BETWEEN ALK_4NBR AND 1,1,3-TRIMETHYL-3-(4-METHYLPHENYL)BUTYL HYDROPEROXIDE ON THE BASE OF NMR 1H INVESTIGATION

YU. V. BERESTNEVA,[1] E. V. RAKSHA,[1] N. A. TUROVSKIJ,[2] and G. E. ZAIKOV[3]

[1]*L.M. Litvinenko Institute of Physical Organic and Coal Chemistry. 70 R. Luxemburg Street, 83-114, Donetsk*

[2]*Donetsk National University, 24 Universitetskaya Street, 83–001 Donetsk, E-mail: elenaraksha411@gmail.com*

[3]*Institute of Biochemical Physics, Russian Academy of Sciences, 4 Kosygin Street, 117–334 Moscow, Russian Federation, E-mail: chembio@sky.chph.ras.ru*

CONTENTS

ABSTRACT

The formation of a complex between 1,1,3-trimethyl-3-(4-methylphenyl) butyl hydroperoxide and tetraalkylammonium bromides (Alk_4NBr) in $CDCl_3$ was experimentally confirmed by NMR ^1H spectroscopy. Thermodynamic parameters for the hydroperoxide-Alk_4NBr complexes were estimated. Effect of the Alk_4N^+ cation on the complex parameters is discussed.

11.1 AIMS AND BACKGROUND

Reactions of peroxide compounds formation and radical decomposition are key to process of hydrocarbons oxidation by molecular oxygen [1, 2], AOP [3], DNA oxidative destruction [4], as well as modification of carbon nanomaterials [5]. Systematic investigations of the catalytic decomposition of organic peroxides (diacyl peroxides [6], cyclohexanone peroxides [2], arylalkyl hydroperoxides [7]) in the presence of a quaternary ammonium salt revealed that a complex formation is the reaction first stage. Results of the molecular modeling of peroxides associative interactions with tetraalkylammonium bromides [7, 8] confirmed that the complexation leads to increased reactivity of the peroxides. The enthalpies of formation of hydroperoxide – tetraalkylammonium bromide complexes estimated on the base of kinetic data were shown to be within (–15 | −22) kJ mol) [7]. NMR ^1H spectroscopy are widely used to investigate the labile supramolecular complexes characterized by non-covalent interactions, that allows to identify structural fragments involved in the intermolecular interactions, as well as to determine the complexation thermodynamic parameters [9].

The aim of this work is investigation of the 1,1,3-trimethyl-3-(4-methylphenyl) butyl hydroperoxide (ROOH) interaction with tetraalkylammonium bromides (Alk_4NBr) by the NMR ^1H spectroscopy. A series of the studied salts included tetraethylammonium, tetrapropylammonium, and tetrabutylammonium bromides.

11.2 EXPERIMENTAL PART

1,1,3-Trimethyl-3-(4-methylphenyl)butyl hydroperoxide was purified according to Ref. [10]. Tetraalkylammonium bromides (Et_4NBr, Pr_4NBr, and Bu_4NBr) were purified by double reprecipitation from an acetonitrile solution with an excess of diethyl ether and stored in a box dried with P_2O_5. Their purity (99.9%) was determined by argentometric titration with the potentiometric fixation of the equivalence point. The NMR ^1H spectra of the hydroperoxide solution and hydroperoxide – tetraalkylammonium mixtures were recorded on a Bruker Avance II 400 (400 MHz). The solvent, chloroform-d (Sigma-Aldrich), was used without additional purification but were stored above molecular sieves for at least 3 days before solution preparation. Tetramethylsilane was an internal standard.

1,1,3-Trimethyl-3-(4-methylphenyl)butyl hydroperoxide, 4-$CH_3C_6H_{4-}$ $C(CH_3)_{2-}CH_{2-}(CH_3)_2COOH$ NMR ^1H (400 MHz, $CDCl_3$, 297 K, δ ppm, *J*/Hz): 1.00 (s, 6H, $-C(CH_3)_2OOH$), 1.39 (s, 6H, $-C_6H_4C(CH_3)_{2-}$), 2.05 (s, 2H, $-CH_{2-}$), 2.32 (s, 3H, $CH_{3-}C_6H_{4-}$), 7.11 (d, *J* = 8.0, 2H, H-aryl), 7.29 (d, *J* = 8.0, 2H, H-aryl), 6.77 (s, 1 H, -COOH).

Tetraethylammonium bromide, Et_4NBr. The decomposition temperature is 285°C. NMR ^1H (400 MHz, $CDCl_3$, 297 K, δ ppm, *J*/Hz): 1.41 (t, 12H, Me, *J* = 8.0 Hz); 3.50 (q, 8H, CH_2, *J* = 8.0 Hz).

Tetrapropylammonium bromide, Pr_4NBr. The decomposition temperature is 270°C. NMR ^1H (400 MHz, $CDCl_3$, 297 K, δ ppm, *J*/Hz): 1.07 (t, 12H, Me, *J* = 8.0 Hz); 1.79 (sextet, 8H, CH2, *J* = 8.0 Hz); 3.37 (t, 8H, CH_2, *J* = 8.0 Hz).

Tetrabutylammonium bromide, Bu_4NBr. The melting point is 103–104°C. NMR ^1H (400 MHz, $CDCl_3$, 297 K, δ ppm, *J*/Hz): 1.01 (t, 12H, Me, *J* = 8.0 Hz); 1.47 (sextet, 8H, CH2, *J* = 8.0 Hz); 1.71 (q, 8H, CH_2, *J* = 8.0 Hz); 3.40 (t, 8H, CH_2, *J* = 8.0 Hz).

11.3 RESULTS AND DISCUSSION

The interaction between 1,1,3-trimethyl-3-(4-methylphenyl)butyl hydroperoxide and tetraalkylammonium bromides was observed by relative change of the chemical shifts in the NMR ^1H spectra. The effect of Alk_4NBr on the signal position of hydroperoxide in the proton magnetic resonance spectrum has been investigated to confirm the complex formation between ROOH and Alk_4NBr at 297–313°K. In the NMR ^1H spectrum of ROOH chemical shift at 6.77 ppm (Figure 11.1) is assigned to proton of the hydroperoxide group (-CO-OH). Spectroscopic studies were carried out in conditions of

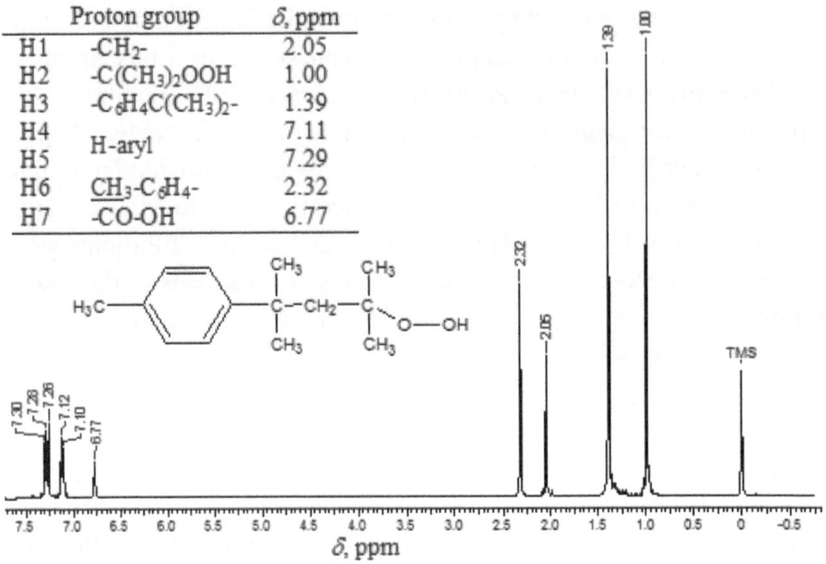

FIGURE 11.1 NMR ^{1}H spectrum of the 1,1,3-trimethyl-3-(4-methylphenyl)butyl hydroperoxide ($[ROOH]_0 = 0.02$ mol dm^{-3}, 297 K, $CDCl_3$).

the quaternary ammonium salt excess at 297–313°K. The concentration of ROOH in all experiments was constant (0.02 mol·dm^{-3}), while the concentration of Alk_4NBr was varied within the range of 0.1–0.6 mol·dm^{-3}.

The monotonous shifting of the NMR signal with increasing of Alk_4NBr concentration without splitting and significant broadening shows fast exchange between the free and bonded forms of the hydroperoxide. Such character of signal changing of the hydroperoxide group proton in the presence of Alk_4NBr (Figure 11.2) indicates the formation of a complex between hydroperoxide and Alk_4NBr in the system. Thus, observed chemical shift of the -CO-OH group proton (δ, ppm) in the spectrum of ROOH – Alk_4NBr mixture is averaged signal of the free (δ_{ROOH}, ppm) and complex-bonded (δ_{comp}, ppm) hydroperoxide molecule.

A nonlinear dependences of the changes of the proton chemical shift $\Delta\delta$ ($\Delta\delta = \delta - \delta_{ROOH}$) of hydroperoxide group on the Alk_4NBr initial concentration (Figure 11.1) were obtained. In conditions of Alk_4NBr excess and formation of the 1:1 complex for the analysis of the experimentally obtained dependence the Foster–Fyfe equation [9] can be used:

$$\Delta\delta/[Alk_4NBr] = -K_C\Delta\delta + K_C\Delta\delta max \quad (1)$$

where K_C – the equilibrium constant of the complex formation between hydroperoxide and Et_4NBr, dm^3 mol^{-1}; $\Delta\delta_{max}$ – the difference between the chemical shift of the -CO-OH group proton of complex-bonded and free hydroperoxide ($\Delta\delta_{max} = \delta_{comp} - \delta_{ROOH}$), ppm.

These dependences of the $\Delta\delta$ on the Alk_4NBr initial concentration are linear (Figure 11.3) in the coordinates of Eq. (1). The equilibrium constants of the complex formation between hydroperoxide and Alk_4NBr and the chemical shift of the -CO-OH group proton of complex-bonded hydroperoxide were determined and listed in Table 11.1.

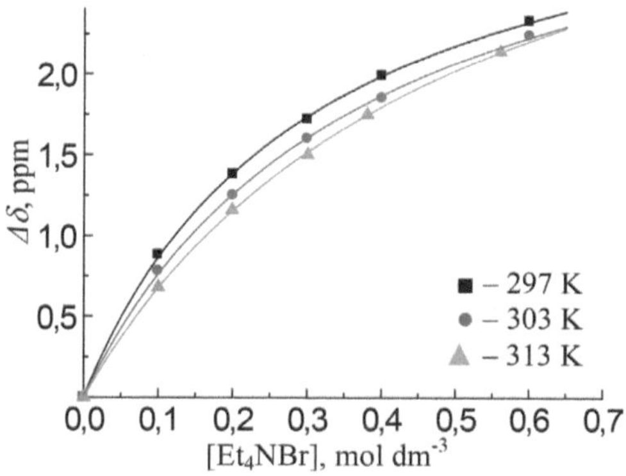

FIGURE 11.2 Change in the –COOH group proton chemical shift of the ROOH *vs* Alk_4NBr concentration ($[ROOH]_0$ = 0.02 mol dm^{-3}, $[Alk_4NBr]$ = 0.1–0.6 mol dm^{-3}, 297 K, $CDCl_3$).

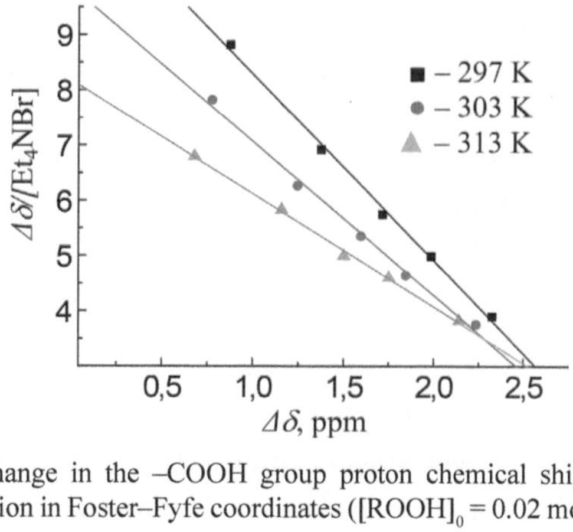

FIGURE 11.3 Change in the –COOH group proton chemical shift of the ROOH *vs* Alk_4NBr concentration in Foster–Fyfe coordinates ($[ROOH]_0$ = 0.02 mol dm^{-3}, $[Alk_4NBr]$ = 0.1–0.6 mol dm^{-3}, 297°K, $CDCl_3$).

TABLE 11.1 The Thermodynamic Parameters of Complex Formation Between the 1,1,3-trimethyl-3-(4-methylphenyl)butyl Hydroperoxide and Alk_4NBr

Alk_4NBr	T, K	K_C, $dm^3\ mol^{-1}$	δ_{comp}, ppm	$\Delta_{comp}H$, $kJ\ mol^{-1}$	$\Delta_{comp}S$, $J\ mol^{-1}\ K^{-1}$	$\Delta^{297}_{comp}G$, $kJ\ mol^{-1}$
Et_4NBr	297	3.37 ± 0.11	10.23 ± 0.05	–24 ± 0.2	–72 ± 0.7	–3.0
	303	2.79 ± 0.12	10.30 ± 0.07			
	313	2.04 ± 0.06	10.74 ± 0.04			
Pr_4NBr	297	2.30 ± 0.4	10.48 ± 0.29	–19.5 ± 2	–58.8 ± 6.2	–2.06
	308	1.70 ± 0.03	10.94 ± 0.03			
	313	1.57 ± 0.04	11.21 ± 0.04			
Bu_4NBr	297	1.81 ± 0.07	11.28 ± 0.06	–11.8 ± 0.7	–34.8 ± 2.4	–1.46
	303	1.67 ± 0.04	11.60 ± 0.04			
	313	1.42 ± 0.06	11.79 ± 0.06			

The chemical shift of the complex-bonded hydroperoxide δ_{comp} slightly increases with temperature for all systems under consideration and is in the range 10.23 ÷ 11.79 ppm (Table 11.1). Values of the binding constants (K_C) characterizing the stability of the ROOH – Alk_4NBr complexes are within 1.42 ÷ 3.37 $dm^3\ mol^{-1}$ and consistent with K_C values observed for complexation of hydroperoxide compounds with metal complexes [11] and quaternary ammonium salts [12–14]. Enthalpies of the hydroperoxide – Alk_4NBr complex formation are within −11 ÷ –24 kJ mol^{-1} (Table 11.1).

Obtained values of the binding constants as well as enthalpies of the hydroperoxide – Alk_4NBr complex formation decrease in the following row of cations: Et_4N^+, Pr_4N^+, Bu_4N^+. Changes of the enthalpy of the hydroperoxide – Alk_4NBr complex formation in the case of different quaternary ammonium cations point out the role of the steric factor at the stage of complex formation. In the simplest case the Van-der-Waals volume of the investigated cations could describe the steric effect. A good correlation between parameters of the complex formation and the Van-der-Waals volumes of cations is observed (Figure 11.4).

Similar effect has been already observed in CD_3CN solution for the hydroperoxide-Alk_4NBr systems [13–15]. The equilibrium constant values also decrease with intrinsic cation volume increasing and this dependence effect remains over the temperature range 297–313°K. Symbate change of the binding constants (Figure 11.5) as well as δ_{comp} values for the hydroperoxide – Alk_4NBr complex determined in CD_3CN and $CDCl_3$ is observed. Thus, the cation effect is not due to the solvent specific properties only but caused by complex structure too.

11.4 CONCLUSIONS

A complex formation between 1,1,3-trimethyl-3-(4-methylphenyl)butyl hydroperoxide and Alk_4NBr has been demonstrated by NMR ^1H spectroscopy investigations. It is shown that the stability of these associates

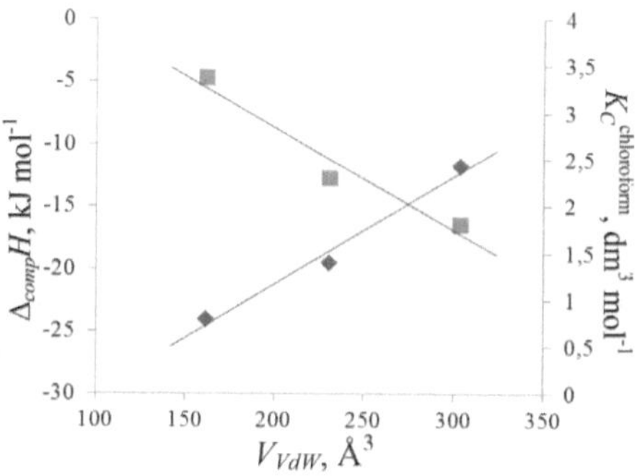

FIGURE 11.4 Enthalpy of the hydroperoxide – Alk_4NBr complex formation *vs* Van-der-Waals volume of the Alk_4N^+ cation. Calculator Plugins were used for structure property prediction and calculation, Marvin 14.7.14.0, 2014, ChemAxon (http://www.chemaxon.com).

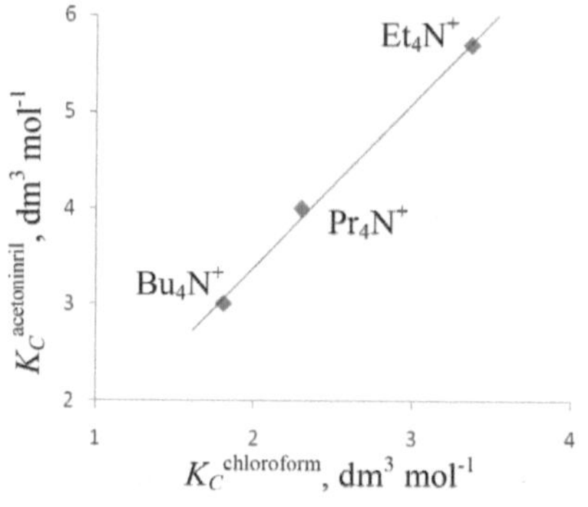

FIGURE 11.5 Symbate change of the binding constants values determined in CD_3CN and $CDCl_3$ (297°K).

decreases with increasing of the salt cation intrinsic volume. Symbate change of the hydroperoxide – Alk_4NBr complex parameters determined in CD_3CN and $CDCl_3$ is observed.

KEYWORDS

- **arylalkyl hydroperoxides**
- **NMR ^1H spectroscopy**
- **tetraalkylammonium bromides**

REFERENCES

1. Matienko, L. I., Mosolova, L. A., Zaikov, G. E., Selective catalytic oxidation of hydrocarbons. New prospects. *Russ. Chem. Rev.*, 78, 221 (2009).
2. Turovskyj, A. M., Opeida, O. I., Turovska, O. M., Raksha, O. V., Kuznetsova, N. O., Zaikov, G. E., Kinetics of radical chain cumene oxidation initiated by α-oxycyclohexylperoxides in the presence of Et_4NBr. *Oxid. Commun.*, 29, 249 (2006).
3. Munter, R. Advanced oxidation processes – current status and prospects. *Proc. Estonian Acad. Sci. Chem.*, 50, 59 (2001).
4. Kuznetsova, A. A., Knorre, D. G., Fedorova, O. S., Oxidation of DNA and its components with reactive oxygen species. *Russ. Chem. Rev.* 78–659 (2009).
5. Ida, Sh., Tsubota, T., Hirabayashi, O., Nagata, M., Ya. Matsumoto, A. Fujishima Chemical reaction of hydrogenated diamond surface with peroxide radical initiators. *Diamond and Related Materials*. 12, 601 (2003).
6. Turovskij, N. A., Pasternak, E. N., Raksha, E. V., Golubitskaya, N. A., Opeida, I. A., E, G., Zaikov Supramolecular reaction of lauroyl peroxide with tetraalkylammonium bromides. Oxid. Commun. 33, 485 (2010).
7. Turovskij, N. A., Raksha, E. V., Yu. Berestneva, V., Pasternak, E. N., Yu, M., Zubritskij, Opeida, I. A., Zaikov, G. E., Supramolecular decomposition of the aralkyl hydroperoxides in the presence of Et_4NBr. *Oxid. Commun.* 38, 1 (2015).
8. Raksha, E. V., Pasternak, E. N., Turovskij, N. A., Golubitskaya, N. A., Ostapenko, E. V., Supramolecular reactions of aralkyl hydroperoxides with tetraethylammonium bromides. Molecular modeling. *Bulletin of the Donetsk National University, A: Natural Sciences*. 1, 108 (2011).
9. Fielding, L., Determination of Association Constants (K_a) from Solution NMR Data. *Tetrahedron*, 56, 6151 (2000).

10. Hock, H., Lang, S., Autoxidation of hydrocarbons (VIII) octahydroanthracene peroxide. *Chem. Ber.*, 77, 257 (1944).
11. Bryliakov, K. P., Talsi, E. P., N, S., Stas'ko, Kholdeeva, O. A., Popov, S. A., Tkachev, A. V., Stereoselective oxidation of linalool with tert-butyl hydroperoxide, catalyzed by a vanadium (V) complex with a chiral terpenoid ligand. *Mol, J., Cat. A: Chem.* 194, 79 (2003).
12. Turovskij, N. A., Raksha, E. V., Yu. Berestneva, V., Yu, M., Zubritskij. Formation of 1,1,3-trimethyl-3-(4-methylphenyl)butyl hydroperoxide complex with tetrabutylammonium bromide. *Russian j. of Gen. Chem.* 84, 16 (2014).
13. Turovskij, N. A., Yu. Berestneva, V., Raksha, E. V., Pasternak, E. N., Yu, M., Zubritskij, Opeida, I. A., Zaikov, G. E., ^{1}H NMR study of the tert-butyl hydroperoxide interaction with tetraalkylammonium. *Polymers Research Journal.* 8, 85 (2014).
14. Turovskij, N. A., Yu. Berestneva, V., Raksha, E. V., Yu, M., Zubritskij, Grebenyuk, S. A., NMR study of the complex formation between tert-butyl hydroperoxide and tetraalkyl ammonium bromides. *Monatsh. Chem.* 145, 1443 (2014).
15. Turovskij, N. A., Yu. Berestneva, V., Raksha, E. V., Opeida, I. A., Yu, M., Zubritskij. Complex formation of hydroperoxides with Alk_4NBr according to NMR spectroscopy data. *Russ. Chem. Bull.* 63, 1717 (2014).

CHAPTER 12

POLYAMIDES AND POLYAMIDOETHER IN MACROMOLECULES CONTAINING TRIPHENYLMETHANE GROUPS

T. A. BORUKAEV, R. Z. OSHROEVA, and N. I. SAMOILYK

Kabardino-Balkarian State University, 173 Chernishevsky Street, Nalchik, Russia, E-mail: boruk-chemical@mail.ru

CONTENTS

ABSTRACT

Polyamides and polyamidoether, containing in macromolecules volume of triphenylmethane group are synthesized. Introduce to macromolecule triphenylmethane fragment results in easily materials are found. The resulting polymer possesses high thermal and strength properties. And polyamidoether has thermotropic liquid crystal properties.

12.1 AIMS AND BACKGROUND

Aromatic polyamides (PA) and polyamidoether (PAE) have unique properties such as thermal stability, impact strength, thermal and chemical stability etc. [1–3]. However, the high melting and softening temperature, limited solubility, significant rigidity of macromolecules complicate the processing of these polymers in the products. Therefore, a great interest of researchers is causing aromatic PA and PAE, which would maintain their inherent high level of physical and mechanical properties, and at the same time were easily processed from solution and melt.

To solve this problem very promising is the use as a starting diamines derivatives of 4,4¢-diaminotriphenylmethane. The last ones due to their chemical structure, bulky substituents, can have a beneficial impact on the process ability of polymers. In addition, these diamines 4,4¢-diaminotriphenylmethane can easily be obtained in one stage from available compounds with a high output.

In the work of the PA and PAE on base of diamines of triarylmethane set have been obtained by means of low-temperature polycondensation and copolycondensation of source monomers in a solution of N-methylpyrrolidone (MP) in an inert atmosphere. As monomers have been used 4,4¢-diaminotriphenylmethane, hydroquinone and dichlorohydrin of tere- and isophthalic acids. The reaction of obtaining PAE we can be represented as follows:

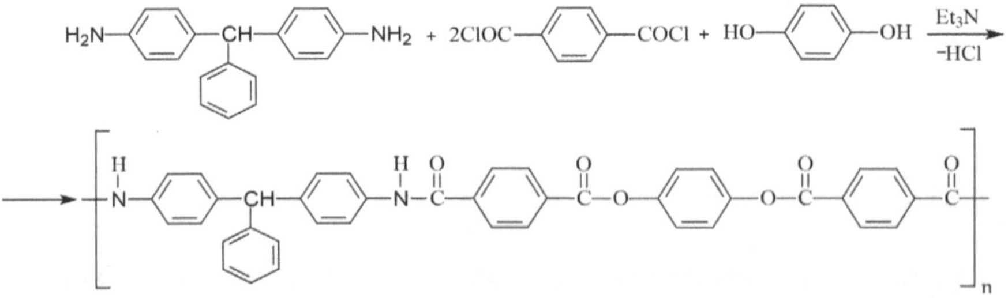

As an HCl acceptor were solvent – MP and triethylamine.

Getting PA with low temperature polycondensation of diamines with dichlorohydrin of tere- and isophthalic acids in solution diethylacetamide (DEA) in an inert gas atmosphere we can be represented by the following scheme:

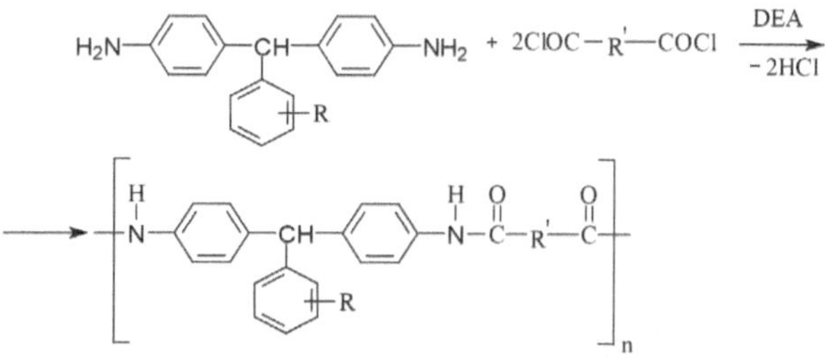

The structure of the obtained polymer was confirmed using IR spectroscopy (IR spectrometer "Specord M-82," the range of 400–4000 cm^{-1}) and elemental analysis. Elemental composition of the products of polycondensation is close to the calculated values.

The analyze polymers on basis 4,4′-diaminotriphenylmethane and dichlorohydrin terephthalic acid spectra is allows to establish the presence of the following groups: monosubstituted benzene ring (stripes 700 and 1110 cm^{-1}), 1,4-disubstituted benzene rings (stripes 785, 1015, 1225 and 1320 cm^{-1}), the stretching vibrations of C-C correspond stripes 1510 and 1560 cm^{-1}, carbonyl group – stripe 1700 cm^{-1}, the amino group – the strip with the inflection 3420 cm^{-1}, and several stripes in the region of 3000–3100 cm^{-1} are responsible stretching vibrations of C-H communication aromatic rings (Figure 12.1).

The introduction of lateral substituents in triphenylmethane, and the use of dichlorohydrin of isophthalic acid to obtain the PA don't lead to significant changes in the IR spectra. In the case of 4,4′-diamino-4″-methyltriphenylamine is observing the disappearance of the bands that correspond to the mono-substituted benzene rings, in General, the nature of the spectrum is preserved

In addition, IR spectra for PAE have been removed. The results of the spectrum analysis based on polymer of 4,4¢-diaminotriphenylmethane, dichlorohydrin terephthalic acid and p-hydroquinone can detect the presence of the following groups: monosubstituted benzene ring (stripes 700, 720 and 1172 cm^{-1}), 1,4-disubstituted benzene ring (stripes 812, 1016, and 1408 cm^{-1}), the stripes 1508–1596 cm^{-1} correspond to stretching vibrations of C-C bond in aromatic rings, ester linkages – stripes 1732 cm^{-1}, carbonyl, amide groups meet stripes 1636 and 1648 cm^{-1}, as the amino group – the strip with the inflection 3400 cm^{-1}, in addition there are several

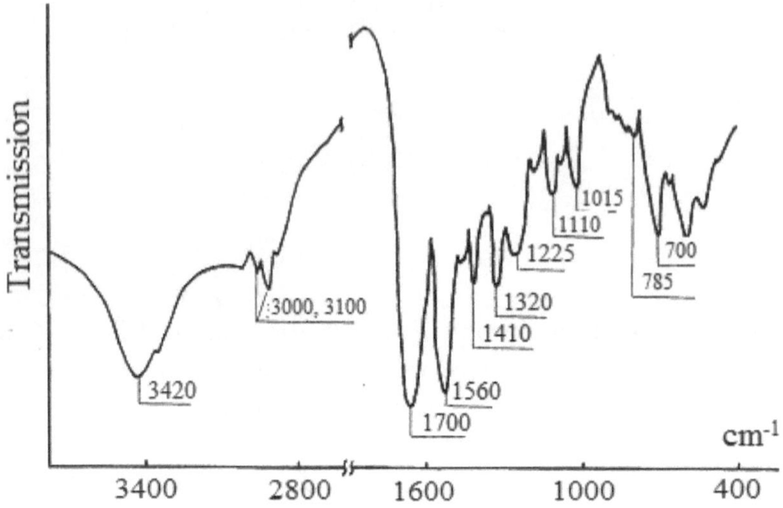

FIGURE 12.1 The IR spectrum of PA on the basis of 4,4′-diaminotriphenylmethane and dichlorohydrin terephthalic acid.

stripes sin the area 3032–3100 cm^{-1} which correspond to stretching vibrations of C-H bonds and aromatic rings (Figure 12.2).

The use of dichlorohydrin isophthalic acid slightly changes the nature of the spectrum. In the spectra (Figure 12.3) are detected stripes, corresponding to 1,3-disubstituted benzene nuclei (700, 876 and 1064 cm^{-1}).

A determining process of thermal decomposition of polymers containing -CO-NH- and -C(O)-O- bonds, is a hydrolytic breakdown of amide and ether groups, catalyzed by terminal carboxyl and amino groups.

Aromatic PA is of great interest as polymers, which possess high thermal stability [4]. However, the high temperature of their melting and softening limited solubility, significant rigidity of macromolecules complicate the processing of these polymers in the product.

We received as PAE through the introduction of ester groups in the main chain. It was interesting to compare the properties of polyamides and polyamidoether based diamines of triarylmethane series.

A study of the thermal stability of PA and PAE have been performed using derivatograph (TGA) Q – 1500 by firm MOM (Hungary) in dynamic mode of heating in the temperature range 20–600°C. The study has been conducted in an inert atmosphere and in air. The heating rate of the samples is 2.5 deg/min. Results of thermal analysis are shown in Table 12.2.

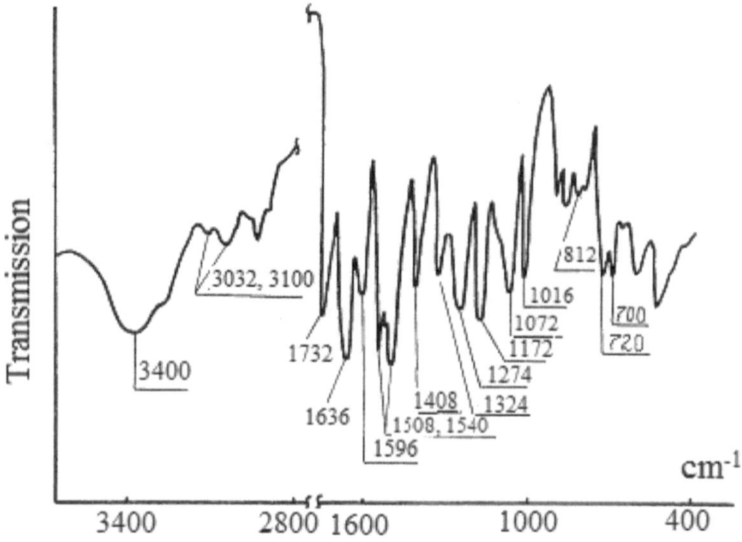

FIGURE 12.2 The IR spectrum of PAE based on 4,4¢-diaminotriphenylmethane, hydroquinone and dichlorohydrin terephthalic acid.

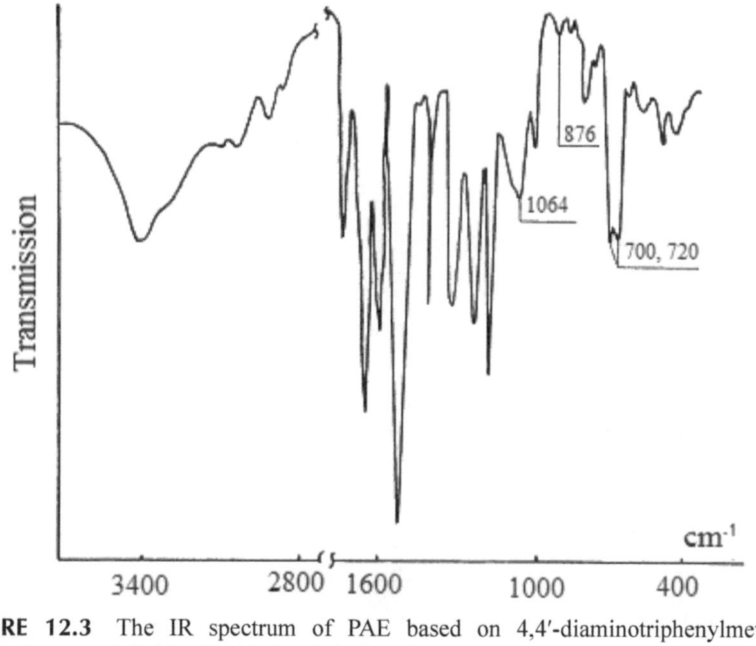

FIGURE 12.3 The IR spectrum of PAE based on 4,4′-diaminotriphenylmethane, hydroquinone and dichlorohydrin isophthalic acid.

TABLE 12.1 The Output and Properties of PA and PAE on Base of Diamines of Triarylmethane Set

№	Polymers	Concentration of monomer, mol/L	ε_y, %	Solvent	Intrinsic viscosity [η], dL/g
I	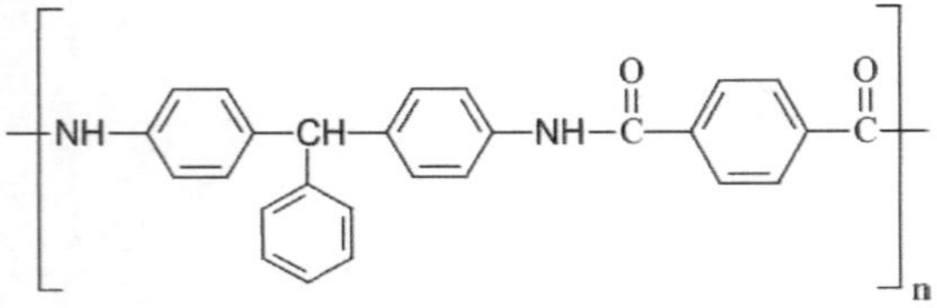	1	98.3	DEA	1.5
II	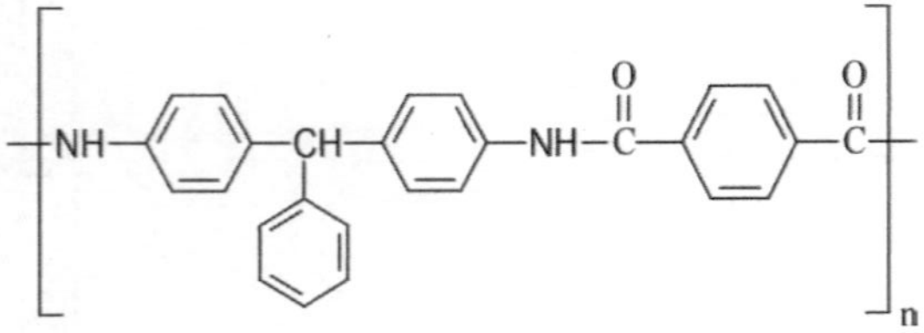	—"—	97.1	—"—	1.2
III	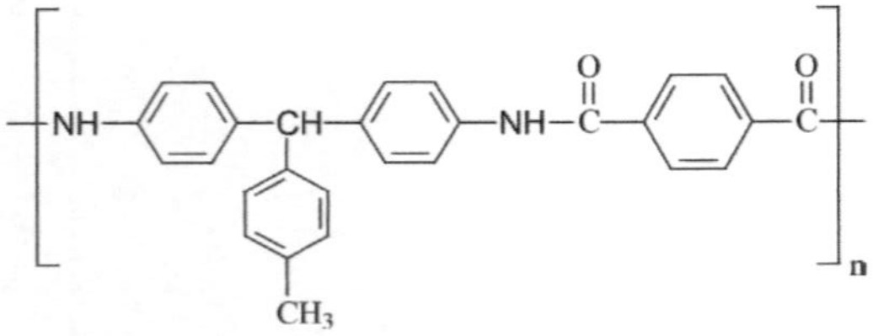	—"—	98.4	—"—	1.5

TABLE 12.1 Continued

№	Polymers	Concentration of monomer, mol/L	ε_y, %	Solvent	Intrinsic viscosity [η], dL/g
IV	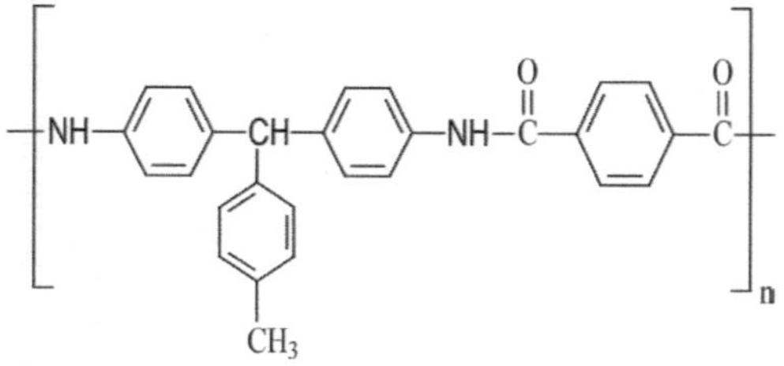	—″—	98.1	—″—	1.3
V	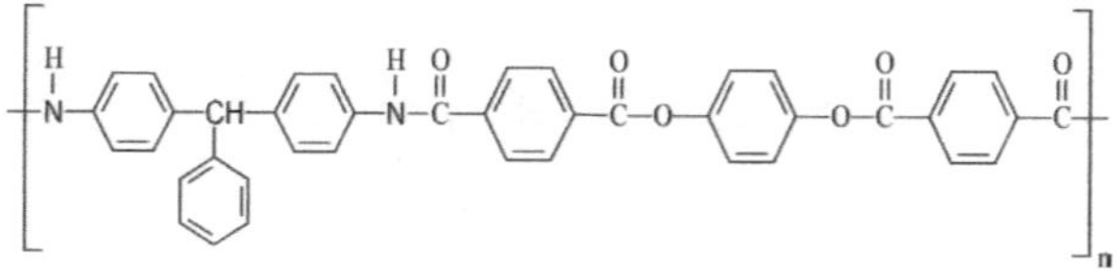	0.6	90.8	MP	0.6
VI	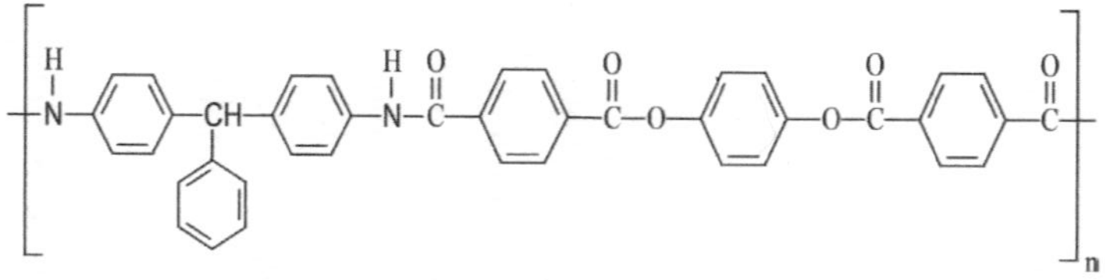	—″—	91.1	—″—	0.6

Note: the Intrinsic viscosity ([η], dL/g) has been determined in a solution of dimethyl-formamide DMF (0.5 g per 100 mL of DMF) on the Ubbelohde type viscometer at the temperature of 20°C.

TABLE 12.2 Thermal Properties of Polyamides and Polyamidoether

Polymers	Intrinsic viscosity [η], dL/g	Weight loss at temperature, °C (on air)		Softening point, *T*, °C
		10%	50%	
I	1,5	425	50%	300
II	1,2	420	>500	290
III	1,5	415	—"—	290
IV	1,3	410	—"—	285
V	0,6	400	—"—	265
VI	0,6	385	—"—	255

As the table shows, the maximum thermal stability is PA based on 4,4′-diaminotriphenylmethane and dichlorohydrin terephthalic acid. In Refs. [5, 6] it is noted that the thermal stability of PA increases in the number of m, m < m, p < p, p, when using the substituted diamines for the synthesis of the PA, the thermal stability of polymers is slightly decreased. The introduction of macromolecules ester groups in the main chain also leads to a reduction of the thermal stability.

The differential thermal analysis (DTA) curve analysis enables to determine the position of the thermal effects associated with oxidative and thermo-oxidative processes. For PA thermal effects on DTA curves associated with oxidative processes we were observed at a temperature as with a maximum at 290°C.

Intensive mass loss that meets of the basic thermal oxidative destruction process, was start at a temperature of 380°C (Figure 12.4).

For PAE thermal effect was observed (DTA) at a temperature of 250°C, associated with the melting of the polymer. Thermal decomposition located in the range of 360–450°C, which is lower than the PA. Low temperature resistant PAE compared to the PA can be explained by the activity of the ester group in the process of heating of the sample and a lower value of the energy gap ester bonds in contrast to the amide. The depth of the transformations when heated in air for PA and PAE are much greater than in an inert atmosphere and the maximum weight of the residue when heated to 500°C, respectively, less than in an inert atmosphere.

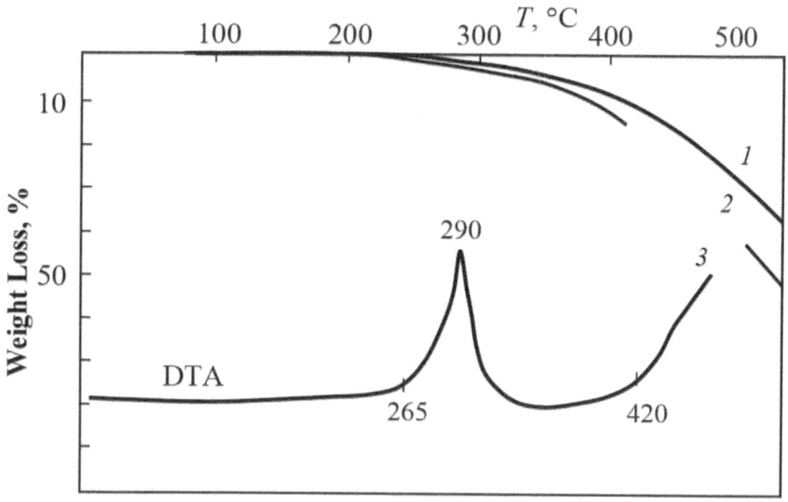

FIGURE 12.4 Dynamic Thermogravimetric analysis (TGA) of PA on the basis of 4,4′-diaminotriphenylmethane, of dichlorohydrin terephthalic acid (1); Dynamic Thermogravimetric analysis of PAE based on 4,4′-diaminotriphenylmethane, hydroquinone and dichlorohydrin terephthalic acid (2); Differential thermal analysis (DTA) of PA (3).

The change in physical properties with temperature, you can determine the operating temperature range of the polymer and its processing. In this regard, we have investigated the thermo mechanical properties of the obtained polymers. Studies were performed on the instrument universal measuring device (UMD-70) [7] in the temperature range 20–300°C in dilatometer mode with a constant heating rate of 5°C.

The results of dilatometer analysis are presented in Table 12.2. From the analysis of experimental data (Table 12.2), we can conclude that the softening point is higher in the PA than PAE. High softening point of the PA can be explained by the presence of hydrogen bonds formed between the amide groups of adjacent molecules. This results in a net of hydrogen bonds, which pervades the whole mass of PA, as shown by Fuller [8]. The energy of hydrogen bonds is less than the main valence bonds (N-C, C-C), however, due to their significant number in each macromolecule total interaction energy can be considerable.

Introduction to basic chain PA ester groups leads to a decrease of the melting temperature of the polymer. The reduction probably occurs by increasing the flexibility of the chain of macromolecules.

For PA and PAE decrease the softening point in the case of using isophthalic acid instead of terephthalic acid (Table 12.2). This is because aromatic PA with m-phenylene groups in the chain unlike poly-p-phenyleneterephthalamide have "like fiber" location of phenylene cycles [6]. Along the fiber axis, alternately arranged benzene nucleus at an angle of 10 and 20°C to the axis. Amide groups are almost perpendicular to the plane of the benzene ring, the rotation of which causes the low symmetry of the aromatic m-polyamides and, consequently, minor orderliness, increased solubility and a lower softening point.

As we noted above, PAE based on 4,4′-diaminotriphenylmethane, dichlorohydrin terephthalic acid and hydroquinone is a partially crystalline polymer (degree of crystallinity of about 35%) with melting point 265°C. For the detection of phase transitions have been used methods of differential scanning calorimetry (DSC), dilatometry and optical microscopy. Observation of PAE (V) in a polarizing microscope with a heating table shows that the polymer melts at 270°C, exhibiting thermotropic liquid crystalline properties. The polymer melt has a nematic liquid crystal state.

The study of polymers using differential scanning calorimetry (DSC we were performed on the device "Metler" at heating rate of 8°C/min) have been shown that when heated sample V (Figure 12.5, curve 1) exothermic effect was being observed at 260°C associated with the crystallization of the polymer, which is then a thermal melting point. After cooling (cooling rate 5°C/min) during the second cycle of heating the endothermic effect at a temperature of 265°C (Figure 12.4, curve 3) was being detected.

Temperature of isotropization (T_i), have been defined using the DSC thermograms, is 285°C. The visual observation of PAE in a polarizing microscope, we could not clearly determine T_i, the polymer acquires a brownish-carbonized color. This area increases in size. The results of DTA (see above) showed that the polymer V losses 5% of the weight in the field of 290–350°C. Temperature isotropization of PAE coincides with the thermal decomposition and it is not thermodynamically controlled phase equilibrium. The observed bending of the thermograms (Figure 12.5, curves 1 and 3) is associated with the transition to the rubbery state of PAE, which is 195°C.

Monitoring PAE based on dichlorohydrin isophthalic acid in a polarizing microscope are showed that the polymer is melted at a temperature of 255°C,

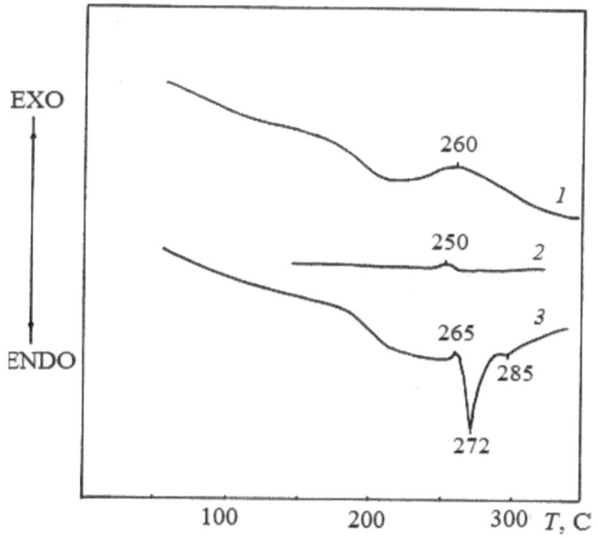

FIGURE 12.5 Differential scanning calorimetry (DSC) of PAE based on 4,4′-diaminotriphenylmethane, hydroquinone and dichlorohydrin terephthalic acid: 1 – the first heating of the sample, 2 – cooling, after the first heating and 3 – second heating.

it is not showing liquid crystal properties, associated with shortness of rotation of the segmental of chain macromolecules. For that reason, and low values of the melting temperature PROBES on the basis of dichlorohydrin isophthalic acid. Thus, the introduction of the nonlinear circuit unit steam-meta leads to lower degree of order of the macromolecules of polymers.

Thus, aromatic polyamides and polyamidoether with triarylmethane fragments in the main chain have been obtained. The obtained polymers are easily processed from solution and melt. It is found that the synthesized polymers show increased thermal stability, films based on polyamides have high strength characteristics. It is shown that polyamidoether based on dichlorohydrin terephthalic acid exhibits liquid crystalline properties in the temperature range of 270–285°C.

ACKNOWLEDGEMENTS

This Work is carried out with the financial support of the Ministry of Education and Science as part of the state task (project code 2199).

KEYWORDS

- **4,4′-diaminodiphenylmethane**
- **liquid-crystalline state**
- **physical and mechanical properties**
- **polyamides**
- **polyamidoethers**

REFERENCES

1. Hsiao Sheng-Huei, Liou Guey-Sheng, Kung Yi-Chun, Pan Hung-Yin, Kuo Chen-Hua. Electroactive aromatic polyamides and polyimides with adamantylphenoxy-substituted triphenylamine units. Eur. Polym. J. – 2009. vol. 45. – №8. 2234–2248.
2. Yu Guipeng, Li Bin, Liu Junling, Wu Shaofei, Tan Haijun, Pan Chunyue, Jian Xigao. Novel thermally stable and organosoluble aromatic polyamides with main chain phenyl-1,3,5-triazine moieties. Polym. Degrad. Stab. 2012. Vol. 97. №9. 1807–1814.
3. Hsiao Sheng-Huei, Chang Yu-Hui. New soluble aromatic polyamides containing ether linkages and laterally attached p-terphenyls. Eur. Polym. J. 2004. Vol. 40. №8. 1749–1757.
4. Hsiao Sheng-Huei, Chang Yu-Min, Chen Hwei-Wen, Liou Guey-Sheng. Novel aromatic polyamides and polyimides functionalized with 4-tret-butyltriphenylamine groups. J. Polym. Sci. A. 2006. Vol. 44. №15. 579–4592.
5. Buhler, K. U. Heat-Resistant and Heat-Stable Polymers [Russian translation]. Moscow: Chemical, 1984. – P.190.
6. Korshak, V. V. Chemical structure and thermal characteristics of polymers. Moscow: Science, 1970. p. 300. [in Russian].
7. Teitelbaum, V. J. Thermal analysis of polymers. Moscow: Science, 1979. p. 236 [in Russian].
8. Fuller, C. S., Baker, W. O., Pape, N. R. Crystalline Behavior of Linear Polyamides. Effect of Heat Treatment. J. Amer. Chem. Soc. 1940. Vol. 62. 3276–3278.

CHAPTER 13

A DETAILED REVIEW ON NANOFIBERS PRODUCTION AND APPLICATIONS

S. PORESKANDAR and SH. MAGHSOODLOU

University of Guilan, Rasht, Iran

CONTENTS

ABSTRACT

When the diameters of fiber materials reduce from micrometers to nanometers, compared to other known form of the material in many research fields, several characteristics are varied. For these cases, researchers become more interested in analyzing the unique properties of these materials. In this chapter, the general and new methods of nanofibers production are studied, at first. After that, the electrospinning process is investigated. Finally, the most important applications of these nanofibers are reviewed.

13.1 INTRODUCTION TO NANOTECHNOLOGY IMPORTANCE

Nanotechnology with unique physical, chemical and biological properties is interested by many scientists in everywhere for novel applications in recent years [1–3]. So, researchers started to analysis these properties [3–4]. When the diameters of polymer fiber materials reduce from micrometers to nanometers, several characteristics are changed compared to other known form of the material in many research fields. These characteristics are [5]:

- a large surface area to volume ratio;
- flexibility in surface functionalities;
- high porosity;
- superior mechanical performance (e.g., Stiffness and tensile force).

Nanofibers can produce from a spacious range of polymers [2, 6–7] (Figure 13.1 and Table 13.1).

These desirable properties make the polymer nanofibers best candidates for many important applications [1–3, 6–23]:

- environmental engineering and Biotechnology
  - medical science
    - tissue engineering
      - wound healing
      - tissue template
    - drug delivery
    - release control
- composites
- defense
  - protective clothing

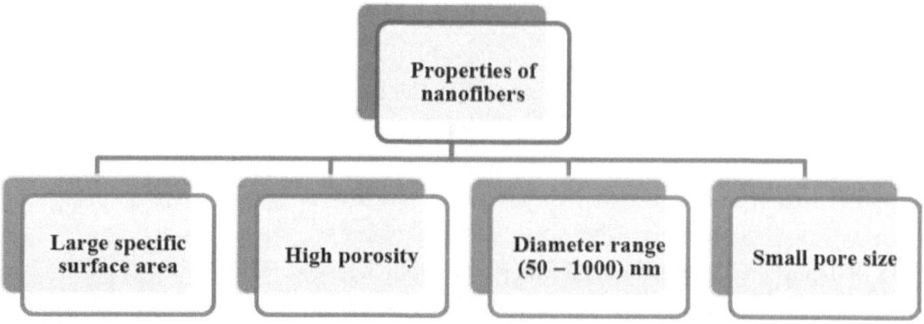

FIGURE 13.1 Properties of nanofibers.

TABLE 13.1 Utilized Polymer Fibers in Electrospinning Process in Different Applications for Tissue Engineering [24]

Fiber diameter	Solvent	Polymer	Application
200–350 nm	2,2,2-trifluoroethanol Water	Poly(ε-caprolactone) (shell) + Poly(ethylene glycol) (core)	Drug Delivery System
1–5 μm	Chloroform DMF Water	Poly(ε-caprolactone) and poly(ethylene glycol) (shell) Dextran (core)	
500–700 nm	Chloroform DMF Water	Poly(ε-caprolactone) (shell) Poly(ethylene glycol) (core	
~4 μm	DCM PBS	Poly(ε-caprolactone-co-ethyl ethylene phosphate)	
260–350 nm	DMF	Poly(D, L-lactic-co-glycolic acid), PEG-b-PLA-PLA	
1–10 μm	DCM	Poly(D, L-lactic-co-glycolic acid)	
690–1350 nm	Chloroform	Poly(L-lactide-co-glycolide) and PEG-PLLA	
2–10 nm	Chloroform methanol	Poly(ε-caprolactone)	General Tissue Engineering
500–900 nm	Chloroform DMF	Poly(ε-caprolactone) (core)+ Zein (shell)	
500 nm	2,2,2-trifluoroethanol	Poly(ε-caprolactone) (core) + Collagen (shell)	
500–800 nm	DMF THF	Poly(D, L-lactic-co-glycolic acid) and PLGA-b-PEG-NH2	
1–4 mm	DMF acetone	Poly(ethylene glycol-co-lactide)	
0.2–8.0 mm	2-propanol and water Water	Poly(ethylene-co-vinyl alcohol)	

TABLE 13.1 Continued

Fiber diameter	Solvent	Polymer	Application
180–250 nm	HFP	Collagen	
0.29–9.10 mm	2,2,2-trifluoroethanol	Gelatin	
120–610 μm	HFP	Fibrinogen	
130–380 nm	HFP	Poly(glycolic acid) and chitin	
0.2–1 nm	Chloroform DMF	Poly(ε-caprolactone)	Vascular Tissue Engineering
200–800 nm	Acetone	Poly(L-lactide-co-ε-caprolactone)	
5 μm	Chloroform	Poly(propylene carbonate)	
300 nm	1,4-dioxane DCM	Poly(L-lactic acid) and hydroxylapatite	
8.77–0.163 m	HFP	Chitin	

13.2 USUAL METHODS OF PRODUCING NANOFIBERS

Various common techniques can be used for preparing polymer nanofibers such as [7, 10, 15]:

a. drawing
b. template synthesis
c. phase separation
d. self-assembly
e. electrospinning

At first these techniques are introduced, and then special ways for producing nanofibers are proceeding.

13.2.1 DRAWING

The drawing technique is associated with evaporating the solvent from viscous polymer liquids directly, leading to solidification of the fiber. In this method, the nanofiber has an order of microns [7, 10, 23]. Figure 13.2

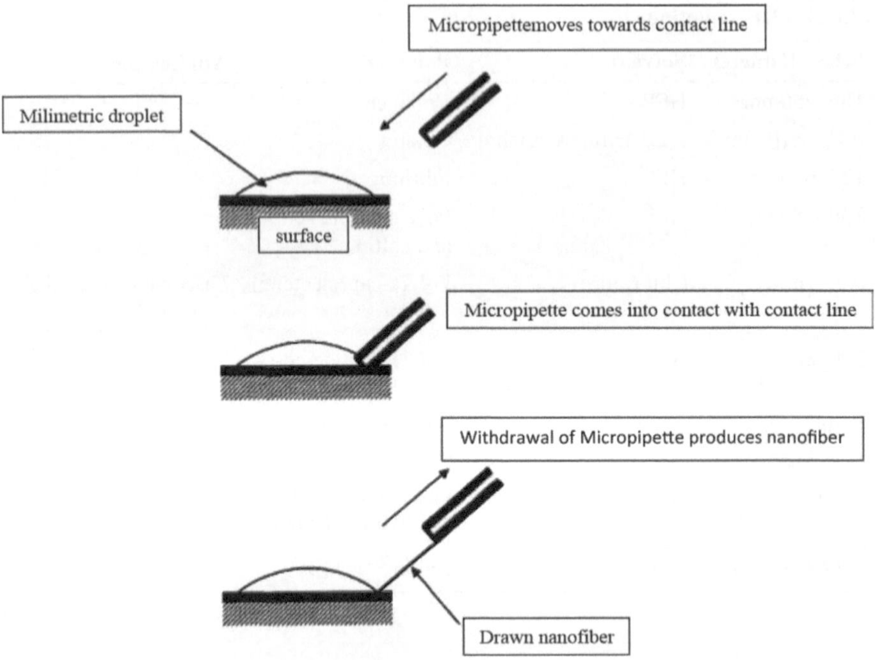

FIGURE 13.2 Schematic of drawing technique (each fiber is drawn from a microdroplet of polymer solution using a micropipette [7, 25].

shows drawing technique, each fiber is made from a micro droplet of polymer solution using a micropipette [7, 25].

13.2.2 TEMPLATE SYNTHESIS

The template synthesis uses templates with pores such as membranes to make solid or hollow form of nanofibers. This technique is similar to the extrusion in manufacturing [7, 25] (Figure 13.3). This method cannot produce one-by-one continuous nanofibers. The most significant advantage of this method is that various materials use for making nanofibers. These materials are [8, 15]:

- conducting polymers
- metals
- semiconductors
- carbons

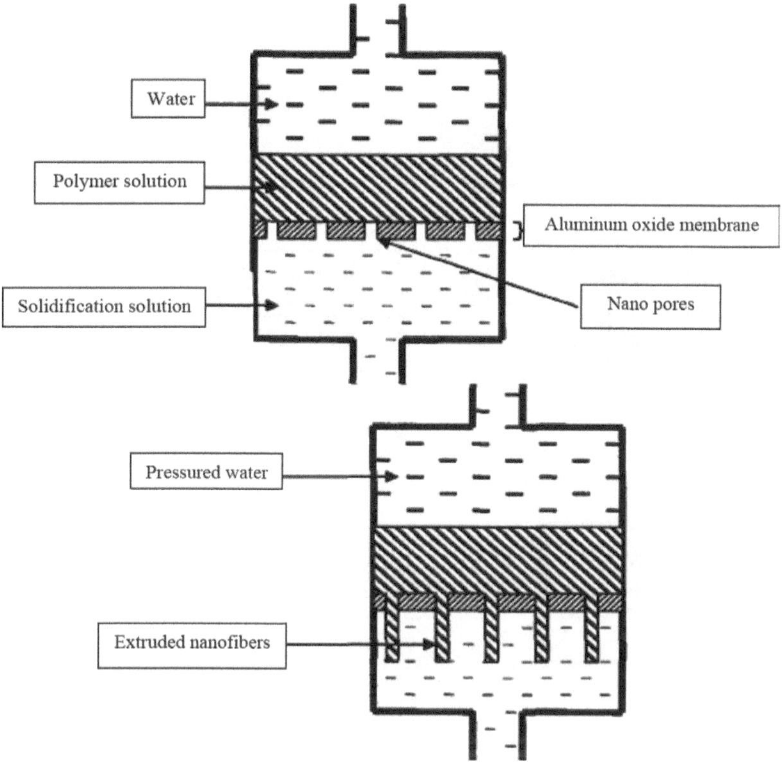

FIGURE 13.3 Schematic of template technique.

Polymer extrudes through a nanoporous template by applying pressure [7, 25].

13.2.3 PHASE SEPARATION

The phase separation involves four levels:

1) Prepare a solution of polymer in solvent
2) Do polymer gelatination with low temperature
3) Get rid of solvent by immersion in water
4) Do freezing and freeze-drying

This technique calls for so long time. Figure 13.4 shows forming nanofiber by phase separation [8, 10, 15, 23, 25].

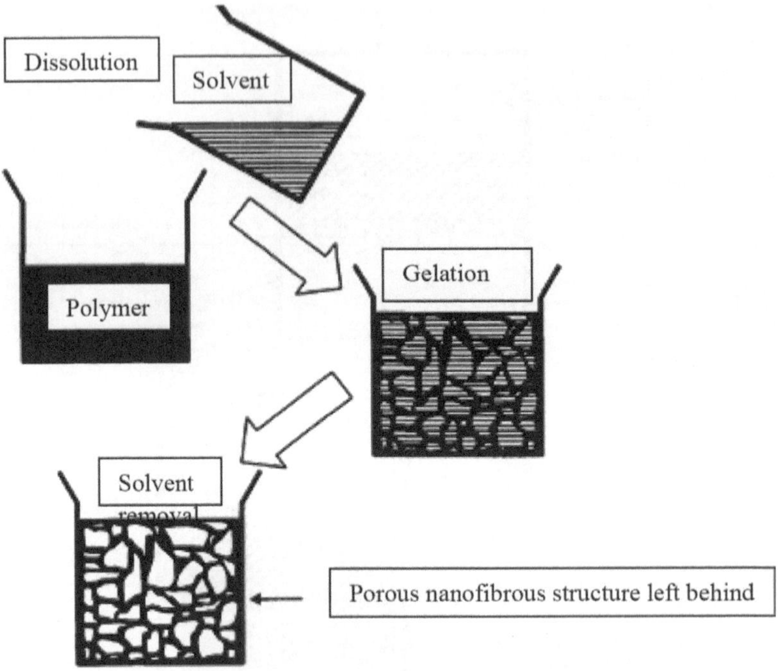

FIGURE 13.4 Formation of nanofiber by phase separation [10, 25]

13.2.4 SELF-ASSEMBLY

The self-assembly is another technique for producing nanofibers. In this technique pre-existing items make up into favorable patterns. Although this technique similar to the phase separation technique. The best feature of this technique is time-consuming for producing continuous polymer nanofibers. In the self-assembly technique, nanofibers are hung up molecule by molecule to bring out specific structures and functions [7, 8, 10, 15, 23, 25] (Figure 13.5).

13.2.5 ELECTROSPINNING

Electrospinning is the most favorite technique for creating more efficient nanofibers [12, 26, 27]. This technique is a simply, cheap and straightforward method to produce nanofibers [5, 6, 28–30]. It:

FIGURE 13.5 (a) Molecular structure, (b) Nanostructure of self-assembling peptide amphiphile nanofiber network [10].

a) creates continuous fibers with diameters in nano range,
b) is applicable for abroad range of materials (e.g., synthetic, natural polymers, metals as well as ceramics and composite), and
c) prepares nanofibers with low cost.

An ordinary electrospinning setup contains three main parts [1, 10]:

- A high power supply voltage
- A syringe with a needle and a pump
- A collector

Likewise, this technique distinguished by four main sections [31]:

- Taylor cone
- Steady jet
- Instability part
- Base part

Nanofibers are formed from polymer solution or melt with a high potential power source. Then this liquid is passed from capillary and collected along the collector [10, 32] (Figure 13.6 and Table 13.2).

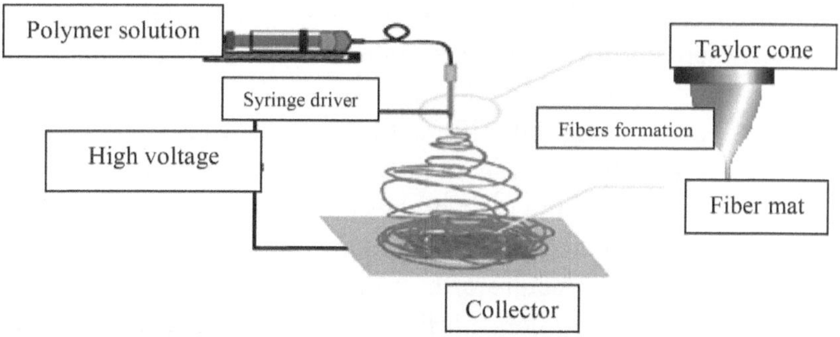

FIGURE 13.6 Standard electrospinning setup.

TABLE 13.2 Compression of Common Technology for Producing Nanofibers

Technique	Technology	Scalability	Simplicity	Controllability	Repeatability
Drawing	Laboratory scale	No	Yes	Yes	No
Template synthesis	Laboratory scale	Yes	Yes	Yes	Yes
Phase separation	Laboratory scale	No	Yes	Yes	No
Self-assembly	Laboratory scale	No	Yes	No	No
Electro spinning	Industrial process	Yes	Yes	Yes	Yes

13.3 NEW METHODS OF PRODUCING NANOFIBERS

13.3.1 GELATION TECHNIQUE

Initially, a gel is made using predetermined amounts of polymer and solvent followed by phase separation and gel formation. Finally, nanofiber forms when the gel is frozen and freeze-dried [7].

13.3.2 BACTERIAL CELLULOSE TECHNIQUE

Cellulose nanofibers produced by bacteria have been long used in diverse applications, including bio-medical [10, 27, 33]. Cellulose synthesis by Acetobacter involves polymerization of glucose residues into chains, followed by the extracellular secretion, assembly and crystallization of the chains into hierarchically comprised ribbons (Figure 13.7) [10, 27, 33].

Networks of cellulose nanofibers with diameters less than 100 nm are readily made. Fibers with different characteristics may be developed by different strains of bacteria. Copolymers have been created by adding polymers to the growth media of the cellulose-producing bacteria [10, 27, 33]. Bacterial cellulose is mixed by the acetic bacterium Acetobacter xylinum. The fibrous structure of bacterial cellulose consists of a three-dimensional network of micro fibrils containing glucanchains bound by hydrogen bonds [27, 33].

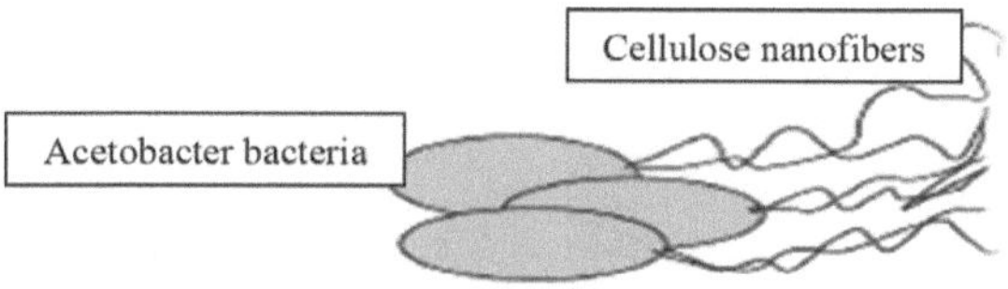

FIGURE 13.7 Acetobacter bacteria cells depositing cellulose nanofibers [10].

13.3.3 EXTRACTION TECHNIQUE

Nanofibers can be extracted from natural materials using chemical and mechanical treatments. Cellulose fibrils can be sorted out from plant cell walls. In one example, cellulose nanofibers were extracted from wheat straw and soy hull with diameters ranging from 10 to 120 nm and lengths up to a few thousand nanometers. Invertebrates have also been utilized as a source for extracting nanofibers[10].

13.3.4 VAPOR-PHASE POLYMERIZATION TECHNIQUE

Polymer nanofibers have also been made from vapor-phase polymerization. Plasma-induced polymerization of vapor phase vinyltrichlorosilane produced organosiloxane fibers with diameters around 25 nm and typical lengths of 400–600 nm and cyanoacrylate fibers with diameters from 100 to 400 nm and lengths of hundreds of microns [10].

13.3.5 KINETICALLY CONTROLLED SOLUTION SYNTHESIS TECHNIQUE

Nanofibers and nano wires have been created in solution using linear aligned substrates as template agents such as iron-cation absorbed reverse cylindrical micelles and silver micelles. PVA – polymethyl methacrylate nanofibers were produced using silver nano particle that was linearly aligned in solution with vigorous magnetic stirring. These nano particle chain assemblies acted as a template for further polymerization of nanofibers with diameters from 10 to 30 nm and lengths up to 60 μm [10].

TABLE 13.3 History of Electrospinning

Name of Researcher	Year	Subject	References
Lord Rayleigh	19th century	Understood the technique of electrospinning	[9]
William Gilbert	1600	Discovered first record of the electrostatic attraction of a liquid	[5]
Zeleny	1914	Introduced one of the earliest studies of electrified jetting phenomenon	[4]
Formhals	1934	Invented the experimental setup for the practical production of polymer filaments with an electrostatic force	[3]
Vonnegut and Neubauer	1952	Produce streams of uniform droplets and invented a simple tool for the electrical atomization	[2]
Drozin	1955	Examine the dispersion of series of liquids into aerosols under high electric potentials	[1]
Simons	1966	Patented a tool for producing non-woven fabrics of ultra thin and weightless	[6]
Taylor	1969	Published his work on the shape of the polymer droplet at the tip of the needle with applying an electric field	[7]
Baumgarten	1971	Made a tool for electrospinning acrylic fibers with a stainless steel capillary tube and a high-voltage DC current.	[15]
		Estimated the jet speed by using energy balance when a critical voltage was applied	
Larrondo and Mandley	1981	Produced polyethylene and polypropylene fibers by melting electrospinning successfully	[8]
Hayati et al.	1987	To study effective factors on jet stability and technique of electrospinning	[28]
Reneker and Chun	1996	Has shown the possibility of electrospinning polymer solutions	[26]

13.3.6 CONVENTIONAL CHEMICAL OXIDATIVE POLYMERIZATION OF ANILINE TECHNIQUE

Chemical oxidation polymerization of aniline is a traditional method for synthesizing poly aniline and during the former stages of this synthesis technique poly aniline nanofibers are formed. Optimization of polymerization conditions such as temperature, mixing speed and mechanical agitation allows the end stage formation of polyaniline nanofibers with diameters in the range of 30–120 nm [10].

13.4 HISTORY OF ELECTROSPINNING AND NANOFIBERS

The word “fiber” has its root from “fibra” and the nano term comes from the definition that has been discussed generously. When the diameter of polymer fiber reduces to the nanoscale, the nanofibers become important in applications [34]. The electrospinning attracts more attention as ultra-fine fibers of varied polymers with lower diameters to nanometers in nanotechnology in the recent years [4, 13, 18]. Employing electrostatic forces to deform materials in the liquid state goes back many centuries [26, 35] but, the origin of electrospinning as fiber spinning technique come back to 100 years ago [9, 36]. Many researchers work on electrospinning set up and effective factor on this technique. Patents characterizing an experimental set-up for producing polymer between 1934 and 1944 [35, 37]. Subjection Formhals work, the focus shifted to developing a better understanding process technique of the electrospinning. Here we get a summary of electrospinning histories in Table 13.3. Several research groups such as Dr. Darrell Reneker and his research group further interest in electrospinning with a series of papers published starting in early to mid 1990s and continuing today up to engage. This renewed interest spread quickly and many secondary academic groups became interested in the field of the electrospinning [1, 14, 23].

The number publications and patents in nanofibers fields and electrospinning have grown significantly in recent years [1, 14, 18]. Secondary academic groups became interested in the subject area of the electrospinning [1, 14, 23] (Figures 13.8 and 13.9).

13.5 THE RULES FOR ELECTROSPINNING OF NANOFIBERS

As it was mentioned before, electrospinning is an efficient and simplest technique for producing of nanofibers with different structures and

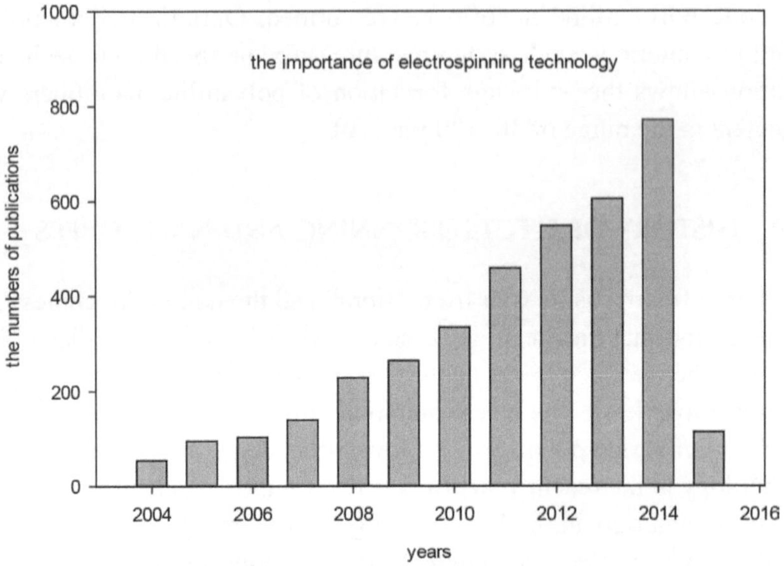

FIGURE 13.8 Numbers of publications about electrospinning.

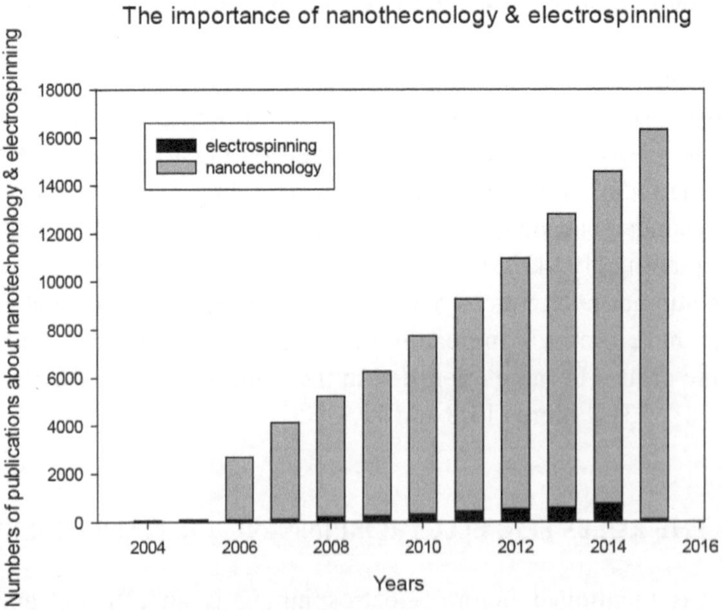

FIGURE 13.9 The importance of nanotechnology and electrospinning.

functionality [22, 34, 38–43]. The advantages of electrospinning technology are [44]:

- producing a high rate of nanofiber
- simplest setup and low costs of production

A common electrospinning set-up includes [11, 39, 43]:

- a high voltage power supply
- a syringe
- a needle
- a grounded collector screen

The technique of electrospinning includes several parts [17, 22, 40]:

- charging of the fluid
- formation of the cone-jet (Taylor cone)
- thinning of the jet with an electric field
- instability of the jet
- collection of the jet on target

Here a simple schematic of the electrospinning technique shows in particular (Figures 13.10 and 13.11). Instead of formal methods of fiber formation (e.g., Dry or wet spinning), electrospinning makes nanofibers by electrostatic forces [17]. Ion migrates in the solution or melts with an electric field. When the potential came into a critical value, stream of jet starts formation to throw. This jet moves straight toward the collector then, bending instability develops into a series of loops expanding with time. The solvent evaporates during the jet moves. At last nanofibers are collected on plate [32, 46–47]. It is important to observe that it is possible to electrospin all polymers into nanofibers, provided the molecular weight of the polymers is enough large and the solvent can be evaporated quickly enough during the technique [8, 44]. The mechanics of this technique deserving a specific attention and necessary to predictive tools or direction for better understanding and optimization and controlling technique. It has been identified that during traveling a solution jet from the tip to collector, the primary jet may show instability during the path. Several videos, graphic and laser light scattering methods for watching over the three-dimensional path of jets in flight, and for seeing the diameter and rate of parts were developed [39]. On the other hand, as in any liquid, the surface

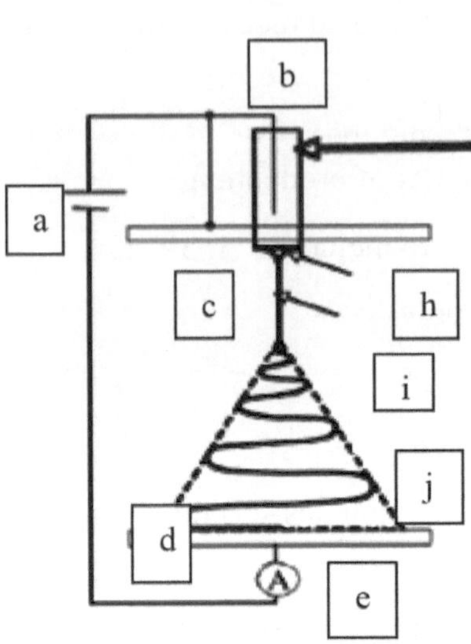

FIGURE 13.10 A simple schematic of electrospinning technique: (a) high voltage power supply; (b) charging devices; (c) high potential electrode (e.g., flat plate); d) collector electrode (e.g., flat plate); (e) current measurement device; (f) fluid reservoir; (g) flow rate control; (h) cone; (i) thinning jet; (j) instability region [45].

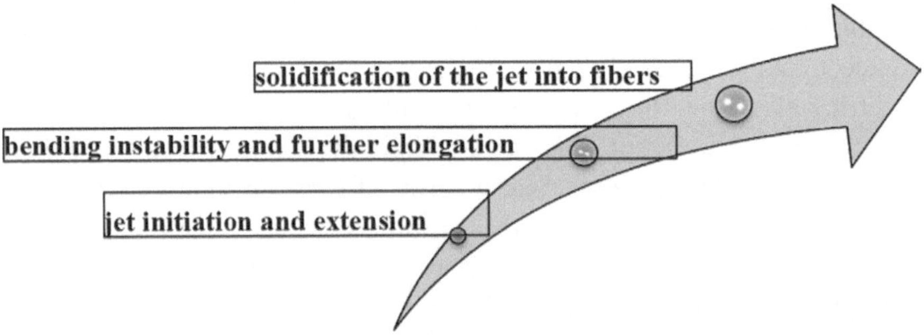

FIGURE 13.11 Basic principle of electrospinning.

tension reduces the entire surface of the jet thus reduces the free energy of the liquid. If the viscosity is not enough to hold the jet as a continuous shape, what usually occurs is an instability that causes the jet to break up into droplets. This effect is known as Rayleigh instability. Which of these two opposing effects prevails depends on the nature of the fluid, especially its viscosity and surface tension. If the viscosity is enough high with good cohesiveness, the charged jet undergoes a straight jet stage and whipping instability takes place, the amplitude depends on the material and solvent, then dry thin fibers are gathered. Although the setup is straightforward, but controlling of electrospinning is complicated. Some studied done by Taylor on the initial jet formation of electrospinning technique. He gained condition for critical electric potential where surface tension is in equipoise with the electrical force [22, 48]:

$$V_c^2 = 4\frac{H^2}{L^2}\left(\ln\frac{2L}{R}-\frac{3}{2}\right)(0.117\pi\gamma R)$$

Although Taylor cone has been viewed in many subjects, the exact shape and the angle of the cone are not fixed and only applicable to slight conducive, monomeric fluids. Researchers studied the initial jet formation through computer simulation and compared with the experimental outcomes. They found that thinning the jet in the initial stage is determined by many features. Viscoelasticity is found to be the key element in the initial jet thinning behavior. Fluid with higher viscoelasticity is thicker. Studies of the whipping motion revealed that in the envelope of the cone it only controls a single jet. The jet undergoes a fast whipping motion and the whipping is so tight that the conventional camera cannot distinguish the splaying with whipping. The bending instability of electrified jet caused by repulsive forces between the charges carried by the jet [13, 32]. The jet remains axisymmetric for some length. Then bending or whipping instability starts. At the onset of this instability, the jet follows a spiral path. As the jet spirals toward the collector, higher order instabilities reveal themselves. This instability makes the jet to loop in spirals with increasing radius [7, 29]. The envelope of this closed circuit is a cone.

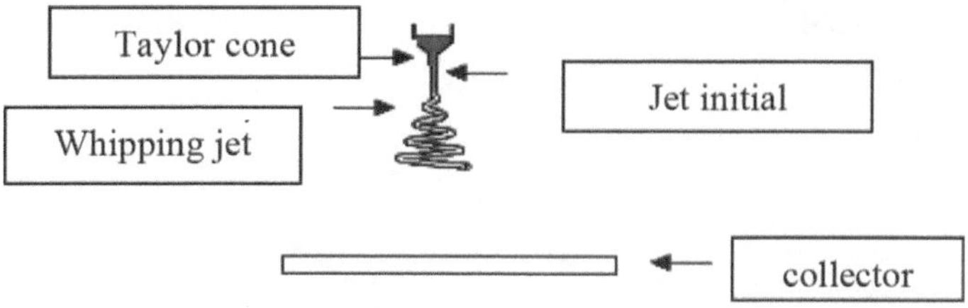

FIGURE 13.12 Whipping instability of jet in electrospinning technique.

Further, the electric field speeds up the jet. So the jet rate increases. This leads to decreasing in the jet diameter. In addition, the electrostatic repulsion between excess charges in the solution stretches the jet. This stretching also decreases the jet diameter [7, 17].

13.6 THE USE OF NANOFIBERS IN VARIOUS SCIENCE SEARCHES

The fine electrospun nanofibers make them useful in a wide range of innovative applications [22, 49]. Many materials are used for electrospinning [8, 44] (Figure 13.13 and Table 13.4).

Also new applications have been explored for these fibers continuously. Main application fields are discussed below [41, 50–51] (Figure 13.14).

For selected applications, it is desirable to control not only the fiber diameter, but as well the internal morphology. Porous fibers are of interest for applications such as filtration or prepare nanotubes by fiber templates

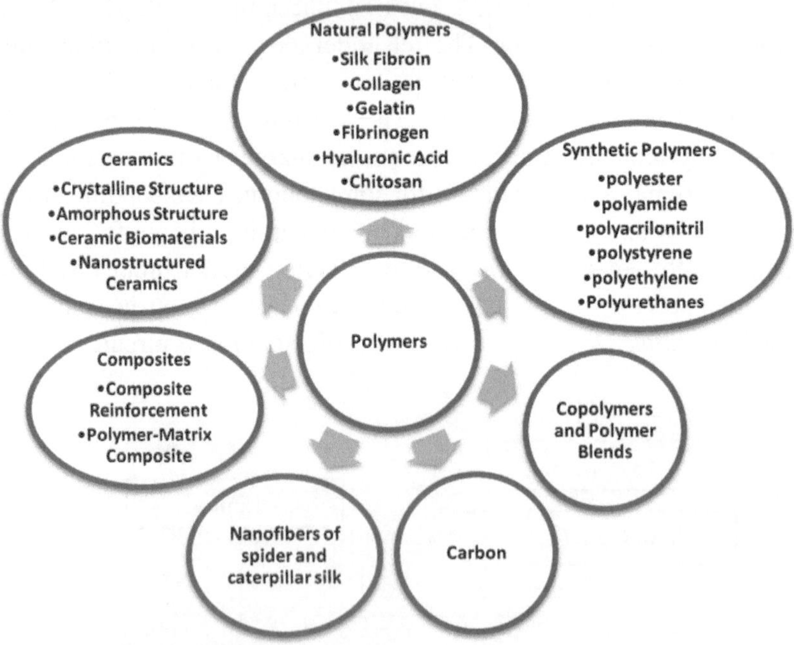

FIGURE 13.13 Verayetis of polymers in electrospinning.

TABLE 13.4 Classes of Polymers with Solvents

Polymer class	Polymer	Solvent
High performance polymers	Polymides	Phenol
	Polyamic acid	m-cresol
	polyetherimide	methylene chloride
Liquid crystalline polymers	Polyaramid	Sulphuric acid
	Polygamma-benyzyl-glumate	dimethylformamide
	Polyp-phenylene terephthalamide	Sulphuric acid
Copolymers	Nylon 6-polyimide	Formic acid
Textile fiber polymers	Polyacrylonitrile	Dimethylformamide
	Polyethylene terephthalate	Trifuoroacetic acid and dichloromethane melt in vacuum
	Naylon	Formic acid
	Polyvinyl alcohol	Water
Electrically conducting polymer	Polyaniline	Sulphuric acid
Biopolymers	DNA	Water
	Polyhydroxy butyrate-valerate	Chloroform
	Polycapro lactone	m-Cresol, Chlorophenol, Formic acid

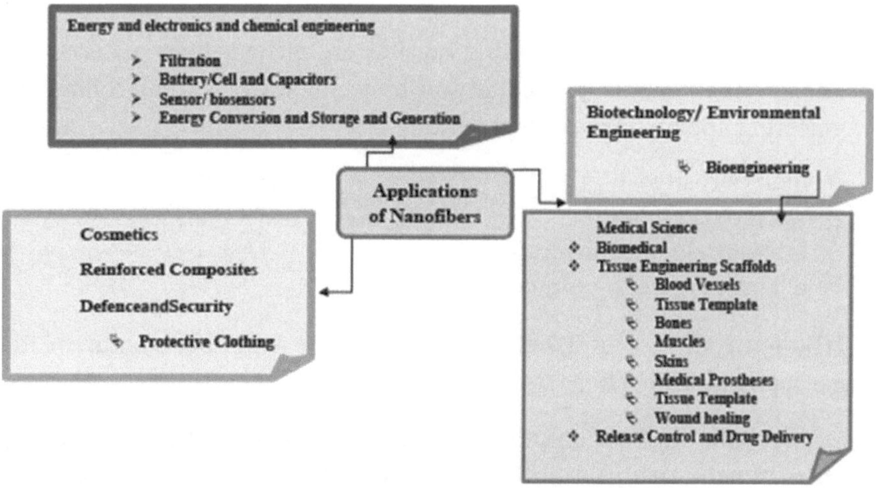

FIGURE 13.14 Potential applications of electrospun fibers.

[37, 40, 47, 52]. Besides, small pore size and high surface area inherent in nanofiber has implications in biomedical applications such as scaffoldings for tissue growth [37, 53]. Also, researchers have spun a fiber from a compound naturally present in the blood. This nanofiber can be used as forms of medical applications such as bandages or sutures that ultimately dissolve into the body. This nanofiber minimizes infection rate, blood loss and is also taken up by the body [44]. One of the most significant applications of nanofibers is to be used as reinforcements in composite developments. With these reinforcements, the composite materials can offer superior properties such as high modulus and strength to weight ratios, which cannot be achieved by other engineered monolithic materials alone. Information on the fabrication and structure-property relationship characterization of such nano composites is believed to be utilitarian. Such continuous carbon nanofiber composite also has possible applications as filters for [54]:

- separation of small particles from gas or liquid
- supports for high temperature catalysts
- heat management materials in aircraft and semiconductor devices
- rechargeable batteries
- super capacitors

13.6.1 BIOTECHNOLOGY/ENVIRONMENTAL ENGINEERING

Non-woven electrospun nanofiber meshes are an excellent material for membrane preparation, particularly in biotechnology and environmental engineering applications for this reason [17]:

- high porosity
- interconnectivity
- micro scale interstitial space
- a large surface to volume ratio

Bio-macromolecules or cells can be tied to the nanofiber membrane for these applications [17]:

- in protein purification and waste water treatment (affinity membranes)
- enzymatic catalysis or synthesis (membrane bioreactors)
- chemical analysis and diagnostics (biosensors)

Electrospun nanofibers can form an effective size exclusion membrane for particulate removal from wastewater [17]. Affinity membranes are a broad class of membranes that selectively captures specific target molecules by immobilizing a specific capturing agent onto the membrane surface. In biotechnology, affinity membranes have applications in protein purification and toxin removal from Bioproducts. In the environmental industry, affinity membranes have applications in organic waste removal and heavy metal removal in water treatment. To be used as affinity membranes, electrospun nanofibers must be surface functionalized with ligands. Mostly, the ligand molecules should be covalently attached to the membrane to prevent leaching of the ligands. Also, Water pollution is now becoming a critical global issue. One important class of inorganic pollutant of great physiological significance is heavy metals, for example, Hg, Pb, Cu, and Cd. Distributing these metals in the environment is mainly applied to the release of metal containing waste-water from industries. For example, copper smelters may release high quantities of Cd, one of the most mobile and toxic among the trace elements, into nearby waterways. It is impossible to eliminate some classes of environmental contaminants, such as metals, by conventional water purification methods. Affinity membranes will play a critical role in wastewater treatment to remove (or recycle) heavy metal ions in the future. Polymer nanofibers functioned with a ceramic nano material, mention in below, could be suitable materials for fabrication of affinity membranes for water industry applications [17]:

- Hydrated alumina hydroxide
- Alumina hydroxide
- Iron oxides

The polymer nanofiber membrane acts as a bearer of the reactive nano material that can attract toxic heavy metal ions, such As, Cr, and Pb, by adsorption or chemisorption and electrostatic attraction mechanisms. Again, affinity membranes provide an alternative access for removing organic molecules from wastewater [17].

13.6.1.1 Bioengineering

In biological viewpoint, almost entirely the human tissues and organs are deposited in nano fibrous forms or structures. Some examples include the following [54]:

- Bone
- Dentin
- Collagen
- Cartilage
- Skin

All of them are characterized by well-organized fibrous structures realigning in nanometer scale. Current research in electrospun polymer nanofibers has focused one of their major applications on bioengineering. We can easily find their promising potential in various biomedical fields [54].

13.6.1.1.1 Medical Science

Nanofibers are used in medical applications, which include, drug and gene delivery, artificial blood vessels, artificial organs, and medical facemasks. For example, carbon fiber hollow Nano tubes, smaller than blood cells, have the potential to transport drugs into blood cells [44].

Biomedical Application

Biomedical field is one of the important application areas among others, using the technique of electrospinning like [11, 55]:

- Filtration material
- Protective material
- Electrical applications
- Optical applications
- Sensors
- Nanofiber reinforced composites

Current medical practice is based almost on treatment regimes. However, it is envisaged that medicine in the future will be based heavily on early detection and prevention before disease expression. With nanotechnology, new treatment will emerge that will significantly reduce medical costs. With recent developments in electrospinning, both synthetic and natural polymers can be produced as nanofibers with diameters ranging from decades to hundreds of nanometers with controlled morphology. The potential of these electrospun nanofibers in human healthcare applications is promising, for example [17]:

- In tissue or organ repair and regeneration
- As vectors to deliver drugs and therapeutics
- As biocompatible and biodegradable medical implant devices
- In medical diagnostics and instrumentation
- As protective fabrics against environmental and infectious agents in hospitals and general surroundings
- In cosmetic and dental applications.

Tissue or organ repair and positive feedback are new avenues for potential treatment, avoiding the need for donor tissues and organs in transplantation and reconstructive surgery. In this advance, a scaffold is usually needed that can be fabricated from either natural or synthetic polymers by many techniques including electrospinning and phase separation. An animal model is utilized to study the biocompatibility of the scaffold in a biological system before the scaffold is introduced into patients for tissue-regeneration applications. Nanofibers scaffolds are suited to tissue engineering. These can be made up and shaped to fill anatomical defects. Its architecture can be designed to supply the mechanical properties necessary to support cell growth, growth, differentiation and motility. Also, it can be organized to provide growth factors, drugs, therapeutics, and genes to stimulate tissue regeneration. An inherent property of nanofibers is that they mimic ECM of tissues and organs. The ECM is a complex composite of fibrous proteins such as collagen and fibronectin, glycoproteins, proteoglycans, soluble proteins such as growth factors, and other bioactive molecules that support cell adhesion and growth. One of the aim is to create electrospun polymer nanofiber scaffolds for engineering blood vessels, nerves, skin, and bone. In the pharmaceutical and cosmetic industry, nanofibers are promising tools for controlled these aims [17, 44, 55]:

1. Delivery of drugs
2. Therapeutics
3. Molecular medicines
4. Body-care supplements

Tissue Engineering Scaffolds

Successful tissue engineering needs synthetic scaffolds to bear similar chemical compositions, morphological, and surface functional groups to their natural counterparts (Figure 13.15). Natural scaffolds for tissue growth

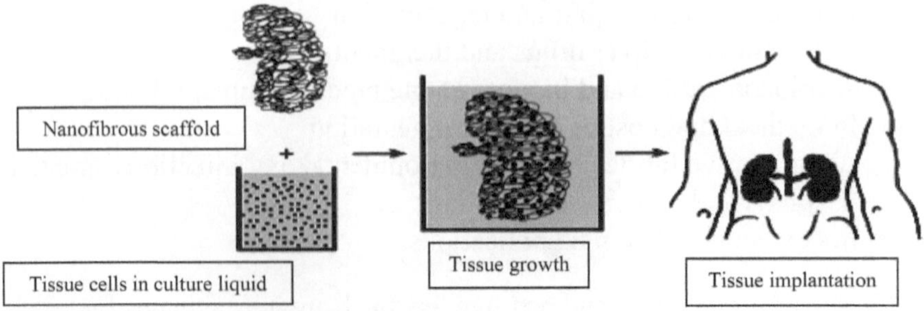

FIGURE 13.15 Principle of tissue engineering.

are three-dimensional networks of nanometer-sized fibers made of several proteins. Non-woven membranes of electrospun nanofibers are well-known for their interconnected, 3D porous structures and large surface areas, which provide a class of ideal materials to mimic the natural ECM needed for tissue engineering. The electrospun nanofibrous support was treated with the cell solution and the nanofiber-cell was cultured in a rotating bioreactor to create the cartilage, which controlled compressive strength similar to natural cartilage. The tissue-engineered cartilages could be applied in treating cartilage degenerative diseases. The scaffold was applied as biomimic ECM, enzyme, gene and medicine to revive skin, cartilage, blood vessel and nerve. The scaffold was helpful in biocompatibility, mechanical property, porosity, degradability in the human physical structure. The electrospun nanofibers showed moderate porosity, excellent mechanical property and biocompatibility, which could be utilized to repair blood vessels, skin and nervous tissue [1]. Tissue engineering is an emerging interdisciplinary and multidisciplinary research study. It involves the utilization of living cells, manipulated through their extracellular environment or genetically to develop biological substitutes for implantation into the body or to foster remodeling of tissues in some active manners. The purpose of tissue engineering is to renovate, replace, say, or improve the function of a particular tissue or organ. For a functional scaffold, a few basic needs have to be satisfied:

- A scaffold should control a high degree of porosity, with a suitable pore size distribution
- A large surface area is needed
- Biodegradability is often needed, with the degradation rate matching the rate of neotissue formation

- The scaffold must control the needed structural integrity to prevent the pores of the scaffold from collapsing during neotissue formation, with the suitable mechanical properties.
- The scaffold should be non-toxic to cells and biocompatible, positively interacting with the cells to promote cell adhesion, growth, migration, and distinguished cell function.

Among all biomedical materials under evaluation, electrospun nanofibrous scaffolds have presented great performances in cell attachment, increase and penetration [56]. One of the most promising potential applications is tissue scaffolding. The non-woven electrospun mat has a high surface area and a high porosity. It contains an empty space between the fibers that is approximately the size of cells. The mechanical property, the topographical layout, and the surface chemistry in the non-woven mat may have a direct effect on cell growth and migration [31]. Ultra-fine fibers of biodegradable polymers produced by electrospinning have found potential applications in tissue engineering because of their high surface area to volume ratios and high porosity of the fibers. However, the flexibility of seeding stem cells and human cells on the fibers makes electrospun materials most suited for tissue engineering applications. The fibers produced can be used systematically to design the structures that they perform not only mimic the properties of ECM, but also control high strength and high toughness. For instance, non-woven fabrics show isotropic properties and support neotissue formation. These mats resemble the ECM matrix and can be applied as a skin-scaffold and wound dressing materials where the materials are needed to be more elastic than stiff. Many natural polymers like collagen, starch, chitin and chitosan and synthetic biodegradable polymers like PCL, PLA, PLGA have been widely investigated for potential applications in developing tissue scaffolds. These results confirm that electrospinning of natural or synthetic polymers for tissue engineering applications are promising [9]. Tissue engineering is one of the most exciting interdisciplinary and multidisciplinary research fields today, and there has been exponential growth in the number of research publications in this area in recent years. It involves the utilization of living cells, manipulated through their extracellular environment or genetically to develop biological substitutes for implantation into the body or to foster remodeling of tissues in some active manners. The purpose is to repair, replace, maintain, or increase the use of a particular tissue or organ.

The core technologies intrinsic to this effort can be organized into three fields [4]:

- Cell technology
- Scaffold frame technology
- Technologies for in vivo integration.

The scaffold frame technology focuses on these objectives [4]:

- Designing
- Manufacturing
- Characterizing three-dimensional scaffolds for cell seeding
- In vitro or in vivo culturing

Blood Vessels

Blood vessels vary in sizes, mechanical and biochemical properties, cellar content and ultra structural organization, depending on their location and specific role. It is needed that the vascular grafts engineered should have wanted characteristics. Blood vessel replacement, a fine blood vessel (diameter<6 mm), has stayed a great challenge. Because the electrospun nanofiber mats can give good support during the initial development of vascular smooth muscle cells, smooth film combining with electrospun nanofiber mat could form a good 3D scaffold for blood vessel tissue engineering [4, 56].

Muscles

Collagen nanofibers were first applied to assess the feasibility of culturing smooth muscle cell. The cell growth on the collagen nanofibers was promoted and the cells were easily integrated into the nanofiber network after 7 days of seeding. Smooth muscle cells also adhered and proliferated well on another polymer nanofiber mats blended with collagen, incorporating collagen into nanofibers was observed to improve fiber elasticity and tensile strength, and increase the cell adhesion. The fiber surface wet ability influences cell attachment. The alignment of nanofibers can induce cell orientation and promote skeletal muscle cell morphogenesis and aligned formation [4].

Medical Prostheses

Polymer nanofibers fabricated by electrospinning have been offered for several soft tissue prosthesis applications such as blood vessel, vascular, breast, etc. In addition, electrospun biocompatible polymer nanofibers can also be deposited as a slender, porous film onto a hard tissue prosthetic device designed to be implanted into the human body. This coating film with a fibrous structure works as an interface between the prosthetic device and the host tissues. It is anticipated to reduce efficiently the stiffness mismatch at the tissue or Interphase and from here prevents the device failure after the implantation[54].

Tissue Template

For treating tissues or organs in malfunction in a human body, one of the challenges in the area of tissue engineering or biomaterials is the design of ideal scaffolds or synthetic matrices. They can mimic the structure and biological functions of the natural ECM. Human cells can attach and organize well around fibers with diameters smaller than those of the cellular phones. Nanoscale fibrous scaffolds can provide an ideal template for cells to seed, migrate, and produce. A successful regeneration of biological tissues and organs calls for developing fibrous structures with fiber architectures useful for cell deposition and cell growth. Of particular interest in tissue engineering is creating reproducible and biocompatible three-dimensional scaffolds for cell growth resulting in biometrics composites for various tissue repair and replacement processes. Recently, people have begun to pay attention to making such scaffolds with synthetic polymers or biodegradable polymer nanofibers. It is believed that converting biopolymers into fibers and networks that mimic native structures will eventually improve the usefulness of these materials as large diameter fibers do not mimic the morphological characteristics of the native fibrils [44, 54].

Wound Healing

Wound healing is a native technique of regenerating dermal and epidermal tissues. When an individual is wounded, a set of complex biochemical actions take place in a closely orchestrated cascade to repair the harm. These events can be sorted into four groups:

1. Inflammatory
2. Proliferative
3. Remodeling phases
4. Epithelialization

Ordinarily, the body cannot heal a deep dermal injury. In full thickness burn or deep ulcers, there is no origin of cells remaining for regeneration, except from the wound edges. Dressings for the wound-healing role to protect the wound, exude extra body fluids from the wound area, decontaminate the exogenous micro-organism, improve the appearance and sometimes speed up the healing technique. For these functions, a wound dressing material should provide a physical barrier to a wound, but be permeable to moisture and oxygen. For a full thickness dermal injury, when an "artificial dermal layer" adhesion and integration consisting of a 3D tissue scaffold with well-cultured dermal fibroblasts will aid there-epithelialization. Nanofiber membrane is a good wound dressing candidate because of its unique properties like [4] (Figure 13.16):

- the porous membrane structure
- well interconnected pores

They are important for exuding fluid from the wound. The small pores and high specific surface area not only inhibit the exogenous micro-organism invasions, but also assist the control of fluid drainage. In addition, the electrospinning provides a simple path to add drugs into the nanofibers for any possible medical treatment and antibacterial purposes[4].

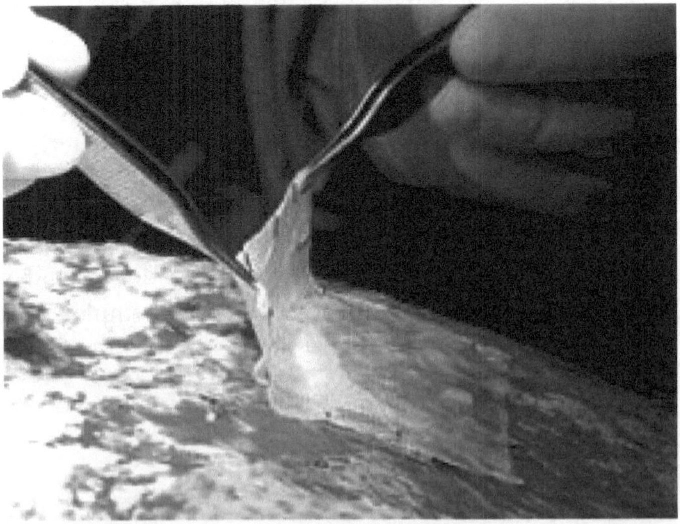

FIGURE 13.16 Nanofiber mats used for medical dressing.

For wound healing, an ideal dressing should have certain features:

1. Hemostatic ability
2. Efficiency as bacterial barrier
3. Absorption ability of excess exudates (wound fluid or pus)
4. Suitable water vapor transmission rate
5. Enough gaseous exchange ability
6. Ability to conform to the contour of the wound area
7. Functional adhesion
8. Painless to patient
9. Ease of removal
10. Low cost

Current efforts using nanofibrous membranes as a medical dressing are still in its early childhood, but electrospun materials meet most of the needs outlined for wound-healing polymer. Because their micro fibrous and nanofibrous provide the non-woven textile with desirable properties [42]. Polymer nanofibers can also be utilized for the treatment of wounds or burns of a human skin, as well as designed for hemostatic devices with some unique characteristics. Fine fibers of biodegradable polymers can spray/spun on to the injured location of the skin to make a fibrous mat dressing. They let wounds heal by encouraging forming a normal skin development and remove form scar tissue, which would occur in a traditional treatment. Non-woven nanofibrous membrane mats for wound dressing usually have pore sizes ranging from 500 nm to 1 mm, small enough to protect the wound from bacterial penetration by aerosol particle capturing mechanisms. High surface area of 5–100 m^2/g is efficient for fluid absorption and dermal delivery [44, 54]. The electrospun nanofibers have been utilized in treating wounds or burns of human skin because of their high porosity which allows gas exchange and a fibrous structure that protects wounds from infection and dehydration. Non-woven electrospun nanofibrous membranes for wound dressing usually have pore sizes in the range of 500 to 1,000 mm, which is low enough to protect the wound from bacterial penetration. High surface area of electrospun nanofibers is efficient for fluid absorption and dermal delivery. Chong invented a composite containing a semi-permeable barrier and a scaffold filter layer of skin cells in wound healing by electrospinning [1]. Electrospinning could create scaffold with more homogeneity besides meeting other needs like oxygen permeation and protection of wound from infection and dehydration for use as a wound-dressing materials. Many other synthetic and natural polymers like carboxyethyl, Chitosan or PVA, collagen or Chitosan, Silk fibroin, have been electrician to advise them for wound-dressing applications [11].

Release Control

Controlled release is an effective technique of delivering drugs in medical therapy. It can balance these features:

1. The delivery kinetics
2. Minimize the toxicity
3. Side effects
4. Improve patient convenience

In a controlled release system, the active substance is loaded into a carrier or device first, and then releases at a predictable rate in vivo when governed by an injected or non-injected route. As a potential drug delivery carrier, electrospun nanofibers have showed many advantages. The drug loading is easy to implement by electrospinning technique, and the high-applied voltage used in the electrospinning technique had little influence on the drug activity. The high specific surface area and short diffusion passage length give the nanofiber drug system higher overall release rate than the bulk material (e.g., film). The release profile can be finely controlled by modulating of nanofiber morphology, porosity and composition. Nanofibers for drug release systems mainly come from biodegradable polymers, such as PLA, PCL, PDLA, PLLA, PLGA and hydrophilic polymers such as PVA, PEG and PEO. Non-biodegradable polymers, such as PEU, were likewise found out [4]. Nanofiber systems for the release of drugs are needed to fill diverse roles. The mattress should be capable to protect the compound from decomposition and should allow for controlled release in the targeted tissue, over a needed period of time at a constant release rate [13]. Drug release and tissue engineering are closely related regions. Sometimes release of therapeutic causes can increase the efficiency of tissue engineering. Various nanostructured materials is applicable in tissue engineering. Electrospun fiber mats provide the advantage of increased drug release compared to roll-films because of the increased surface area [11, 37].

Drug Delivery and Pharmaceutical Composition

Delivery of drug or pharmaceuticals to patients in the most physiologically acceptable manner has always been an important concern in medicine. In general, the smaller the dimensions of the drug and the coating material wanted to encapsulate the drug, the better the drug to be assimilated by human being. Drug delivery with polymer nanofibers is based on the rule that the dissolution rate of a particulate drug increases with increasing surface area of both the drug and the similar carrier if needed. As the drug and carrier materials can be mixed for electrospinning of nanofibers, the likely modes of the drug in the resulting nano structure products are:

- Drug as particles attached to the surface of the carrier, which is in the form of nanofibers;
- Both drug and carrier are nanofiber-form, therefore the product will be the two kinds of nanofibers interlaced together;
- The blend of drug and carrier materials integrated into one fiber containing both sections;
- The carrier material is electrospun into a tubular frame in which the drug particles are encapsulated.

However, as the drug delivery in the form of nanofibers is still in the early stage exploration, a real delivery mode after production and efficiency has yet to be determined in the future [53]. Drug delivery with electrospun nanofibers is based along the principle that drug releasing rate increases with increasing surface area of both the drug and the similar carrier used. The increased surface area of drug improved the bioavailability of the poor water-soluble drug. Various drugs such as avandia, eprosartan, carvedilol, hydrochloridethiazide, aspirin, naproxen, nifedipine, indomethacin, and ketoprofen were entrapped into PVP to form pharmaceutical compositions, which provided controllable releasing. Not only synthetic polymers, but also natural polymers can be applied for modeling drug delivery system [5]. Controlled drug release over a definite period of time is possible with biocompatible delivery matrices of polymers and biodegradable polymers. They mostly used as drug delivery systems to deliver therapeutic agents because they can be well designed for programed distribution in a controlled fashion. Nanofiber mats applied as drug carriers in drug delivery system because of their high functional characteristics. The drug delivery system relies on the rule that the dissolution rate of a particulate drug increases with increasing surface area of both the drug and the similar carrier. Importantly, the large surface area associated with nano spun fabrics allows for quick and efficient solvent evaporation, which provides the incorporated drug limited time to recrystallize which favors forming amorphous dispersions or solid solutions. Depending on the polymer carrier used, the release of pharmaceutical dosage can be designed as rapid, immediate, delayed, or varied dissolution. Many researchers successfully encapsulate drugs within electrospun fibers by mixing the drugs in the polymer solution to be electrospun. Various solutions containing low molecular weight drugs have been electrospun, including lipophilic drugs such as ibuprofen, cefazolin, rifampin, paclitaxel and Itraconazole and hydrophilic drugs such as mefoxin and tetracycline hydrochloride. However, have encapsulated proteins in electrospun polymer fibers. Besides the normal electrospinning process, another path to develop drug-loaded polymer nanofibers for controlling drug release is to use coaxial electrospinning and research has successfully encapsulated two kinds of medicinal pure drugs through this process [42]. Electrospinning affords great flexibility in selecting materials for drug delivery applications. Either biodegradable or non-degradable materials can be utilized to control whether drug release occurs by diffusion alone or diffusion and scaffold degradation. Also, because of the flexibility in material selection many drugs can be delivered including:

- Antibiotics
- Anticancer drugs
- Proteins
- DNA

Using the various electrospinning techniques, many different drug loading methods can also be applied:

1. Coatings
2. Embedded drug
3. Encapsulated drug (coaxial and emulsion electrospinning

However, as the drug delivery in the form of nanofibers is still in the early stage exploration, a real delivery mode after production and efficiency has yet to be found in the future [43].

13.6.2 COSMETICS

The current skin masks applied as topical creams, lotions or ointments. They may include dusts or liquid sprays and more likely than fibrous materials to migrate into sensitive areas of the body, such as the nose and eyes where the skin mask is being utilized to the face. Electrospun polymer nanofibers have been tried as a cosmetic skin care mask for treating skin healing, skin cleaning, or other therapeutic or medical properties with or without various additives. This nanofibrous skin mask with small interstices and high surface area can make easy far greater utilization and speed up the rate of transfer of the additives to the skin for the fullest potential of the additive. The cosmetic skin mask from the electrospun nanofibers can be applied gently and painlessly as well as directly to the three-dimensional topography of the skin to provide healing or cure treatment to the skin [53]. Electrospun nanofibers have been aimed for use in cosmetic cares such as treating skin healing and skin cleaning with or without various additives in recent years. Despite the growth in the number of electrospun polymer nanofiber publications in the recent years, there is a rare work, including scientific papers and patents, in the cosmetic field about the use of electrospun nanofibers. Developing nanofibers in this field have been focused on skin treatment applications, such as care mask, skin healing and skin cleaning, with active agents (cosmetics) with controlled release from time to time. The cosmetic application be included in the biomedicine application, which admits the drug delivery system employing the active agents used in cosmetics, body care supplements. Therefore, the cosmetic and drug delivery are closely interrelated areas. The electrospun nanofibers provide the advantage of the increasing drug release when compared to cast-films because of the increased surface area. Besides, it's agreeable to processing different polymers such as natural, synthetic and blends, according to their solubility or melting point. It is significant that although most of the researches in polymeric nanofiber by electrospinning consider the technique simple, cost-effective and easily scalable from laboratory to commercial production, only a limited number of companies have commercially performed electrospun fibers [5].

13.6.3 ENERGY AND ELECTRONICS AND CHEMICAL ENGINEERING

The demand for energy use of goods and services has been increasing every year throughout the world. However, it was reported the estimated reserve amounts of petroleum and natural gas in the world are only 41 years and 67 years, respectively. To solve this problem, new, clean, renewable and sustainable energies have to be ground and used to replace the current non-sustainable energies. Wind generator, solar power generator, hydrogen battery, and polymer battery are among the most popular alternatives to produce new energies. In recent years, electrospun nanofibers have presented their potential in these applications:

- Super capacitors
- Lithium cells
- Fuel cells
- Solar cells
- Transistors

Further, electrospun nanofibers with electrical and electro-optical have also got much interest recently because of their potential applications in creating nanoscale electronic and optoelectronic devices [5] (Figure 13.17).

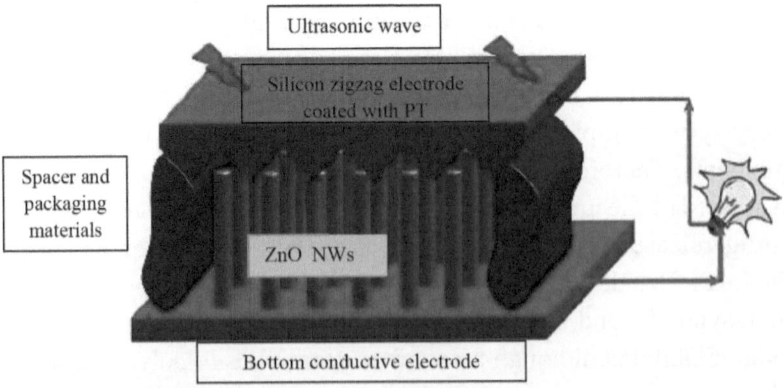

FIGURE 13.17 ZnO nanofibers in energy application.

13.6.3.1 Filtration Application

Filtration is necessary in many engineering fields. It was estimated that future filtration market would be up to US 700 billion US dollars by the year 2020 [1, 15, 44] (Figure 13.18).

Fibrous materials used for filter media provide advantages of high filtration efficiency and low air resistance. Filtration efficiency, which is closely related with the fiber fineness, is one of the most important concerns for the filter performance. One direct way of developing high efficient and effective filter media are by using nanometer sized fibers in the filter structure [15, 54]. With outstanding of polymeric nanofibers properties such as high specific surface area, high porosity, and excellent surface adhesion, they are suited to be made into filtering media for filtering out particles in the sub micron range. Also, this filter system could be used for processing waste water containing active sludge [1, 9]. Electrospun fibers are being widely studied for aerosol filtration, air-cleaning applications in industry and for particle collection in clean rooms. The advantage of using electrospun fibers in the filtration media is the fiber diameters can be easily controlled and can produce an impact in high efficiency particulate air

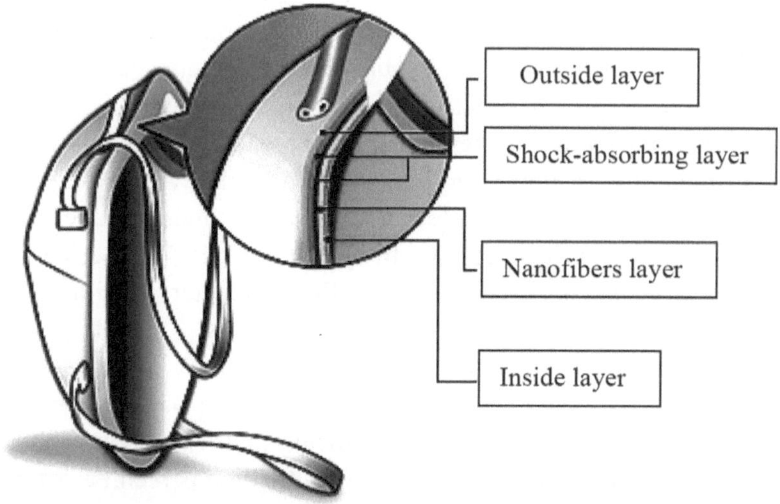

FIGURE 13.18 Applications of Nanofibers in Filtration.

filtrations [9]. The filtration efficiency is commonly influenced by these parameters [4]:

- The filter physical structure like
  - ✎ Fiber fineness
  - ✎ Matrix structure
  - ✎ Thickness
  - ✎ Pore size
  - ✎ Fiber surface electronic properties
  - ✎ Its surface chemical characteristic
- Surface free energy

The particle collecting capability is also associated with the size range of particles being collected. Besides the filtration efficiency, other properties such as pressure drop and flux resistance are also important factors to be assessed for a filter media [4]. Filter efficiency increases linearly with the decrease of thickness of filter membrane and applied pressure increase [42].

13.6.4 REINFORCED COMPOSITES/REINFORCEMENT

Although electrospun fiber reinforced composites have significant potential for development of high intensity or high toughness materials and materials with good thermal and electrical con conductivity. Few studies have found out the use of electrospun fibers in composites. Traditional reinforcements in polymer matrices can create stress concentration sites because of their irregular shapes and cracks spread by burning through the fillers or traveling up, down and around the particles. However, electrospun fibers have various advantages over traditional fillers. The reinforcing effects of fibers are influenced essentially by fiber size. Smaller size fibers give more efficient support. Fibers with finer diameters have a preferential orientation of polymer chains along the fiber axis. The orientation of macromolecules in the fibers improves with decreasing in diameter, making finer diameter fibers strong. Therefore, the use of nanometer sized fibers can significantly raise the mechanical integrity of polymer matrix compared to micron-sized fibers. However, the high percentage of porosity and irregular pores between fibers can contribute to an interpenetrated structure when spread in the matrix, which also improves the mechanical strength because of the interlocking mechanism. These characteristic features of nanofibers enable the transfer of applied stress to the fiber-matrix in a more serious fashion than most of the commonly used filler materials. Current issues related to the use of electrospun nanofibers as reinforcement materials are the control of dispersion and orientation of the fibers in the polymer matrix. To achieve better reinforcement, electrospun nanofibers, may require to be collected as an aligned yarn instead of a randomly distributed felt so the post-electrospinning stretching process could be applied to further improve the mechanical properties. Further, if crack growth is transverse to the fiber orientation, the crack toughness of the composite can be optimized. So, the interfacial adhesion between fibers and matrix material needs to be controlled such the fibers are deflecting the cracks by fiber-matrix interface debonding and fiber pullout. The interfacial adhesion should not be excessively strong or too weak. Ideal control can only be obtained by careful selective fiber surface treatment. Spreading electrospun mats in the matrix can be improved by cutting down the fibers to shorter fragments. This can be accomplished, if the electrospun fibers are collected as aligned bundles (instead of non-woven network), which can then be optically or mechanically trimmed to get fiber fragments of several 100 nm in length [9]. Early studies on electrospun nanofibers also included reinforcement of polymers. As electrospun nanofiber mats have a large specific surface area and an irregular pore structure, mechanical interlocking among the nanofibers should occur [4]. One of the most significant applications of traditional fibers, especially engineering fibers such as carbon, glass, and Kevlar fibers, is to be used as reinforcements in composite developments. With these reinforcements, the composite materials can provide superior structural properties such as high modulus and strength to weight ratios, which cannot be attained by other engineered monolithic materials alone. Nanofibers will also eventually find important applications in making nano composites. This is because nanofibers can have even better mechanical properties than micro fibers of the same materials, and therefore the superior structural properties of nano composites can be anticipated. However, nanofiber reinforced composites may control some extra merits which cannot be shared by traditional (microfiber) composites. For instance, if there is a difference in refractive indices between fiber and ground substance, the resulting composite becomes opaque or nontransparent because of light scattering. This limit, however, can be avoided when the fiber diameters become significantly smaller than the wavelength of visible illumination [44] (Figure 13.19).

13.6.5 DEFENSE AND SECURITY

Military, firefighter, law enforcement, and medical personal need high-level protection in many environments ranging from combat to urban, agricultural, and industrial, when dealing with chemical and biological threats like [17]:

- Nerve agents,
- Mustard gas,
- Blood agents such as cyanides,
- Biological toxins such as bacterial spores, viruses, and rickettsiae.

Nanostructures with their minuscule size, large surface area, and light weight will improve, by orders of magnitude, our capability to [17]:

- Detect chemical and biological warfare agents with sensitivity and selectivity;
- Protect through filtration and destructive decomposition of harmful toxins;
- Provide site-specific naturally prophylaxis.

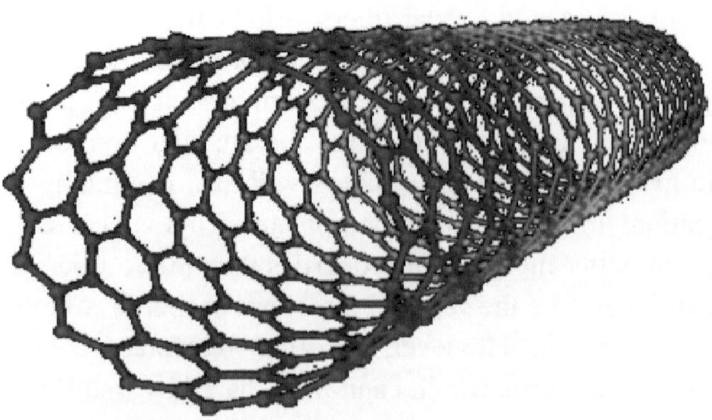

FIGURE 13.19 Carbone nanotubes composite.

Polymer nanofibers are considered as excellent membrane materials owing to their lightweight, high surface area, and breathable (porous) nature. The high sensitivity of nanofibers toward warfare agents makes them excellent candidates as sensing of chemical and biological toxins in concentration levels of parts per billion. Governments across the globe are investing in strengthening the protection levels offered to soldiers in the battlefield. Various methods of varying nanofiber surfaces to improve their capture and decontamination capacity of warfare agents are under investigation. Nanofiber membranes may be employed to replace the activated charcoal in adsorbing toxins from the atmosphere. Active reagents can be planted in the nanofiber membrane by chemical functionalization, post-spinning variation, or through using nano particle polymer composites. There are many avenues for future research in nanofibers from the defense perspective. As well as serving protection and decontamination roles, nanofiber membranes will also suffer to provide the durability, wash ability, resistance to intrusion of all liquids, and tear strength needed of battledress fabrics [17] (Figures 13.20 and 13.21).

13.6.5.1 Protective Clothing

The protective clothing in the military is largely expected to help increase the suitability, sustainability, and combat effectiveness of the individual soldier system against extreme climate, ballistics, and NBC warfare. In peace ages, breathing apparatus and protective clothing with the particular role of against chemical warfare agents such as sarin, soman, tabun

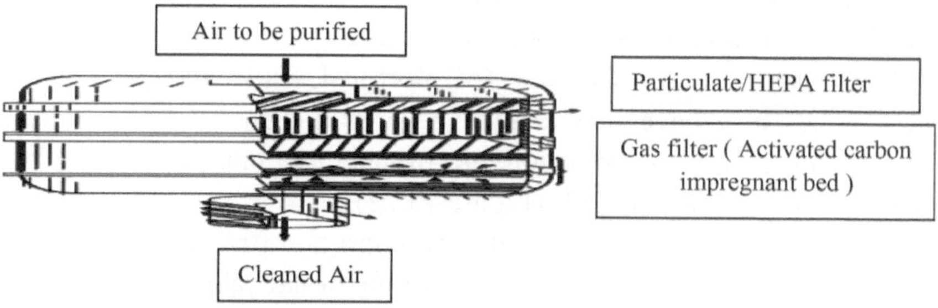

FIGURE 13.20 Cross-sectioning of a facemask canister used for protection from chemical and biological warfare agents.

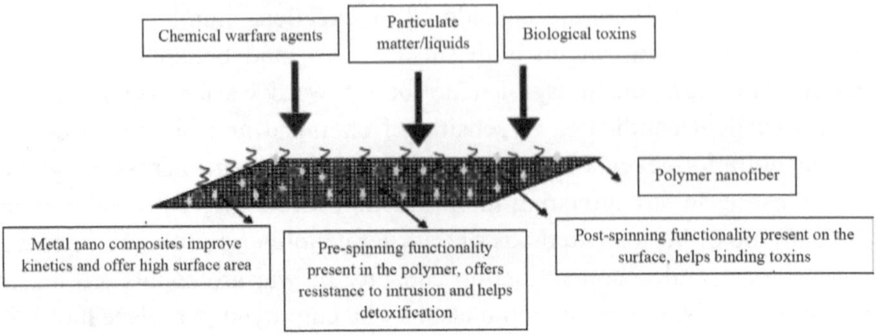

FIGURE 13.21 Incorporating functional groups into a polymer nanofiber mesh.

become a special concern for combatants in conflicts and civilian populations in terrorist attacks. Current protective clothing containing charcoal absorbents has its terminal points for water permeability, extra weight-imposed to the article of clothing. A lightweight and breathable fabric, which is permeable to both air and water vapor, insoluble in all solvents and reactive with nerve gasses and other deadly chemical agents, is worthy. Because of their large surface area, nanofiber fabrics are neutralizing chemical agents and without impedance of the air and water vapor permeability to the clothing. Electrospinning results in nanofibers lay down in a layer that has high porosity but small pore size, offering good resistance to penetrating chemical harms agents in aerosol form. Preliminary investigations indicate that compared to conventional textiles the electrospun nanofibers present both small impedance to moisture vapor diffusion and efficiency in trapping aerosol particles, as well as show strong promises as ideal protective clothing. Conductive nanofibers are expected to be utilized in fabricating tiny electronic or machines such as Schottky junctions, sensors and actuators. Conduct (of electrical, ionic and photoelectric) membranes also have potential for applications including electrostatic dissipation, corrosion protection, electromagnetic interference shielding, photovoltaic device, etc. [53] (Figure 13.22).

Electrospun nanofibers can play an important part in textile applications as protective clothing and other functional fabric materials. The electrospun nanofibrous membranes are capable of neutralizing chemical agents without impedance of the air and water vapor permeability to

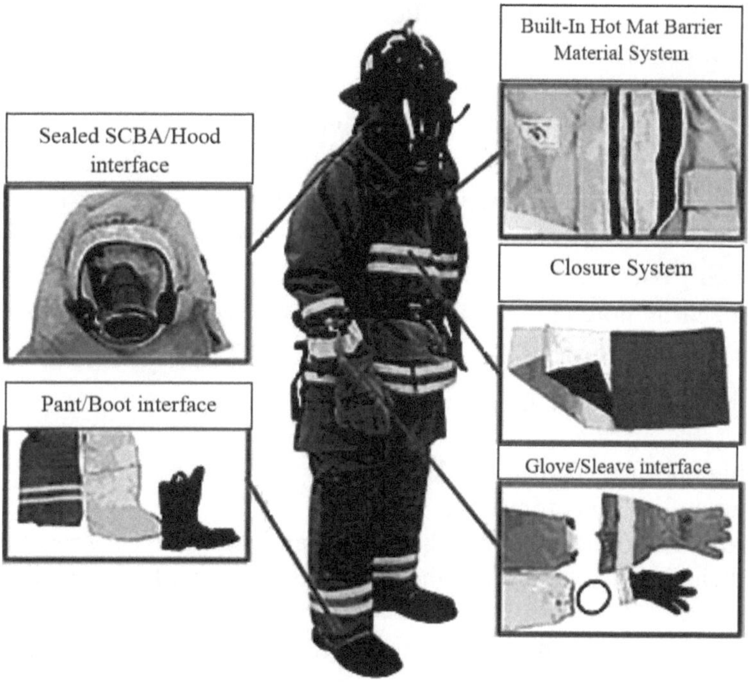

FIGURE 13.22 Protective clothing applications.

the clothing because of their high specific surface area and high porosity but small pore size. Preliminary investigations suggest the electrospun nanofibers control both minimal impedance to moisture vapor diffusion and efficiency in trapping aerosol particles compared with conventional textiles. Smith prepared a fabric comprising electrospun PEI nanofibers as lightweight protective clothing which was captured and neutralizing chemical warfare agents. This formed fabric also could be used in protective breathing apparatuses because PEI provides multiple amine sites for the nucleophilic decomposition of mustard gasses and fluorophosphates nerve gasses. A protective mask was constructed by attaching PC/PS electrospun nanofibrous layer to one side of a moist fabric composed of cellulose and wool. The diameter of the nanofibers in the protective layer was in the range of 100–10,000 nm. A non-woven fabric composed of a submicrosized fiber, which receives a PC shell and a polyurethane core was made by co-axial electrospinning. The resultant fabric combines the filtration efficiency of the PC and the mechanical effectiveness of polyurethane, which is useful in exposure suits and aviation clothing. A water-resistant and air-permeable laminated fabric was manufactured by utilizing hot-melt polyester as dots onto the surface of the electrospun nylon non-woven fabric [5]. Ideally, protective clothing should have close to essential properties such as, lightweight, breathable fabric, air and water vapor permeability, insoluble in all solvents and improved toxic chemical resistance. Electrospun nanofiber membranes recognized as potential candidates for protective clothing applications for these causes:

- they're lightweight;
- large surface area;
- high porosity (breathable nature);
- great filtration efficiency;
- resistant to penetration of harmful chemical agents in aerosol form;
- their ability to neutralize the chemical agents without impedance of the air;
- water vapor permeability to the clothing.

Various methods for variation of nanofiber surfaces have been examined to improve protection against toxins. One protection method that has been used includes chemical surface variation and attachment of reactive groups such as axioms, Cyclodextrins, and chloramines that bind and detoxify warfare agents [42].

13.7 CONCLUSION

Nanotechnology with unique properties is interested by many scientists in everywhere for novel applications in recent years. Various common techniques can be used for preparing polymer nanofibers. Also, special ways such as gelation and bacterial cellulose were utilized for producing nanofibers. Among these methods, electrospinning has been widely used as a novel technique for generating nano scale fibers. Therefore, electrospun nanofibers are utilized in a wide range of applications. Also, this work was analyzed the recent advances of this technology in tissue engineering, drug delivery, etc.

KEYWORDS

- **electrospinning**
- **nanofibers**
- **nanofibers applications**
- **nanofibers production**
- **nanotechnology**

REFERENCES

1. Lu, and B. Ding, *Applications of Electrospun Fibers.* Recent patents on nanotechnology, 2008. 2(3), 169–182.
2. Reneker, D. H., et al., *Electrospinning of Nanofibers from Polymer Solutions and Melts.* Advances in applied mechanics, 2007. 41, 43–346.
3. Vonch, J., A. Yarin, and C. M. Megaridis, *Electrospinning: A Study in the Formation of Nanofibers.* Journal of Undergraduate Research, 2007. 1, 1–6.
4. Fang, J., et al., *Applications of Electrospun Nanofibers.* Chinese Science Bulletin, 2008. 53(15), 2265–2286.
5. RAFIEI, S., et al., *Mathematical Modeling in Electrospinning Process of Nanofibers: A Detailed review.* Cellulose Chemistry And Technology, 2013. 47(5–6), 323–338.
6. Fang, J., X. Wang, and T. Lin, *Functional Applications of Electrospun Nanofibers.* Nanofibers-production, properties and functional applications, 2011, 287–326.
7. Karra, S., *Modeling Electrospinning Process and a Numerical Scheme using Lattice Boltzmann Method to Simulate Viscoelastic Fluid Flows.* 2007, Texas A&M University.
8. Angammana, C. J., *A Study of the Effects of Solution and Process Parameters on the Electrospinning Process and Nanofiber Morphology.* 2011, University of Waterloo.
9. Baji, A., et al., *Electrospinning of Polymer Nanofibers: Effects on Oriented Morphology, Structures and Tensile Properties.* Composites Science and Technology, 2010. 70(5), 703–718.
10. Beachley, V. and X. Wen, *Polymer Nanofibrous Structures: Fabrication, Biofunctionalization, and Cell Interactions.* Progress in Polymer Science, 2010. 35(7), 868–892.
11. Agarwal, S., J. H. Wendorff, and A. Greiner, *Use of Electrospinning Technique for Biomedical Applications.* Polymer, 2008. 49(26), 5603–5621.
12. Fridrikh, S. V., et al., *Controlling The Fiber Diameter during Electrospinning.* Physical review letters, 2003. 90(14), 144502–144502.
13. Garg, K. and G. L. Bowlin, *Electrospinning Jets and Nanofibrous Structures.* Biomicrofluidics, 2011. 5(1), 013403–1 – 013403–19.

14. Haghi, A. K., *Electrospun Nanofiber Process Control.* Cellulose Chemistry & Technology, 2010. 44(9), 343–352.
15. Huang, Z. M., et al., *A Review on Polymer Nanofibers by Electrospinning and their Applications in Nanocomposites.* Composites science and technology, 2003. 63(15), 2223–2253.
16. Kowalewski, T. A., S. NSKI, and S. Barral, *Experiments and Modelling of Electrospinning Process.* Technical Sciences, 2005. 53(4), 385–394.
17. Ramakrishna, S., et al., *Electrospun Nanofibers: Solving Global Issues.* Materials Today, 2006. 9(3), 40–50.
18. Reneker, D. H. and I. Chun, *Nanometer Diameter Fibres of Polymer, Produced by Electrospinning.* Nanotechnology, 1996. 7(3), 216–223.
19. Wang, H. S., G. D. Fu, and X. S. Li, *Functional Polymeric Nanofibers from Electrospinning.* Recent patents on nanotechnology, 2009. 3(1), 21–31.
20. Zhang, C., X. Ding, and S. Wu, *The Microstructure Characterization and the Mechanical Properties of Electrospun Polyacrylonitrile-Based Nanofibers.* p. 177–196.
21. Zhang, S., *Mechanical and Physical Properties of Electrospun Nanofibers.* 2009.
22. Zhou, H., *Electrospun Fibers from Both Solution and Melt, processing, Structure and Property*. 2007, Cornell University.
23. Zanin, M. H. A., N. N. P. Cerize, and A. M. de. O, *Production of Nanofibers by Electrospinning Technology: Overview and Application in Cosmetics*, in *Nanoco-smetics and Nanomedicines*. 2011, Springer. p. 311–332.
24. Khan, N., *Applications of Electrospun Nanofibers in The Biomedical Field.* Studies by Undergraduate Researchers at Guelph, 2012. 5(2), 63–73.
25. Ramakrishna, S. and K. Fujihara, *An Introduction to Electrospinning and Nanofibers*. 2005. 1–383.
26. Zeng, Y., et al. *Numerical Simulation of Whipping Process in Electrospinning*. in *WSEAS International Conference. Proceedings. Mathematics and Computers in Science and Engineering*. 2009: World Scientific and Engineering Academy and Society.
27. Ciechańska, D., *Multifunctional Bacterial Cellulose/Chitosan Composite Materials for Medical Applications.* Fibres & Textiles in Eastern Europe, 2004. 12(4), 69–72.
28. Stanger, J. J., et al., *Effect of Charge Density on the Taylor Cone in Electrospinning.* International Journal of Modern Physics B, 2009. 23(06).
29. Chronakis, I. S., *Novel Nanocomposites and Nanoceramics based on Polymer Nanofibers using Electrospinning Process—A Review.* Journal of Materials Processing Technology, 2005. 167(2), 283–293.
30. Wu, Y., et al., *Controlling Stability of the Electrospun Fiber by Magnetic Field.* Chaos, Solitons & Fractals, 2007. 32(1), 5–7.
31. Zhang, S., *Mechanical and Physical Properties of Electrospun Nanofibers.* 2009, 1–83.
32. Li, W. J., et al., *Electrospun Nanofibrous Structure: A Novel Scaffold for Tissue Engineering.* Journal of biomedical materials research, 2002. 60(4), 613–621.
33. Brown, E. E. and M. P. G. Laborie, *Bioengineering Bacterial Cellulose/Poly (Ethylene Oxide) Nanocomposites.* Biomacromolecules, 2007. 8(10), 3074–3081.
34. De. V, S., et al., *The Effect of Temperature and Humidity on Electrospinning.* Journal of materials science, 2009. 44(5), 1357–1362.

35. Tao, J. and S. Shivkumar, *Molecular Weight Dependent Structural Regimes during The Electrospinning of PVA.* Materials Letters, 2007. 61(11), 2325–2328.
36. Kowalewski, T. A., S. B. Ł. O. NSKI, and S. Barral, *Experiments and Modelling of Electrospinning Process.* Technical Sciences, 2005. 53(4).
37. Zong, X., et al., *Structure and Process Relationship of Electrospun Bioabsorbable Nanofiber Membranes.* Polymer, 2002. 43(16), 4403–4412.
38. Lyons, J., C. Li, and F. Ko, *Melt-Electrospinning Part I, processing Parameters and Geometric Properties.* Polymer, 2004. 45(22), 7597–7603.
39. Reneker, D. H. and A. L. Yarin, *Electrospinning Jets and Polymer Nanofibers.* Polymer, 2008. 49(10), 2387–2425.
40. Bognitzki, M., et al., *Nanostructured Fibers via Electrospinning.* Advanced Materials, 2001. 13(1), 70–72.
41. Deitzel, J. M., et al., *The Effect of Processing Variables on the Morphology of Electrospun Nanofibers and Textiles.* Polymer, 2001. 42(1), 261–272.
42. Bhardwaj, N. and S. C. Kundu, *Electrospinning: A Fascinating Fiber Fabrication Technique.* Biotechnology Advances, 2010. 28(3), 325–347.
43. Sill, T. J. and H. A. von. R, *Electrospinning: Applications in Drug Delivery and Tissue Engineering.* Biomaterials, 2008. 29(13), 1989–2006.
44. Patan, A.k., et al., *Nanofibers-A New Trend in Nano Drug Delivery Systems.* International Journal of Pharmaceutical Research & Analysis, 2013. 3, 47–55.
45. Rutledge, G. C. and S. V. Fridrikh, *Formation of Fibers by Electrospinning.* Advanced Drug Delivery Reviews, 2007. 59(14), 1384–1391.
46. Sawicka, K. M. and P. Gouma, *Electrospun Composite Nanofibers for Functional Applications.* Journal of Nanoparticle Research, 2006. 8(6), 769–781.
47. Yousefzadeh, M., et al., *A Note on The 3D Structural Design of Electrospun Nanofibers.* Journal of Engineered Fabrics & Fibers (JEFF), 2012. 7(2), 17–23.
48. Yarin, A. L., S. Koombhongse, and D. H. Reneker, *Bending Instability in Electrospinning of Nanofibers.* Journal of Applied Physics, 2001. 89(5), 3018–3026.
49. Keun. S, W., et al., *Effect of PH on Electrospinning of Poly (Vinyl Alcohol).* Materials Letters, 2005. 59(12), 1571–1575.
50. Feng, J. J., *The Stretching of An Electrified Non-Newtonian Jet: A Model for Electrospinning.* Physics of Fluids (1994-present), 2002. 14(11), 3912–3926.
51. Maleki, M., M. Latifi, and M. Amani. T, *Optimizing Electrospinning Parameters for Finest Diameter of Nano Fibers.* World Academy of Science, Engineering and Technology, 2010. 40, 389–392.
52. Thompson, C. J., *An Analysis of Variable Effects on A Theoretical Model of the Electrospin Process for Making Nanofibers.* 2006, University of Akron.
53. Luo, C. J., M. Nangrejo, and M. Edirisinghe, *A Novel Method of Selecting Solvents for Polymer Electrospinning.* Polymer, 2010. 51(7), 1654–1662.
54. Huang, Z.-M., et al., *A Review on Polymer Nanofibers by Electrospinning and Their Applications in Nanocomposites.* Composites Science and Technology, 2003. 63, 2223–2253.
55. Frenot, A. and I. S. Chronakis, *Polymer Nanofibers Assembled by Electrospinning.* Current opinion in colloid & interface science, 2003. 8(1), 64–75.
56. Fang, J., X. Wang, and T. Lin, *Functional Applications of Electrospun Nanofibers.* Nanofibers-production, properties and functional applications, 287–326.

PART II

BIOCHEMICAL PHYSICS

CHAPTER 14

COMPOSITION OF BIOREGULATOR OBTAINED FROM GARLIC *Allium sativum* L.

O. G. KULIKOVA,[1] A. P. ILYINA,[1] V. P. YAMSKOVA,[2] and I. A. YAMSKOV[1]

[1]*Nesmeyanov Institute of Organoelement Compounds, Russian Academy of Sciences, St. Vavilova 28, Moscow, 119991 Russia*

[2]*Koltsov Institute of Developmental Biology, Russian Academy of Sciences, St. Vavilova 26, Moscow, 119334 Russia, E-mail: yamskova-vp@yandex.ru*

CONTENTS

ABSTRACT

This study devote to extracting and investigating bioregulator of MGTBs group from garlic Allium sativum L. Garlic Allium sativum L. are widely spread medicine plant which can be used for prophylaxis and treatment different kind of diseases. Bioregulator obtained consist of biological activity peptide component with molecular weight 2–9 kDa and protein component 56–70 kDa, which modulate activity of peptide containing fraction. Using this peptide-protein complex can be very perspective for investigation biological activity in the models in vivo and in vitro.

14.1 AIM AND BACKGROUND

Aim of this work is obtaining bioregulator of MGTBs group and study its composition for further its using as medical drug. *Allium sativum* L. is commonly used medical plant. Throughout history, many different cultures have recognized the potential use of garlic for prevention and treatment of different diseases. Recent studies support the effects of garlic and its extracts in a wide range of applications. These studies raised the possibility of revival of garlic therapeutic values in different diseases. Different compounds in garlic are thought to reduce the risk for cardiovascular diseases, have anti-tumor and anti-microbial effects, and show benefit on high blood glucose concentration. However, the exact mechanism of all ingredients and their long-term effects are not fully understood. Further studies are needed to elucidate the pathophysiological mechanisms of action of garlic as well as its efficacy and safety in treatment of various diseases [1].

14.2 INTRODUCTION

Plants are the traditional source of biologically active compounds, most of them are not studied and characterized enough [2,3]. Compounds influencing important biological processes (migration, adhesion, proliferation, and cell differentiation) are among them. In this aspect, we should pay attention to new biologically active membranotropic homeostatic tissue specific bioregulators (MHTBs), which were found in different tissues of mammal, plants and fungi [4]. It was established that MHTBs isolated from mammalian tissues are the complex of peptides whose molecular weight does not exceed 11 kDa and protein-modulators. These MHTBs are located in the intercellular space in tissues [5–9]. Small peptides (1000–11000 Da), which are responsible for biological activity of MHTBs are bound with protein modulated their activity [4, 10]. These bioregulators are active at concentration 10^{-8}–10^{-15} mg/mL and characterized by the presence of tissue specificity and the absence of species specificity [4, 10, 11]. MHTBs also stimulate reparative processes in pathologically altered tissues, facilitating the repair of their disturb structures by additional activating cell regeneration sources [5–7].

All members of this group bioregulators have similar physical and chemical properties. Thus for isolating these MHTBs from different tissue sources and for their purification we use experimental approach including different methods of their extraction from tissues and the range of traditional biochemistry methods [4].

14.3 MATERIALS AND METHODS

Garlic (*Allium sativum* L.) was obtained from the nursery of the N.V. Tsitsin Main Botanical Garden, RAS. In this study we used purified water (18 MOhm), acetonitrile, isopropanol, acetic acid and trifluoroacetic acid, mercaptoethanol, hydrochloric acid, ethanol, acetone, KCl, KH_2PO_4, and ammonium sulfate (Reahim, Russia), Tris, dodecyl sulfate (SDS-Na), dithiothreitol and trypan blue (Serva, Germany) acrylamide, methylene bis-acrylamide, ammonium persulfate, N,N,N,'N'-tetramethylethylene diamine (TEMED), Na_2HPO_4, Tween-20 (Sigma, USA).

The biological activity of fractions obtained was tested in F1-hybrid mice line C57BL/CBA (males weighing 18–20 g).

Centrifugation was performed with a T32A centrifuge (Janetzki, Germany). We also used the Mini vertical gel system EC 120 (BioRad, United States), PowerPacBasic electrophoresis power supply

(BioRad, UNITED STATES), Ultrospec 1100 pro spectrophotometer (Biochrom Ltd., UK), Jenaval light microscopes (Carl Zeiss, Germany) and Olympus Vanox ABVTZ (Japan), Olympus U-PMTVC digital camera (Japan), Oalcton 2100 pH meter (EUTECH Instruments, Singapore), and a Transsonic Digital ultrasonic bath (Elma, Switzerland).

14.3.1 BIOREGULATOR ISOLATION AND PURIFICATION

For extraction, we used garlic bulbs cut into small (1–1.5 cm) fragments, which were incubated at 8–10°C for 5–6 h in an extraction solution containing 2.06×10^{-2} M NH_4NO_3, 1.88×10^{-2} M KNO_3, 10^{-3} M$CaCl_2 \bullet 2H_2O$, 1.5×10^{-3} M $MgSO_4 \oplus 7H_2O$, 1.25×10^{-3}M KH_2PO_4. Thus obtained extract was filtered through several layers of cheesecloth and centrifuged at 3000 g for 30 min; the pellet was discarded. Then, dry ammonium sulfate was added to the plant extract under stirring to a concentration of 780 g/L until a saturated salt solution formed; pH was adjusted at 7.5–8.0 with ammonium hydroxide. The obtained protein mixture was incubated at 4°C for 95–100 h. The precipitated proteins were separated by centrifugation at 25000 g and 4–8°C for 30 min. Thus obtained supernatant and pellet were dialyzed against water or physiological solution until complete removal of ammonium sulfate and then concentrated at 37–40°C in a vacuum rotor evaporator.

The *protein concentration* was determined according to the Lowry method.

14.3.2 ELECTROPHORESIS OF PROTEINS IN POLYACRYLAMIDE GEL

Electrophoresis of proteins in 12.5% polyacrylamide gel (PAGE) under denaturing conditions was carried out according to the Laemmli method. Protein markers were used as molecular weight markers with the following (Sigma Aldrich, USA): α-lactalbumin –14200 Da, soybean trypsin inhibitor – 20,000 Da, trypsinogen from bovine pancreas – 24,000 Da, carbonic anhydrase – 29,000 Da, glyceraldehyde-3–phosphate dehydrogenase – 36000 Da, valbumin –45,000 Da, albumin – 66,000 Da. Gels were stained with Coomassie G-250 [12].

14.3.3 MASS-SPECTROMETRIC ANALYSIS

Molecular weights of peptides were determined by the matrix-assisted laser desorption/ionization (MALDI-TOF) technique in an UltraFlex 2 time-of-flight mass spectrometer (Germany). Time-of-flight mass spectra were recorded in a linear model and reflecting mode. Samples for mass spectrometric analysis were obtained by evaporation until dry with subsequent dilution in 70% acetonitrile containing 0.1% trifluoroacetic acid with α-cyano-4-hydroxycynamic acid being used as a matrix.

14.3.4 ADHESIOMETRIC METHOD

We used a previously developed method for MHTB identification, which includes short-term organ cultivation of mouse liver in vitro (bioassay method), to investigate the membrano-trophic activity of the obtained fractions [13]. The experiment was performed at least 3 times for each fraction. Statistical data processing was performed by variation statistics using Student's t-test (Origin 6.0 Profession Program).

14.3.5 EXCLUSION CHROMATOGRAPHY

Fractions of pellets were analyzed using a high-pressure chromatograph (Kromasil, US) equipped with a HPLC Gel Filtration Column Bio-Sil TSK-125 300×7.5 мм (Japan). Elution was conducted in phosphate buffer 0.05 M pH 6.8 with velocity 0.4 mL/min for 25 min. Detection was carried out at 280 nm.

14.3.6 TRYPTIC HYDROLYSIS

Tryptic hydrolysis was performed as follows: gel pieces sized about 2 × 2 mm^2 washed twice in 100 mL of 40% acetonitrile in 0.1 M NH_4HCO_3 for 20 min at room temperature to remove the dye. After removing the solution 100 mL of acetonitrile was added. After removing acetonitrile and dried piece of the gel 4 mL of modified trypsin (Promega) in 0.05 M NH_4HCO_3 at a concentration of 15 mg/mL was added. The hydrolysis was conducted for 18 hours at 37°C, then 7 mL of 0.5% TFA in 10%

acetonitrile solution in water was added to a solution of and stirred thoroughly. Nadgelevy solution was used for MALDI-mass spectra.

Identification of proteins was performed using the program Mascot (www.matrixscience.com). The search was conducted in the NCBI protein among all organisms with the same accuracy with the possibility of methionine oxidation by atmospheric oxygen and possible modification of cysteines acrylamide. Candidate proteins having parameter reliability score > 83 in the NCBI database is considered definitely reliable ($p < 0.05$).

14.4 RESULTS AND DISCUSSION

A small fragments of bulbs were extracted with saline solution at a lower temperature. Then obtained extract was treated with ammonium sulfate that resulted to the two fractions formation: pellet and supernatant. The adhesiometry bioassay method has shown that only the fraction of supernatant had membranotropic activity characteristic for MGTBs (Figure 14.1). Figure 14.1 demonstrate dose dependence of biological activity that had

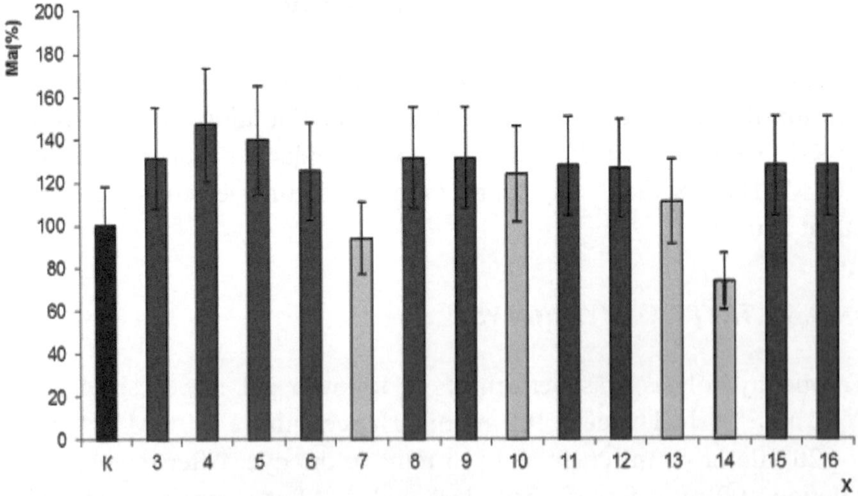

FIGURE 14.1 Dose dependence activity (%) of supernatant fraction of garlic extract (*Allium sativum* L.). The abscissa axis shows the degree of tenfold sequential dilution of the fraction (the initial concentration 0.1 mg protein/ ml); K- control; ($p < 0.05$). Biological active concentrations are gray ($p < 0.05$).

polymodal character with positive extremum. If fraction was biologically active the meaning of membranotropic effects is up to 157% that is significantly different from control. In case of extract and pellet, which are not biologically active the meaning of membranotropic effect had not reached 124% (Table 14.1).

Because of fact that supernatant had only membranotropic effect it can be supposed that after procedure of treatment with ammonium sulfate extract separated into biologically active component (supernatant) and biologically inactive component (pellet).

Using MALDI-TOF mass-spectrometry it was shown that small peptides with molecular weights about 8 kDa are included in fraction of supernatant (Figure 14.2).

Then fraction of supernatant will be called as peptide containing fraction. On the assumption of previously stated about membranotropic activity distribution we can suppose that pellet fraction in its composition have substance inhibited activity of peptide containing fraction. To prove this statement we have tried to combine two separated fractions – supernatant and pellet in ratio 1:10 to mass in the presence of Ca^{2+} and Mn^{2+} ions (10^{-2}M). This combined fraction we studied by adhesion metrical method that have showed that combined fraction are not active. This result allowed us to conclude the presence of inhibiting component in pellet fraction (Figure 14.3).

Thus, pellet fraction was studied by PAGE in the presence of SDS-Na (Figure 14.4). The range of proteins with different molecular weights were found.

To determine the inhibiting component in pellet we used exclusion chromatography. According to this method the fraction of pellet was separated into several fractions, which differ molecular weights. Figure 14.5 demonstrates four purified zones that were analyzed if they provide inhibiting activity in relation to peptide containing fraction. To this end we attempt

TABLE 14.1 Investigation of Biological Activity of Obtained Plant Fractions

Fraction	Initial concentration, mg/mL	Activity
Extract	0.1	Inactive
Pellet	0.1	Inactive
Supernatant	0.1	Active

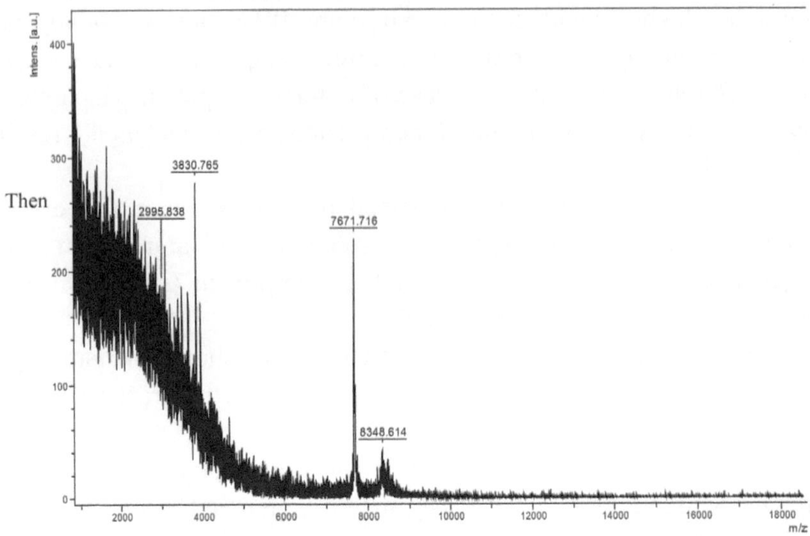

FIGURE 14.2 MALDI-TOF spectrum of supernatant fraction of garlic extract (*Allium sativum L.*).

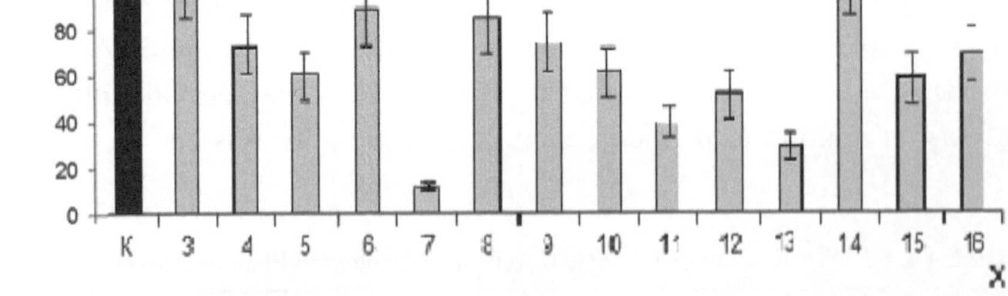

FIGURE 14.3 Dose dependence activity (%) of complex fraction (supernatant+pellet) *Allium sativum* L. [The abscissa axis shows the degree of tenfold sequential dilution of the fraction (the initial concentration 0.1 mg protein/mL); K-control ($p < 0.05$)].

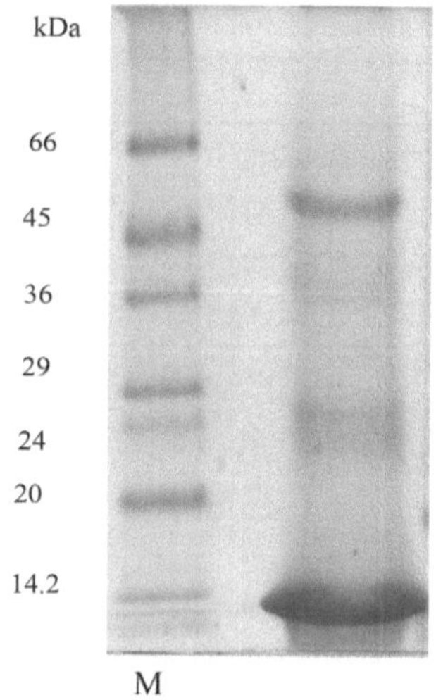

FIGURE 14.4 PAGE of pellet of *Allium sativum* L. extract. Markers: α-lactalbumin –14,200 Da, soybean trypsin inhibitor – 20,000 Da, trypsinogen from bovine pancreas – 24,000 Da, carbonic anhydrase – 29,000 Da, glyceraldehyde-3–phosphate dehydrogenase – 36,000 Da, valbumin – 45,000 Da, albumin – 66,000 Da. Gel was stained with Coomassie G-250.

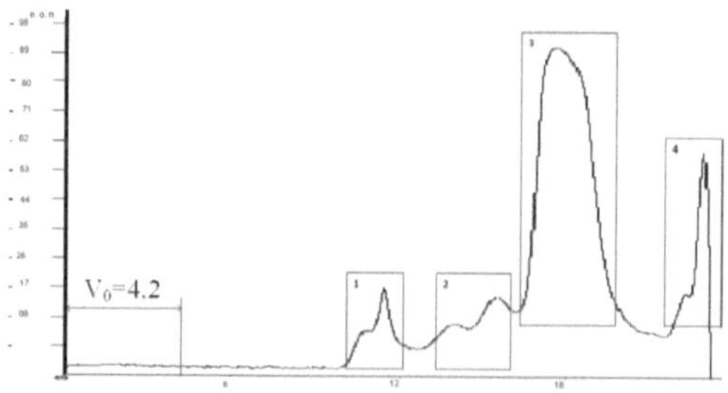

FIGURE 14.5 Separation of pellet fraction by exclusion chromatography HPLC Gel Filtration Column Bio-Sil TSK-125 300x7.5 MM (Japan). Elution was conducted in phosphate buffer 0.05 M pH 6.8 with velocity 0.4 ml/min for 25 min. Detection was carried out at 280 nm.

to create four combined fractions between peptide containing fraction and each of zone after exclusion chromatography. The results of adhesiometric analyzes of these combined fractions are presented in Table 14.2. So, the complex formation was observable in case of fraction 1. This means that fraction 1 inhibited activity of peptide containing fraction.

Then we carried out the preparative exclusion chromatography for separating fraction 1 (Figure 14.6).

TABLE 14.2 Membranotropic Activity of Combined Fraction

Fraction	Activity
Fraction 1+ peptide containing fraction	inactive
Fraction 2 + peptide containing fraction	active
Fraction 3 + peptide containing fraction	active
Fraction 4 + peptide containing fraction	active

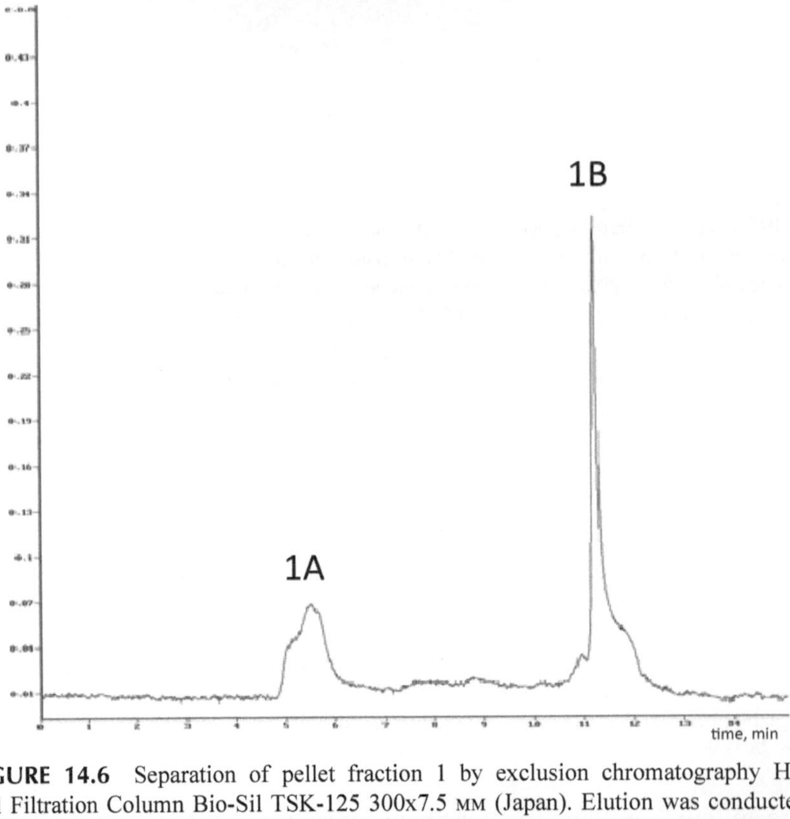

FIGURE 14.6 Separation of pellet fraction 1 by exclusion chromatography HPLC Gel Filtration Column Bio-Sil TSK-125 300x7.5 MM (Japan). Elution was conducted in phosphate buffer 0.05 M pH 6.8 with velocity 0.4 ml/min for 25 min. Detection was carried out at 280 nm.

Figure 14.6 demonstrates that fraction 1 separated into two fraction – 1A and 1B, which differ molecular weights. These fractions were preparative obtained by chromatography and analyzed by adhesiometric method. Two pairs combined fractions between 1A and peptide containing fraction and 1B and peptide containing fraction were created. Study biological activity of these two combined fraction have shown that only fraction 1A inhibited activity of peptide containing fraction (Table 14.3).

To study protein composition of fraction 1A and 1B we carried out PAGE under denaturing condition. The result of experiment shows that protein with molecular weights 6.4 kDa are included in both fractions 1A and 1B composition. Moreover, fraction 1A includes high molecular component which molecular weights are 56 kDa and 70 kDa (Figure 14.7)

Because of formation complex between fraction 1B and peptide containing fraction were not observed we can exclude role of protein with molecular weights 6.4 kDa as inhibitor. On the basis of biological activity of combined fraction and electrophoresis composition investigation we can suppose that high molecular component (56 and 70 kDa) are responsible for inhibiting activity of fraction 1A.

Then we have tried to identified electrophoretic obtained protein components – 6.4, 56 and 70 kDa. For this purpose we used cryptic hydrolysis of Coomassie stained zones corresponding to electroforegram (Figure 14.7). Tryptic hydrolysis accompanied with MALDI-TOF mass spectrometry analysis that revealed that protein with molecular weights 6.4 kDa have high homology with A chain mannose-specific agglutinin (lectin) of *Allium sativum* L. Proteins with molecular weights were identified as unknown proteins (NCBI data).

14.5 CONCLUSIONS

Thus, in present study from garlic *Allium cepa* L. we obtained biologically active component of plant bioregulator which are performed by the peptide complex with molecular weight range 2000–9000 Da. From pellet fraction of plant extract, high molecular protein complex was obtained. It was shown that this protein complex inhibits activity of peptide containing

TABLE 14.3 Investigation of Biological Activity of Combined Fraction

Combined fraction	Initial concentration, mg/mL	Activity X (10^x)
Fraction 1A + peptide contained fraction	0.1	Inactive
Fraction 1B + peptide contained fraction	0.1	6, 12, 15–16

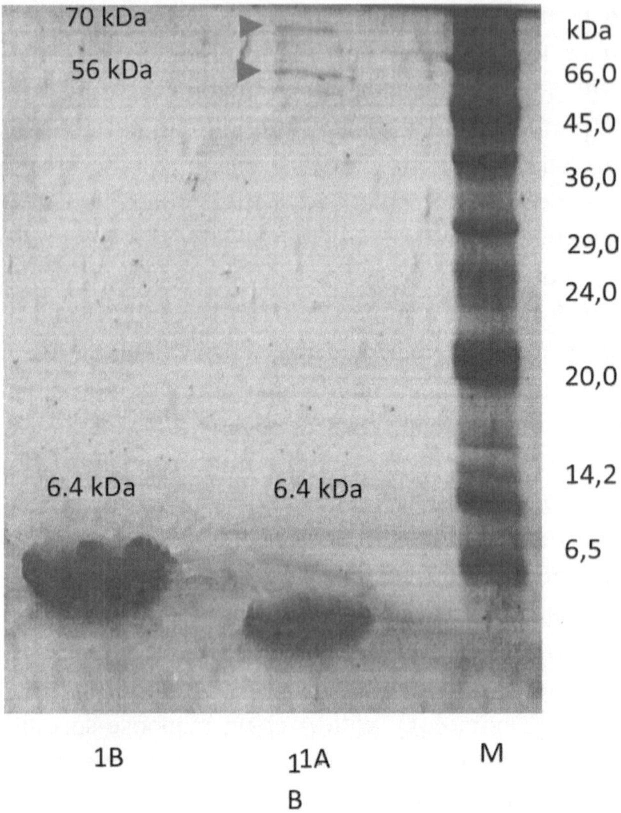

FIGURE 14.7 PAGE of pellet fractions 1A and 1B. Markers: aprotinin from lung bovine – 6500 Da, α-lactalbumin – 14,200 Da, soybean trypsin inhibitor – 20,000 Da, trypsinogen from bovine pancreas – 24,000 Da, carbonic anhydrase – 29,000 Da, glyceraldehyde-3-phosphate dehydrogenase – 36,000 Da, valbumin – 45,000 Da, albumin – 66,000 Da. Gel was stained with Coomassie G-250.

fraction. Obtained results are correlate with results for other bioregulators of MGTBs group [4].

KEYWORDS

- ***Allium sativum*** **L.**
- **biological activity**
- **peptide**

REFERENCES

1. Bayan, L., Koulivand, P. H., Gorji, A. Garlic: a review of potential therapeutic effects. Avicenna J Phytomed. 2014, V. 1, №1, p. 1–14.
2. Olennikov, D. N., Tankhaeva, L. M., Mikhailova, T. M., and Samuelsen, A. B., Khim. Prir. Soedin., 2005, no. 4, pp. 354–355.
3. Masoomeh, S. G., Shokoohamiri, M. R., Amirrajab, N., Behnaz, M., Ali, G., Farideh, Z., Golnar, S., Mehdi, R. A., Fitoterapia, 2006, v. 77, no. 3 P, pp. 321–323.
4. V.P. Yamskova, M.S. Krasnov, I.A. Yamskov. New experimental and theoretical aspects in bioregulation. The mechanism of action of tissue-specific homeostatic membranotrop nyh bioregulators. Saarbrucken: Lambert Academic Publishing, 2012, p. 136.
5. A.V. Borisenko, V.P. Yamskova, M.S. Krasnov, I.V. Blagodatskikh, V.V. Vecherkin, I.A. Yamskov. Regulatory proteins from the mammalian liver that display biological activity at ultra low doses, pp. 35–45. Book "Biochemical Physics Frontal Research," Ed. by S.D. Varfolomeev, E.B. Burlakova, A.A. Popov, G.E. Zaikov, N.Y. Hauppauge. Nova Science Publishers Inc., 2007, pp. 35–46.
6. D.V. Margasyuk, M.S. Krasnov, I.V. Blagodatskikh, E.N. Grigoryan, V.P. Yamskova, I.A. Yamskov. Regulatory Protein from Bovine Cornea: Localization and Biological Activity, pp. 47–59. Book "Biochemical Physics Frontal Research," Ed. by S.D. Varfolomeev, E.B. Burlakova, A.A. Popov, G.E. Zaikov, N.Y. Hauppauge. Nova Science Publishers Inc., 2007, pp. 47–60. Yamskova et al. 2007.
7. M.S. Krasnov, V.P. Yamskova, D.V. Margasyuk, O.G. Kulikova, A.P. Il'ina, E.Yu. Rybakova, I.A. Yamskov. Study of a new group of bioregulators isolated from the greater plantain (*Plantago major* L.). Applied Biochemistry and Microbiology, 2011, Vol. 47, No. 2, pp. 128–135
8. O.G. Kulikova, V.P. Yamskova, A.P. Il'ina, D.V. Margasyuk, A.A. Molyavka, I.A. Yamskov. Identification of a new bioregulator acting in ultralow doses in bulb onion (*Allium cepa* L.). Applied Biochemistry and Microbiology, 2011, Vol. 47, No. 4, pp. 356–360.
9. P.A. Nazarova, V.P. Yamskova, M.S. Krasnov, A.G. Filatova, I.A. Yamskov. Regulatory proteins biologically active in ultralow doses from mammalian glands and their secretions, pp. 78–88. Book "Biochemical Physics Frontal Research," Ed. by S.D. Varfolomeev, E.B. Burlakova, A.A. Popov, G.E. Zaikov, N.Y. Hauppauge. Nova Science Publishers Inc., 2007, pp. 73–82.
10. V. P. Yamskova, A. V. Borisenko, M. S. Krasnov, A. P. Il'ina, E. Yu. Rybakova, D. I. Malcev, I. A. Yamskov, On Mechanisms Underlying Regeneration and Reparation Processes in Tissues. Bulletin of Experimental Biology and Medicine, 2010, Vol. 149, No. 1, pp. 140–143.
11. M.S. Krasnov, E.N. Grigoryan, V.P. Yamskova, D.V. Boguslavskiy, I.A. Yamskov. Regulatory proteins obtained from vertebrates eye tissues. Radiation Biology and Radioecology, 2003, No. 3, pp. 265–268.
12. Laemmli, U.K., Nature, 1970.

CHAPTER 15

MORPHOLOGICAL AND BIOENERGETICAL CHARACTERISTICS OF MITOCHONDRIA

V. I. BINYUKOV,[1] E. M. MIL,[1] I. V. ZHIGACHEVA,[1]
I. P. GENEROZOVA,[2] and M.M. RASULOV[3]

[1]*Emanuel Institute of Biochemical Physics, Russian Academy of Sciences, Street Kosygina, 4, Moscow, 119334 Russia; Fax: +7 (499) 137-41-01; E-mail: elenamil2004@mail.ru, zhigacheva@mail.ru*

[2]*Timiryazev Institute of Plant Physiology, Russian Academy of Sciences, ul. Botanicheskaya 35, Moscow, 127276, Russia, Tel.: (495)903-93-40, E-mail: igenerozova@mail.ru*

[3]*Research Institute of Chemistry and Technology Organoelement Compounds, Enthusiasts Highway 38, Moscow 111123, Russia, Tel.: +7(495)673-13-78; Fax: +7 (495) 783–6444; E-mail: maksud@bk.ru*

CONTENTS

ABSTRACT

The comparing of the effects of different concentrations of 1-(germatran-1-yl)-1-hydroxyethyl amine (germatrane) on mitochondrial bioenergetic characteristics of animal and plant origin was done. Introduction of the drug 10^{-5}–10^{-11} M into the incubation medium resulted in increased efficiency of oxidative phosphorylation and the maximum rate of NAD-dependent substrates oxidation by mitochondria. In these same concentrations germatrane has reduced the intensity of LPO up to the control values in the membranes of rat liver mitochondria and the 6-day etiolated pea seedlings (*Pisum sativum L.*) at the model system of "aging" of the mitochondria. A statistically significant change in the shape of mitochondria pea seedling had detected by the method Atomic force microscopy (AFM) at a pea seedlings that were exposed to insufficient watering with moderate to 14°C cooling. A significant increase volume (swelling) of mitochondria compared to mitochondria isolated from the control group of seedlings was found. The treatment with a solution of 10^{-5} M germatrane has prevented the swelling of mitochondria. That can evidence of possible anti-stress of the drug properties

15.1 AIM AND BACKGROUND

Germany is one of the necessary compounds for living organisms, the lack of which causes the development of a number of pathological conditions. The biological activity of this compound was discovered in 1928 but intensive study of their biological activity have been started only in the middle 60s of last century [5]. At the same time were synthesized tricyclic esters of triethanolamine and germanium with the general formula $XGe(OCH_2CH_2)_3N$, or germatrane, and began the study of their biological activity in the laboratory of Voronkov [6, 7]. Studies conducted on animals have shown that germatranes have high biological activity (neurotropic, analgesic, anti-viral) [8]. However, the study of the effect of these compounds on the growth and development of plants has not been done. The growth and development of plants and their resistance to stress factors, primarily dependent on energy metabolism, in this connection the aim of our work was to study the influence of the germatrane (gross formula $N(CH_2CH_2O)_3Ge(OCH_2CH_2)NH_2$) per the functional state of 6-day etiolated pea seedlings mitochondria in control and stress conditions.

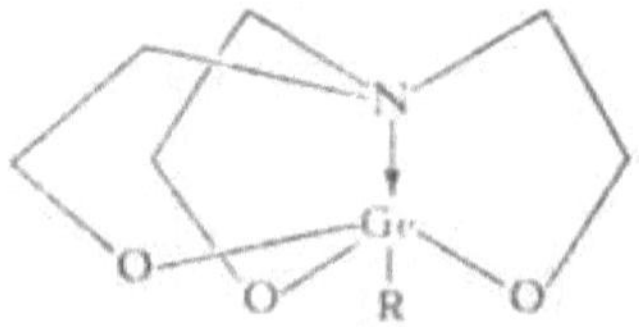

II

$R\text{-}(OCH_2CH_2)NH_2$

We investigated the effects of combined action of insufficient watering and moderate cooling to 14°C (conditions typical of early spring period after snowless winter as stress condition.

15.2 INTRODUCTION

In 1963 Academician M.G. Voronkov and colleagues synthesized tricyclic esters of triethanolamine silicon and germanium of general formula XM $(OCH_2CH_2)_3N$, where M = Si, Ge, and started a study of their biological activity [6]. It was shown that their biological activity is determined by the unique structure and nature substituent X at the atom of silicon or germanium. This discovery in subsequent years has stimulated the rapid development of investigation of silatranes and germatranes both in Russia and abroad. It was found that tricyclic chelate compounds of silicon and germanium, respectively silatranes germatrane usually have nearly identical biological activity [9], probably due to the similarity of the elements Si and Ge regarding their atomic radius and electronegativity. However germatranes having the same or higher biological activity, are less toxic than their silicon counterparts [6, 8, 10]. Unlike silatranes, which are widely used in medicine and agriculture, germatranes, has been paid much less attention. However, animal studies have shown the germatranes are drugs with a broad spectrum of biological activity. They activate macrophages system and B-cell immunity, and enhances the body's natural resistance [4, 11, 12]. Less studied properties of these compounds as regulators of plant growth and development (PGRs). Although it is shown that treatment of strains of tissue culture plants of tropical *Polyscias Filicifolia LX-5* and *dalistsia Polyscias filicifolia* most studied germatranolom-1 increases the intracellular content nucleic acids and proteins, as well as increases the activity of antioxidant enzymes [13]. In addition, this drug exhibits anti-stress properties, increasing the resistance of plants (wheat germ) at the temperature stress [14]. It can be assumed that germatranes may have adaptogenic properties. Moreover, that the results of field trials, silatranes (the preparations which similar the biological activity with germatranes) has intensified growth, reparative processes, the formation and maturation of tissues in different plant crops [7, 15].

It is known that stress factors lead to a shift of antioxidant–prooxidant relationship towards increasing the ROS generation and leads to the development of pathological states [16], which is associated with damage to cell components. Under stress conditions, the main sources of ROS are mitochondria of animal [17] and the mitochondria and chloroplasts of plants [18]. On this basis, it could be assumed that drugs having adaptogenic properties should primarily affect the generation of ROS with these organelles. In our work we mainly paid attention on the mitochondria, as these organelles in plants and in animals play a major role in the body's response to the action of stressors. In particular, it was found that the change in the ambient temperature leads to a change in the lipid composition of the mitochondrial membrane. At the same time there were a change in the number free fatty acids and in degree of saturation of the free fatty acids that probably indicate on the factor in stress [18]. In particular, it was found that the ambient temperature changes lead to changes in the lipid composition of mitochondrial membranes. At the same time there is a change in the number and degree of saturation of free fatty acids, which is probably a sign of the stress factor [19]. Increase the amount of free fatty acids (FFA) changes the redox state of the inner mitochondrial membrane. Depending on the redox state of the mitochondrial respiratory chain expression changes of stress genes. Under the effect of extreme temperatures, excessive salt or lack of moisture in some degree increased ROS. Thus mitochondria are extremely functional-dependent organelles. In animal cells and yeast, these organelles are combined into an extensive network, referred to as "mitochondrial reticulum" [20] in higher plants, the mitochondria singly and have either a spherical or cylindrical shape [21, 22]. Under stress conditions, mitochondria form branched networks (anoxia) [23] or the aggregate and form dense clusters around chloroplasts or in other regions of cytosol (heat shock, UV irradiation, effect of strong oxidants) [24]. Creation of a "giant mitochondria" is accompanied by an increase in the generation of ROS [25, 26]. Antioxidants prevent the formation of a "giant mitochondria" and increase the generation of ROS by these organelles [24, 27, 28]. The standard procedure for selection of the mitochondria in a sucrose solution leads to the complete destruction intermitochondrial contacts. For this reason, the mitochondria are presented in separate vesicles before 1.2 μm in diameter. The morphology of isolated mitochondria possibly reflects their functional state [29], consequently, the degree of the plants adaptation to changing environmental conditions.

The aim of the work was also to study the bioenergetic characteristics of mitochondria under stress conditions and the effect of the germatrane (gross formula $N(CH_2CH_2O)_3Ge(OCH_2CH_2)NH_2$) on these characteristics. The mitochondrial respiratory chain of plants and animals has a general plan of organization and the main differences relate to CN-resistant electron transport and a region of NADH-dehydrogenase of the respiratory chain [30]. In this regard the basic mechanisms of functioning of germatrane we investigated on plant mitochondria (6 day etiolated pea seedlings) and animal mitochondria (rat liver mitochondria). An effect of germatrane, as well as other biologically active substances which exerting their activity in a dose dependent manner, apparently depends on their concentration [31]. Therefore, using the model of "aging" of the mitochondria, we researched the effect of different concentrations of the drug on the fluorescence intensity of the final products of LPO – Schiff bases. At the same time studied the effect of the drug on the bioenergetic characteristics of mitochondria. And finally, exploring anti-stress properties of the drug, studied the effect of water deficit with moderate cooling to 14°C (conditions appropriate early spring after a snowless winter) and seed treatment of pea *(Pisum Sativum L.)* with 1-(germatran-1-yl)-1-hydroxy-ethyl amine on lipid peroxidation in the membranes of mitochondria and on morphology of mitochondria isolated from 6-day etiolated seedlings.

15.3 MATERIAL AND METHODS

15.3.1 SPROUTING SEEDS

Pea (Pisum Sativum L., cv. Flora 2) seeds were washed with soapy water and 0.01% $KMnO_4$. Control seeds were then soaked in water, experimental seeds – in 10^{-5} M germatrane for 30 min. Thereafter, seeds were transferred into covered trays on moistened filter paper in darkness for 2 days. After 2 days, half of control (water deficit + moderate cooling) (WD+MC) and germatrane (water deficit + moderate cooling + germatrane) (WD+MC+GER) treated seeds were transferred in the open trays on dry filter paper where they were kept at 14°C for 2 days. After two days of water deficit treatment with moderate cooling, seeds were transferred to covered trays on wet filter paper, where they were kept for next two days at 22°C. Another half of control plants were retained in closed trays on wet filter paper, where they were kept for 6 days at 22°C. On the sixth day, mitochondria were isolated from seedling epicotyls.

15.3.2 ISOLATION OF MITOCHONDRIA

Isolation of mitochondria from 6-day-old epicotyl of pea seedlings (*P. sativum*) performed by the method [32] in our modification. The epicotyls having a length of 1.5–5 cm (20–25 g) were placed into a homogenizer cup, poured with an isolation medium in a ratio of 1:2, and then were rapidly disintegrated with scissors and homogenized with the aid of a press. The isolation medium comprised: 0.4 M sucrose, 5 mM EDTA, and 20 mM KH_2PO_4 (pH 8.0), 10 mM KCl, 2 mM 1, 4-Dithio-di-theiritol, and 0.1% fatty acids-free (FA-free) BSA. The homogenate was centrifuged at 25,000 g for 5 min. The precipitate was re-suspended in 8 mL of a rinsing medium comprised: 0.4 M sucrose, 20 mM KH_2PO_4, 0.1% FA-free BSA (pH 7.4) and centrifuged at 3000 g for 3 min. The supernatant was centrifuged for 10 min at 11,000 g for mitochondria sedimentation. The sediment was re-suspended in 2–3 mL of solution contained: 0.4 M sucrose, 20 mM KH_2PO_4 (pH 7.4), 0.1% FA-free BSA and mitochondria were precipitated by centrifugation at 11000 g for 10 min.

The isolation of liver mitochondria was carried out by a method of differential centrifugation [33]. The first centrifugation was carried out at 600 g for 10 min.; the second, at 9000 g for 10 min. The precipitate was resuspended in the isolation medium. The tissue: medium ratio was 1:0.25. The isolation medium was: 0.25 M sucrose, 10 mM HEPES, pH 7.4.

The respiration rate of mitochondria from rat liver, and pea germs were recorded using a Clark type electrode on an LP-7 polarograph (Czechia). The incubation medium of liver mitochondria contained 0.25 M sucrose, 10 mM Tris-HCl, 2 mM $MgSO_4$, 2 mM KH_2PO_4, 10 mM KCl (pH 7.5) (28°C). The incubation medium of pea germs mitochondria contained 0.4 M sucrose, 20 mM HEPES-Tris-buffer (pH 7.2), 5 mM KH_2PO_4, 4 mM $MgCl_2$, 0.1% BSA (28°C).

Protein was measured with a biuret method.

Lipid peroxidation (LPO) activity. LPO activity was assessed by fluorescent method [34]. Lipids were extracted by the mixture of chloroform and methanol (2:1). Lipids of mitochondrial membranes (3–5 mg of protein) were extracted in the glass homogenizer for 1 min at 10°C. Thereafter, equal volume of distilled water was added to the homogenate, and after rapid mixing the homogenate was transferred into 12 mL centrifuge tubes. Samples were centrifuged at 600 g for 5 min. The aliquot (3 mL) of the chloroform (lower) layer was taken, 0.3 mL of methanol was added, and fluorescence was recorded in 10 mm quartz cuvette with a spectrofluorometer (FluoroMax Horiba Yvon, Germany). Background fluorescence was recorded using a mixture of 3 mL chloroform and 0.3 mL methanol. The excitation wavelength was 360 nm, the emission wave length was 420–470 nm. The results were expressed in arbitrary units per mg protein.

15.3.3 ATOMIC FORCE MICROSCOPY (AFM)

Mitochondrial morphology was investigated by atomic force microscopy. Mitochondrial samples were fixed with 2% glutaraldehyde for 1 h, followed by washing with water by the centrifugation. Precipitation mitochondria were applied to the polished surface a silicon wafer and dried in air. The study was performed on a SOLVER P47 SMENA at a frequency of 150 kHz in tapping mode. NSG11 used cantilever with a radius of curvature of 10 nm. An analysis of mitochondria AFM images under study permits us to determine the volume of individual mitochondria. The volume of a mitochondrial image is equal to the product of the sectional area of the mitochondrial image and the average height of the image in the region of section and is calculated by an Image Analysis program to the coordinate data and scanning pitch. The section was made at a height of 30 nm. Volume of image drugs mitochondria corresponded to the product of the area of the cross section image of mitochondria multiplied on the medium altitude-of this image in area of section. In the analysis and processing the data file, there was used Statistica 6. In the analysis there were used individual mitochondria.

15.3.3.1 Reagents

Sucrose, Tris, FCCP (carbonyl cyanide-*p*-trifluoromethoxyphenylhydrazone), malate, glutamate, succinate, EDTA, ADP (Sigma, USA), BSA (Sigma, USA), HEPES (MP Biomedicals, Germany).

15.4 RESULTS AND DISCUSSION

In stress conditions one of the main source of ROS are the mitochondria in this regard t we had to develop a model, which simulating a stress, notably have to find conditions under which will increased ROS production by mitochondria, and thus will be activated LPO. We solved this problem by having developed a model of "aging" (the incubation of mitochondria isolated from 6 day etiolated pea seedlings in a hypotonic medium at room temperature). The incubation of mitochondria in a hypotonic solution of sucrose caused a weak swelling of mitochondria and growth of the ROS generation that resulted in a 3 to 4-fold increase in the intensity of fluorescence of LPO products [14], which is consistent with the data of Earnshaw [35]. The introduction of germatrane into the mitochondria incubation medium resulted in decreasing the LPO intensity that exhibited dose dependence. Germatrane decreased the intensity of fluorescence of LPO products in membranes of "aged" mitochondria in concentrations of 10^{-5} and 10^{-11} M almost to control levels (Figures 15.1a and 15.1b), which may be indicative of the presence of the drug have anti-stress properties. Note that the most effective concentration was 10^{-5} M germatrane.

The addition of germatrane in these concentrations to the incubation medium for the mitochondria of rat liver or mitochondria isolated from 6 day etiolated pea seedlings changed the mitochondria energy. The preparation increased the rates of oxidation of NAD+-dependent substrates and succinate in the presence of the uncouple (FCCP) (carbonyl cyanide-*p*-trifluoromethoxyphenylhydrazone) by liver mitochondria by 53–54% (10^{-11} M) and 26–32% (10^{-5} M), respectively (Table 15.1). The rates of oxidation of NAD+-dependent substrates in the presence of ADP increased by 1.6–2.3 and 1.4 times Table 15.1).

Incubation medium containing: 0.2–5 M sucrose, 10 mM Tris-HCl, 2 mM KH_2PO_4, 5 mM MgSO4, 10 mM KCl, pH 7.5. Further additives: 200 μM ADP, 10^{-6} M FCCP (carbonyl cyanide-*p*-trifluoromethoxyphenylhydrazone), 4 mM glutamate, 1 mM malate or 5 mM succinate.

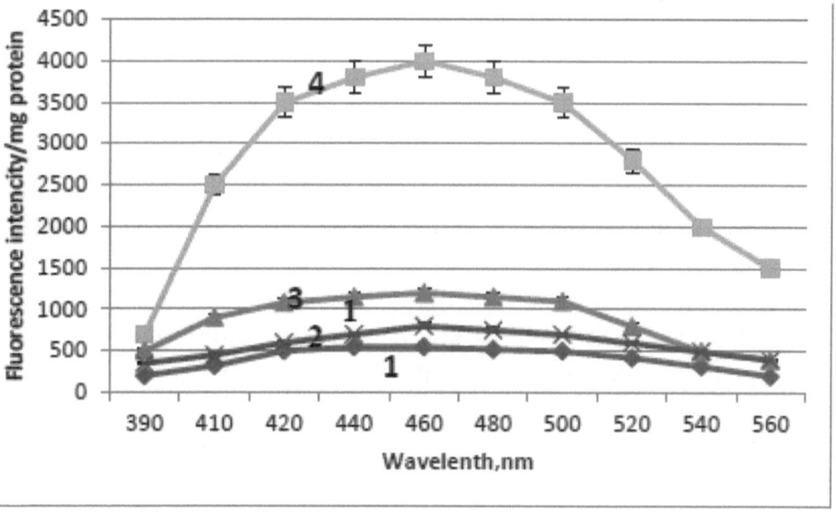

FIGURE 15.1A The fluorescence intensity of LPO products after introduction of different concentrations of germatrane into the incubation medium of the mitochondria isolated from the 6 day etiolated pea seedlings Y-axis: fluorescence intensity, arbitrary units/mg protein; X-axis: the concentration of germatrane.

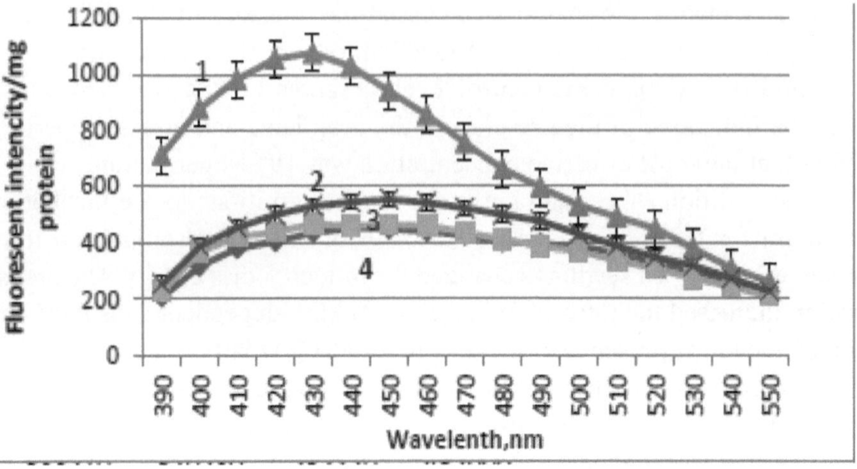

FIGURE 15.1B The fluorescence intensity of LPO products after introduction of different concentrations of germatrane into the incubation medium of the rat liver mitochondria. Y-axis: fluorescence intensity, arbitrary units/mg protein; X-axis: the concentration of germatrane.

Notation: Vo – the rate of oxidation of substrates; V_3 – rate of substrate oxidation in the presence of ADP; V_4 – oxidation rate at rest (the rate of oxidation of the substrate after the exhaustion of ADP).

The respiratory control rate (RCR) increased by 33–50% (for oxidation NAD-dependent of substrates) and 16–20% (by oxidation succinate). In the respiratory chain of pea seedling mitochondria, the rates of oxidation of NAD+dependent substrates increased by 27.7% (10–5 M) – 42.7% (10–11 M) in the presence of FCCP and by 17–47% in the presence of ADP (Table 15.2).

Incubation medium: 0.4 M sucrose, 20 mM HEPES-Tris buffer (pH 7.2), 5 mM KH_2PO_4, 4 mM $MgCl_2$, 0,1% BSA. Further additives: 5 mM malate, glutamate or 10 mM succinate, 5 mM, 125 μM ADP, 0.5 mM FCCP (carbonyl cyanide-*p*-trifluoromethoxyphenylhydrazone).

The respiratory control rate increased from 2.30±0.01 to 2.90±0.02 and 3.25±0.05. The differences between plant and animal mitochondria were observed when the oxidation substrate used was succinate. All studied concentrations of germatrane had no effect on the maximum rates and the respiratory control rate in oxidation of succinate by pea seedling mitochondria, which is an evidence for the adaptive character of the germatrane

TABLE 15.1 The Effect of Different Concentrations of Germatrane on the Kinetics of Oxygen Consumption of the Rat Liver Mitochondria (ng. mol O_2/min × mg protein) (data from 10 experiments)

Substratum	Germatrane, M	V_0	V_3	V_4	FCCP	RCR
Glutamate+malate	0	6.18±1.21	18.3±2.08	7.50±0.38	25.48±4.34	2.44±0.02
Glutamate+malate	10^{-5}	8.72±0.84	43.0±2.50	11.62±1.30	38.95±3.31	3.70±0.03
Glutamate+malate	10^{-9}	13.3±2.24	44.0±3.08	16.92±1.50	64.24±4.82	2.60±0.03
Glutamate+malate	10^{-11}	8.35±0.15	29.20±3.60	8.98±0.58	39.22±4.61	3.25±0.03
Glutamate+malate	10^{-18}	9.89±2.20	36.36±2.28	11.68±2.46	46.81±1.10	2.77±10.02
Succinate	0	14.88±1.55	36.48±2.10	14.40±1.84	62.0±3.21	2.53±0.03
Succinate	10^{-5}	17.95±2.00	51.40±1.83	16.85±2.00	78.32±2.00	3.05±0.02
Succinate	10^{-9}	15.10±1.65	48.73±2.54	18.18±3.50	74.00±5.30	2.68±0.04
Succinate	10^{-11}	18.10±2.86	50.58±3.22	17.15±1.60	82.03±5.54	2.95±0.02

TABLE 15.2 The Effect of Different Concentrations of Germatrane on the Kinetics of Oxygen Consumption of the Mitochondria Isolated from the 6-day Etiolated Pea Seedlings (ng. atom O_{2}/min × mg protein) (Data from 10 Experiments)

Substratum	**Germatrane, M**	**V_0**	**V_3**	**V_4**	**FCCP**	**RCR**
Glutamate+malate	0	11.2±2.0	58.1±5.0	25.3±2.0	62.0±4.6	2.30±0.01
Glutamate+malate	10^{-5}	16.24±2.6	85.5±3.0	26.0±1.0	79.2±5.2	3.28±0.05
Glutamate+malate	10^{-9}	20.0±3.1	82.2±4.3	31.0±1.2	128.0±2.4	2.65±0.02
Glutamate+malate	10^{-11}	15.1±2.8	68.4±3.0	23.6±1.3	88.5±4.0	2.90±0.02
Glutamate+malate	10^{-18}	17.9±2.3	80.2±3.0	30.7±1.0	94.8±4.2	2.61±0.02
Succinate	0	32.4± 3.8	105.0± 2.5	42.3±3.1	120.0± 7.2	2.48± 0.10
Succinate	10^{-5}	33.5±2.4	108.4±2.4	46.1±2.6	129.0±4.0	2.35±0.06
Succinate	10^{-9}	38.7 ± 4.1	102.3± 3.4	42.3±2.5	123.0± 4.0	2.42±0.03
Succinate	10^{-11}	35.2±3.2	110.0±4.6	44.0±3.2	119.4±3.7	2.50±0.03
Succinate	10^{-18}	32.8±1.4	106.5±2.8	43.3±3.7	118.4±5.4	2.46±0.01

effect. As the germinating seeds are characterized by rather low speed oxidation of NAD-dependent substrates, the stimulating of activity of NAD-dependent dehydrogenases germatrane apparently contributes to the activation of energy processes in the cell and increases the resistance of seedlings to the action stressors [35].

Checking protective properties of the drug were performed using mitochondria isolated from a 6 day etiolated pea seedling suffered from insufficient watering at a temperature of 14°C. The choice of such a model of stress due to the fact that plants grow in constantly changing environmental conditions and undergoes combined effects of various abiotic and biotic natural factors. In this case, quite often there are situations when in low water availability (after the snowless winter) there is even a slight drop in temperature, that normal for the early spring season. Since 10^{-5} M germatrane has a significant effect on the bioenergetic characteristics and effectively reduce the intensity of the fluorescence of LPO products in membranes of rat liver mitochondria and pea seedlings, in our studies of anti-stress properties of drug we used this concentration. A combined effect of insufficient watering and moderate cooling (14°C) (WD+MC) resulted in changes in the mitochondria morphology of pea (*P. sativum*), grade Flora-2, isolated from 6-day etiolated seedlings as compared with the control (22°C) and with the seedlings whose seeds were pretreated with germatrane (10^{-5}M) (WD+MC+GER). In the insufficient watering and moderate cooling group (WD+MC) there was observed an increase in the volume of AFM images (Figure 15.2c). The figure shows the two-dimensional image of mitochondria [x (mkm), y (mkm), z (nm)].

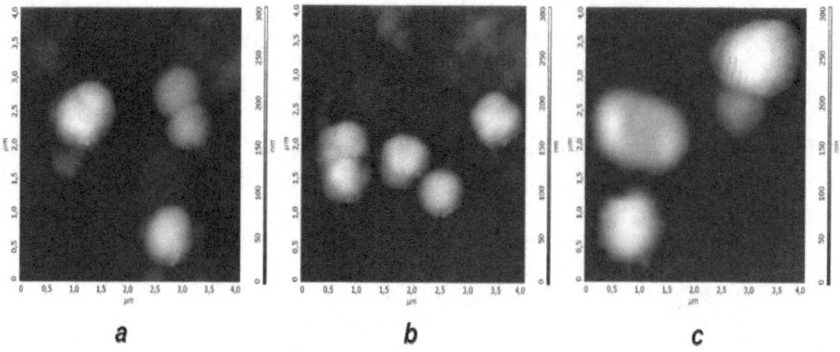

FIGURE 15.2 AFM images of the pea seedlings mitochondria in the control (a); WD+MC+GER) (b); WD+MC (c).

Statistical analysis of the volume of pre-fixed with glutaraldehyde mitochondria indicates the appearance in this group single of mitochondria with a large volume (Vav. = 115.1 $(mkm)^2 \times nm$) compared with the control group (Vav. = 80.7 $(mkm)^2 \times nm$). A comparison of the published data with the obtained results can be assumed that a combined effect of moderate cooling and insufficient watering of pea seedling mitochondria promotes the increase of generation of ROS and subsequent swelling of mitochondria [28]. Indeed, simultaneous cooling and deficit of moisture led to LPO activation in the mitochondrial membranes of pea seedlings. In this case, the fluorescence intensity of LPO products increased 3 and 2.5 times, respectively (Figure 15.3). Similar data were obtained in the study of a combined effect of drought and heat

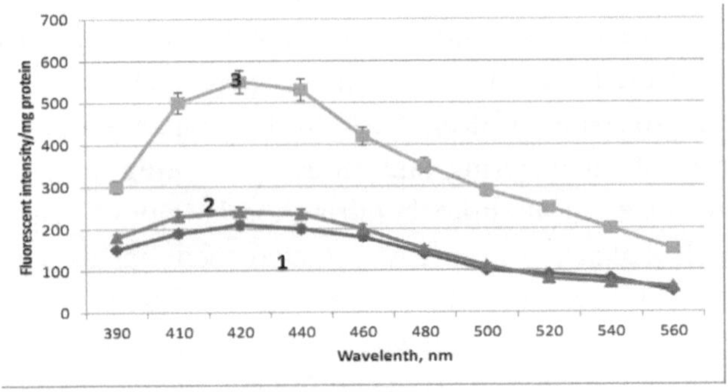

FIGURE 15.3 The fluorescence spectra of LPO products in the mitochondrial membranes of pea seedlings exposed to combined action of insufficient moisture and moderate cooling and in the mitochondrial membranes from pea seedlings treated by germatrane and exposed to combined action of insufficient moisture and moderate cooling. Y-axis: fluorescence intensity, arbitrary units/mg protein; X-axis: wavelength, nm [1 – Control; 2 – Ger+water deficit + moderate cooling (10^{-5} M germatrane); 3 –water deficit + moderate cooling].

TABLE 15.3 The Impact of the Combined Action of Water Deficit, Moderate Cooling and Germatrane (10^{-5} M) on the Average Volume of Mitochondria Pea Seedlings AFM Measurements

Treatments	Vav. $(mkm)^2 \times nm$,	+95% $(mkm)^2 \times nm$	–95% $(mkm)^2 \times nm$
Control	80.7	70.8	90.6
WD+MC+GER	59.4	53.4	65.4
WD+MC	115.1	103.0	127.2

shock on the ultrastructure of mitochondria and chloroplasts of leaves of *Triticum aestivum* [36].

Soak the seeds in a 10^{-5}M germatrane resulted in a decrease of LPO products in the membranes of the mitochondria: the fluorescence intensity of lipid peroxidation products decreased almost to the control level. Such treatment prevents changes to the morphology of mitochondria, approaching the size of mitochondria towards control.

Thus the average volume of mitochondria was lower control values (Vav. = 59.4 (mm) × 2 nm), which is consistent with previous data for the drug "melaphen" [37].

15.5 CONCLUSION

It can be assumed that the protective effect of germatrane is due to its antioxidant properties. Reducing the intensity of free radical oxidation is reflected in the low intensity of lipid peroxidation. Preventing activation of lipid peroxidation, with germatrane, apparently contributes to the preservation of the functional state of mitochondria, which is reflected in the preservation of the morphological characteristics of mitochondria. Protective properties of the drug, apparently bound to its ability to activate NAD – dependent dehydrogenases. It should be kept in mind that complex I of mitochondrial electron-transport chain is a main site for enter a reducing agent of the mitochondrial matrix. Since simultaneous proton translocation by complex I is coupled with mitochondrial oxidative phosphorylation generating ATP, activation of NAD-dependent dehydrogenases activates energy metabolism in the cell and this improves plant tolerance to varying environmental conditions. This is especially important for germinating seeds, where the rate of NAD-dependent substrates is low [38].

KEYWORDS

- **atomic force microscopy (AFM)**
- **germatrane**
- **lipid peroxidation (LPO)**
- **mitochondria**

REFERENCES

1. Kaars Sijpesteijn, A., Rijkens T., Kerk, G. J. M. Antimicrobial Activity of Organigermanium Derivatives. Nature. 1964. V. 201. P. 736–737.
2. Voronkov, M. G., Zelchan, G. I., Mironov, V. F., Kemme, A.A, Bleydelis Ya. Ya. Atran. Organilgermatrany. Chemistry of Heterocyclic Compounds 1968, 2, 227–229
3. Lukevics, E.Ya, L. Ignaovich. Comparative study of the biological activity of organosilicon and organogermanium compounds. Appl. Organomet. Chem. 1992, 6, 113.
4. Lukevics, E. Ya. Gar, T. K., Ignatovich, L. M., Mironov, V. F. The biologically activity of the compounds of germanium. E.Ya. Lukevics, – Riga: Publishing house "Zinatne." 1990, 191pp.
5. Menchikov, L. G., Ignatenko, M. A. The biological activity of organic compounds of germanium. Chem-Pharm. J. 2012.46 (11). 3–6. Russ.
6. Voronkov, M.G, Baryshok, V. P. Atran -new generation of biologically active substances. Herald of the RAS 2010.80 (11). 985–992. Russ.
7. Voronkov, M. G., Baryshok, V. P. Silatrane in medicine and agriculture. Novosibirsk: Publishing house of SB RAS. 2005, 258 pp. Russ.
8. Gar, T. K., Mironov, V. F. The biological activity of the compounds of germanium. M. Research Institute of feasibility studies of the chemical industry, 1982, 26 pp. Russ.
9. Voronkov, M. G., Baryshok, V. P. The Silatranes in medicine and agriculture. Novosibirsk: Publishing House of SB RAS. 2005, 258 pp. Russ.
10. Baryshok, V. P., L. N. Thuy Zhang, Voronkov, M. G. 1-Acyloxygermatranes. Russian Journal of General chemistry. 2013, 83 (8), 1267–1269
11. Ignatenko, M. A. Antitumor activity of organic compounds of silicon and germanium (Review). Chem Pharm. J. 1987, 4, 402–408
12. Yablonskaya, O. V. Germatranyl as immunostimulant at growing calves. Bulletin of the agroecological University. Zhitomir. 2002.1, 56–62. Russ.
13. Spacenav, A. I. Protein synthetical ability and stability of two strains of tissue culture *Polysciasfilicifolia* under stress. Abstract. PhD. Biol. Sciences. St. Petersburg, 2006, 24 pp.
14. Shigarova AM, Borovsky, G. B., Thuy Le Nhat Zhang, Baryshok, V. P. Factors plant resistance in extreme climatic conditions and man-made environment. Proceedings of the Scientific Conference June 10–13, 2013. Irkutsk 2013, 294–297. Russ.
15. Voronkov, M. G., Leviticus, T. H., Kirillov, A. F., Baryshok, V. P., Kozmik, R. A. et al. Action of germatranol upon cold resistance of grapes. Doklady Akademii Nauk. 1988, 299 (2). 509–512. Russ.
16. Tailor, N. L., Day, D. A., Millar, A. H. Targets of stress-induced oxidative damage in plant mitochondria and their impact on cell carbon/nitrogen metabolism. J. of Exp. Botany. 2003, 55(394). 1–10.
17. Plotnikov, E., Chupyrkina, A., Vasileva, A. Kazachenko, A., Zorov, D. The role of reactive oxygen and nitrogen species in the pathogenesis of acute renal failure. BBA 2008, 1777. S58–S59.
18. Rodríguez, M., E. Canales, O. Borrás-Hidalgo. Molecular aspects of abiotic stress in plants. Biotecnología Aplicada 2005, 22 (1). 1–10.
19. Voynikov, V. K. A Nuclear-mitochondrial interaction per the redox regulation of expression of genes of plants under stress. Plant and stress. Russian Symposium. Abstracts 9–12 November 2010 M. 2010, 90–91 Russ.

20. Bakeeva, L. E., Chentsov Yu.S., Skulachev, V. P. Mitochondrial framework (reticulum mitochondriale) in rat diaphragm muscle. Biochim. Etbiophys. Acta. 1978, 501(3). 349–369.
21. Logan, D. C., Leaver, C. J. Mitochondria-targeted GFP highlights the heterogeneity of mitochondrial shape, size and movement within living plant cells. J. Exp. Bot. 2000, 51, 865–871.
22. Logan, D. C. The mitochondrial compartment. J. Exp. Bot. 2006, 57.1225–1243.
23. Vartapetian, B. B., Andreeva, I. N., Kozlova, G. I., Agapova, L. P. Mitochondrial Ultrastructure in Roots of Mesophyte and Hydrophyte at Anoxia and after Glucose Feeding. Protoplasma. 1977, 93, 243–256.
24. Gao, C., Xing, D., Li, L., Zhang, L. Implication of reactive oxygen species and mitochondrial dysfunction in the early stages of plant programed cell death induced by ultraviolet-C overexposure. Planta. 2008, 227(4), 755–672.
25. Gesten, K. V., Verbelen, J. P. Giant mitochondria are response to low oxygen pressure in cells of tobacco (*Nicotianatabacum, L.*). J. Exp. Botany. 2002, 53(371), 1215–1218.
26. Yu Hua Bi, Wen Li Chen, Wei Na Zhang, Quan Zhou, Li Juan Yun, Da Xing. Production of reactive oxygen species, impairment of photosynthetic function and dynamic changes in mitochondria are early events in cadmium-induced cell death in *Arabidopsis thaliana.* Biol. Cell. 2009, 101, 629–643.
27. Scott, I., Logan, D. C. Mitochondrial morphology transition is an early indicator of subsequent cell death in *Arabidopsis.* New Phytologist. 2008, 177, 90–101.
28. Zhang L, Yinshu Li, Da-Xing, Caiji Gao. Characterization of mitochondrial dynamics and subcellular localization of ROS reveal that HsfA2 alleviates oxidative damage caused by heat stress in *Arabidopsis.* J. Exp. Bot. 2009, 60, 2073–2091.
29. Claypool, S. M., Mc Caffery, J. M., Koehler, C. M. Mitochondrial mis-localization and altered assembly of a cluster of Barth syndrome mutant tafazzins. The Journal of Cell Biology. 2006, 174(3). 379–90.
30. Sweetlove, L. J., Heazlwood, J. L., Hearld, V., Holtzapffel, R., Day, D. A., Leaver, C. J., Millar, A. H. The impact of oxidative stress on *Arabidopsis* mitochondria. The Plant, J. 2002, 32, 891–904.
31. Burlakova, E. B., Konradov, A. A., Maltseva, E. L. Effect of ultralow doses of biologically active substance and low-intensity physical factors. M.-Izhevsk. 2007, 390–423. Russ
32. Popov, V. N., Ruge, E. K., Starkov, A. A. Effect of electron transport inhibitors on the formation of reactive oxygen species in the oxidation of succinate by pea mitochondria. Biochemistry. 2003, 68(7). 910–916. Russ.
33. Zhigacheva, I. V., Kaplan, E. Ya., Pahomov, V. Yu, Rozantseva, T. V. Hristianovich, D. S., Burlakova, E. B. Drug "Anphen" and energy status of the liver. Doklady Akademii Nauk. 1995, 340(4). 547–550. Russ.
34. Fletcher, B. I., Dillard, C. D., Tappel, A. L. Measurement of fluorescent lipid peroxidation products in biological systems and tissues. Anal. Biochem. 1973, 52, 1–9.
35. Koster, K. L., Reisdorph, N., Ramsau, J. L. Changing Desiccation Tolerance of Pea Embryo Protoplasts during Germination. J. Exp. Bot. 2003, 54, 1607–1614.
36. Stupnікova, I., Benamar, A., Tolleter, D., Grelet, J., Borovskii, G., Dorne, A. J., Macherel, D. Pea Seed Mitochondria Are Endowed with a Remarkable Tolerance to Extreme Physiological Temperatures. Plant Physiol. 2006, 140, 326–335.

37. Zhigacheva, I., Mill, E., Binukov, V., Generozova, I., Shugaev, A. Fatkullina L. Combined Effect of Insufficient Watering, Moderate Cooling, and Organophosphorous Plant Growth Regulator on the Morphology and Functional Properties of Pea Seedling Mitochondria. Annual Research & Review in Biology 2014, 4(19), 3007–3025.
38. Generozova, I. P., Shugaev, A. G. Respiratory metabolism of mitochondria of pea seedlings of different age under conditions of water deficiency. Plant Physiol. 2012, 59, 262–273. Russ.

CHAPTER 16

HALOPHILIC MICROORGANISMS FROM SALINE WASTES OF STAROBIN POTASH DEPOSIT

N. I. NAUMOVICH, A. A. FEDORENCHIK, and
Z. M. ALESCHENKOVA

Institute of Microbiology, NAS Belarus, Kuprevich Str. 2, 220141 Minsk, Belarus, E-mail: microbio@mbio.bas-net.by, Tel./Fax: +375-17-267-77-66

CONTENTS

ABSTRACT

Three cultures of halophilic phosphate-solubilizing bacteria isolated from solid saliferous wastes of Starobin potash mines run by Belaruskali concern are capable to grow in the presence of 10–15% sodium chloride and to promote germination and growth of alfalfa (*Medicago sativa*).

16.1 INTRODUCTION

Stressful environmental factors, like salinization, drought, xenobiotics, heavy metals, radiation etc. are extremely unfavorable for agricultural practice. In most cases contamination of environmental media in industrial regions is distinguished by a complex pattern. Added to increased levels of pollutants represented by mono- and polyaromatic hydrocarbons, the soil is subjected to extreme action of other factors, for example, enhanced salt background [1].

Belaruskali company exploiting the mines since the sixties of XX century has dumped on formerly fertile lands over 730 million tons of solid clay-saline wastes across overall area about 2 thousand hectares [2].

Salt stress negatively affects growth and productivity of farm crops. Despite adverse salinization impact, soils with high technogenic load generate microbiota capable to withstand hypersaline exposure [3]. A promising strategy for remediation of salinized soils envisages application of legume varieties, like alfalfa to upgrade soil fertility via atmospheric nitrogen fixation. Alfalfa salt tolerance is promoted by inoculation of selected rhizobial strains and PGPR bacteria.

Aim of the study was isolation of microbial cultures resistant to soil salinization and stimulating germination and growth of alfalfa (*Medicago sativa*).

16.2 EXPERIMENTAL PART

Salt-resistant phosphate-solubilizing microorganisms were recovered from saline wastes of Starobin deposit on solid nutrient medium in compliance with Muromtsev guidelines [4]. Tolerance of microbial isolates to sodium-, potassium- and calcium chlorides in concentrations 3, 5, 10, 15, and 20% was tested on TY medium [5]. Nitrogen-fixing *Ensifer meliloti* strains able to grow on TY medium containing 300 mM NaCl were isolated from nodules of alfalfa, nifH gene encoding a small subunit of nitrogenase complex was used as nitrogen fixation marker. A couple of primers nifH-1F and nifH-1R proved suitable for all examined samples of alfalfa-specific nodulating bacteria [6].

Growth-stimulating action of bacterial isolates under salinization conditions was evaluated using alfalfa seeds. Tested plant seeds were steeped in 2% aqueous bacterial suspension, control plant seeds were dipped into sterile tap water.

The seeds placed into vials were incubated at 28°C during 2 hours and transferred into sterile Petri plates on damp filter paper impregnated with sodium chloride in 100 mM concentration. The treated seeds were germinated in incubation chamber for 2–3 days at 28°C.

16.3 RESULTS AND DISCUSSION

Unique microflora capable to survive under extreme environmental conditions is generated in soil with high salinization and technogenic load. In this respect special emphasis is focused on microorganisms adapted to elevated mineralization background (7). Such bacteria appear attractive due to increased biotechnological potential. They may be applied to design biopreparations remediating saline soils and effluents with complex multichemical composition of pollutants (8).

Isolation of halophilic microorganisms showing resistance to chlorides of sodium, potassium, calcium was performed from solid saliferous wastes of Starobin potash deposit. The enrichment culture was carried out by inoculating homogenated solid saliferous wastes into selective liquid nutrient media and subsequent fermentation during 1 week. Submerged culture was conducted in 250 mL flasks containing 100 mL of selective medium on laboratory shaker (agitation rate 200–220 rpm) at temperature +28°C. Glucose-asparagine medium proposed by Muromtsev served as selective medium for enrichment culture of phosphate-solubilizing microorganisms. Isolates of phosphate-solubilizing bacteria were recovered by plating samples of enrichment culture on agar selective medium of the same composition.

At the next research stage isolates were screened according to criteria of phosphate-solubilizing activity and resistance to enhanced salt concentrations. Four cultures were sorted out of phosphate-solubilizing bacterial colonies dominating on glucose-asparagine medium with calcium phosphate. All isolates were Gram-positive rods showing catalase activity, while variant 4F displayed oxidase activity. Based on physiological-biochemical characterization using automatic system VITEK-2 (bioMerieux) and taxonomic nomenclature of Bergey's Manual of Determinative Bacteriology (1994) all four phosphate-solubilizing strains were affiliated to genus *Bacillus*.

Survival of selected bacilli exposed to the action of sodium-, potassium- and calcium chloride in different concentrations was assessed in a series of model experiments (Table 16.1).

In accordance with the aim of our investigation strains capable to grow on media containing elevated chloride levels should be chosen as promising objects for further studies. It may be deduced from our data that three strains – *Bacillus sp.* 2F, *Bacillus sp.* 4F and *Bacillus sp.* 8F suit perfectly to meet the set research objectives. They may grow on media with 15% sodium-, potassium- and calcium chloride and form halo zones on glucose-asparagine medium with calcium orthophosphate, indicating conversion of added salt into water-soluble form.

Four nitrogen-fixing strains of bacteria *E.meliloti* able to grow on TY medium comprising 300 mM NaCl were isolated from nodules of alfalfa (*M. sativa*).

Strain capacity to fix nitrogen mediated by nitrogenase enzyme complex was confirmed by presence of nifH gene. So far 10 nifH genes have been identified in Rhizobia, including nifH encoding a small subunit of enzyme complex. Amplification of nifH gene fragment with a couple of primers nifH-1F and nifH-1R revealed one specific zone sized approximately 430 bp (Figure 16.1). Partial nifH gene amplification involving primers nifH-1F and nifH-1R demonstrated that specific PCR product available in all 4 strains of nodulating bacteria *E. meliloti* is identical to that of nitrogen-fixing reference strain *Rhizobium loti.*

A series of model experiments was conducted to evaluate growth-stimulating effect of selected halophilic phosphate-solubilizing bacterial strains on seed germination and development of alfalfa sprouts in saline media (NaCl concentration 100 mM) (Figure 16.2).

The obtained result points out the leading role of strain Bacillus sp. 8F promoting germination of alfalfa seeds by 50% and elongation of seedlings by 16%. The other halophilic phosphate-solubilizing bacteria *Bacillus sp* 2F and *Bacillus sp* 4F did not cause any favorable impact on germination and length of alfalfa seedlings.

Mixed influence of nitrogen-fixing strain *E. meliloti* S3 and phosphate-mobilizing culture *Bacillus sp.* 8F on growth and development of alfalfa crop (*M. sativa*) in salinized environment (NaCl concentration 100 mM) was tested in model experiment (Figure 16.3).

TABLE 16.1 Growth of Halophilic Phosphate-Solubilizing Bacteria on TY Medium in the Presence of Sodium-, Potassium- and Calcium Chloride

Isolate N	Control (salt free)	Concentration, (%)														
		NaCl					KCl					$CaCl_2$				
		3,5	5	10	15	20	3,5	5	10	15	20	3,5	5	10	15	20
Bacillussp. 2F	++	+++	+++	++	+	—	+++	+++	++	+	—	+++	+++	+	+	—
Bacillussp. 4F	++	+++	+++	++	+	—	+++	+++	++	+	—	+++	+++	+	+	—
Bacillussp. 5F	++	+++	+++	+	—	—	+++	+++	+	—	—	+++	+++	—	—	—
Bacillussp. 8F	++	+++	+++	++	+	—	+++	+++	++	+	—	+++	+++	+	+	—

Note: +++ – abundant growth, ++ – good growth, + – satisfactory growth, — – no growth

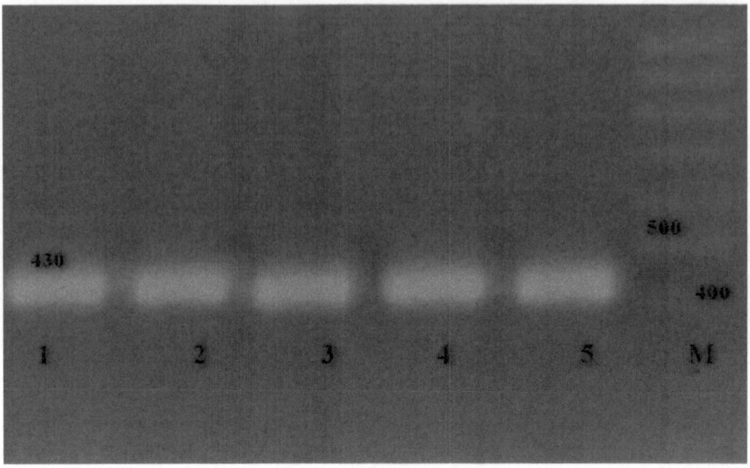

FIGURE 16.1 Amplification of nifH gene fragment with primers nifH-1F and nifH-1R using DNA isolated from pure culture [1 – *Rhizobium loti* L3 (positive control); 2 – *Ensifer meliloti* S3; 3 – *Ensifer meliloti* S5; 4 – *Ensifer meliloti* IS0; 5 – *Ensifer meliloti* 3583; M – DNA molecular weight marker (100 bp)].

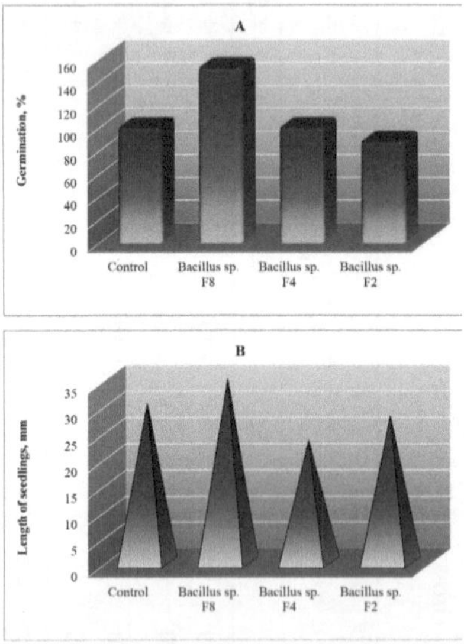

FIGURE 16.2 Effect of halophilic strains of phosphate-solubilizing bacteria on seed germination (A) and length of alfalfa seedlings (*M. sativa*) (B) under saline conditions.

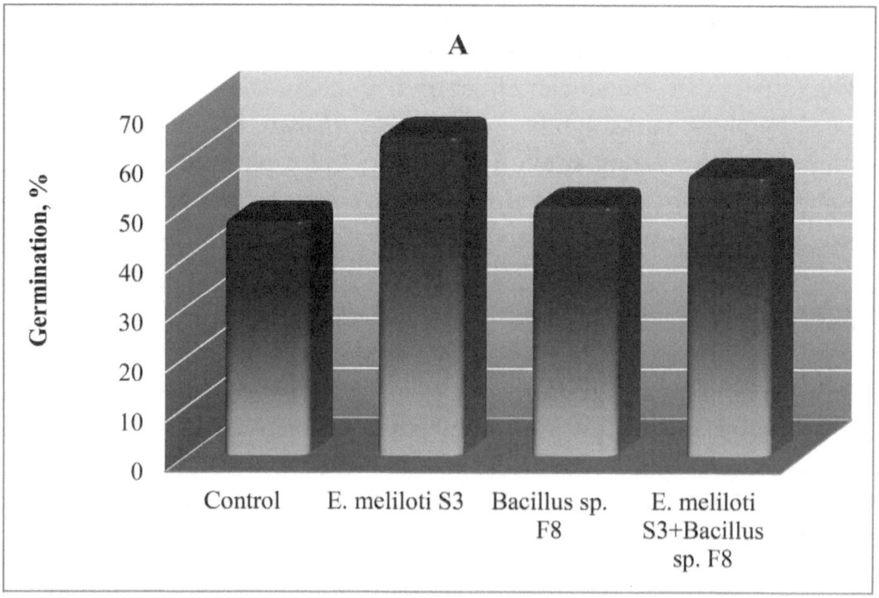

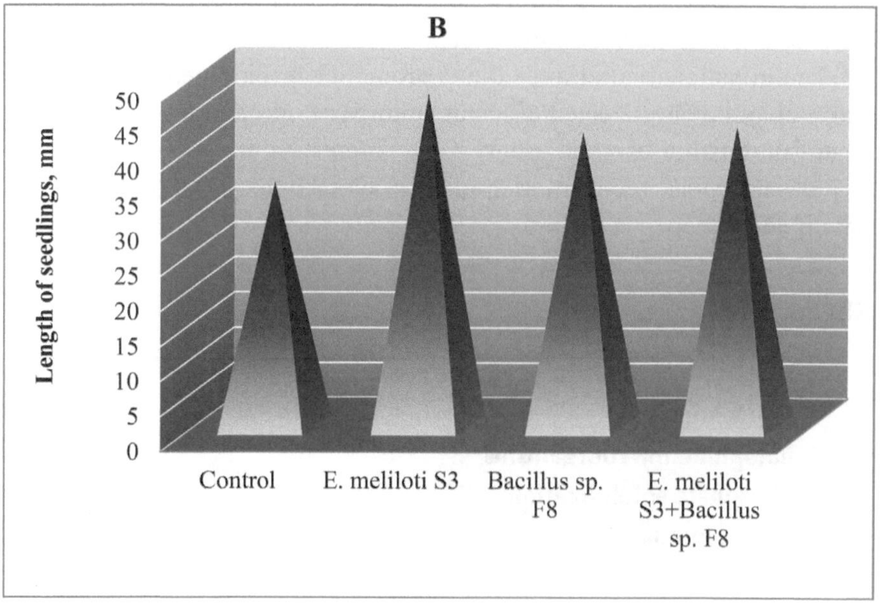

FIGURE 16.3 Combined effect of strains *E. meliloti* S3 and *Bacillus sp.* 8F on seed germination (A) and length of alfalfa (*M. sativa*) seedlings (B) under the salinized conditions.

Analysis of results evidences stimulating action of strain *E. meliloti* S3 on alfalfa seed germination and development of seedlings – 17% and 34%, respectively. Beneficial effect on both parameters was also demonstrated by phosphate-solubilizing strain *Bacillus sp.* 8F. Dual application of nitrogen-solubilizing strain *E. meliloti* S3 plus phosphate-solubilizing strain *Bacillus sp.* 8F exposed to excessive salinity ensured increase in seed germination rate by 9% and length of seedlings by 21%.

16.4 CONCLUSION

Three bacterial strains representing genus *Bacillus* and distinguished by the maximum tolerance to soil salinization and active ability to grow on media in the presence of 15% concentrations of sodium-, potassium- and calcium chloride were isolated from solid saliferous wastes of Starobin potash mines managed by Belaruskali concern.

Screening of three isolates for phosphate-solubilizing activity in pure culture demonstrated the ability of all selected variants to mobilize calcium orthophosphate. Strains *E. meliloti* S3 and *Bacillus sp.* 8F stimulated germination of alfalfa seeds and promoted development of *Medicago sativa* sprouts. The selected efficient cultures of nitrogen-fixing and phosphate-solubilizing bacteria arouse vivid interest as potential components of plant-microbial association applied in biotechnology of salinized soil phytoremediation.

KEYWORDS

- **biotechnology**
- **halophilic microorganisms**
- **phosphate solubilization**
- **phytoremediation**
- **salinization**

REFERENCES

1. Boonchan, S., Britz, M. L., Stanley, G. A., Appl. Environ. Microbiol, 66 (3), 1007 (2000).
2. Zhumar, P. V., Chertko, N. K., Proc. Int. Conf., Belarus State Agricultural Academy, Gorky 2007, 124 p (In Russian).
3. Hincha, D. K., Hagemann, M., Biochem. J. 383, 277 (2004).
4. Main methods of microbiological soil investigations, All-Union Research Institute of Agricultural Microbiology, Leningrad, 1987. 33 p. (in Russian).
5. Ibrahimova, M. V., et al. Microbiology. 75 (1), 94 (2006). (in Russian).
6. Soltani, A. A., Toolarood et al.: Intern. Res. J. Appl. and Basic Sci., 3(7), 1470 (2012).
7. Margesin, R., Schinner, F., Extremophiles, 5, 73(2001).
8. Holt, J. G., et al. (Eds.): Bergey's manual of determinative bacteriology. Williams & Wilkins, Baltimore, Maryland, 1994. 787p.

CHAPTER 17

BIOCHEMICAL CHARACTERISTICS OF INSECTS HERMETIA ILLUCENS

R. V. NEKRASOV,[1] I. V. PRAVDIN,[2] L. Z. KRAVTSOVA,[2] A. I. BASTRAKOV,[3] L. A. PASHKOVA,[1] and N. A. USHAKOVA[3]

[1]*All Russian Research Institute of Animal Husbandry After Academy Member L.K. Ernst, 142132 Russia, Moscow Region, Podolsk District, Settlement Dubrovitsy, Russia, E-mail: nek_roman@mail.ru*

[2]*The "NTC BIO" LLC, 309292 Russia, Belgorod Region, Shebekino Town, Dokuchayev Str., Russia, E-mail: ntcbio@mail.ru*

[3]*A.N. Severtsov Institute of Ecology and Evolution, Russian Academy of Sciences (RAS), Moscow, Leninski Avenue, 119071 Russia, E-mail: naushakova@gmail.com*

CONTENTS

ABSTRACT

The prospects of use of the *Hermetia illucens* larvae for feeding young pigs have been discussed. At rearing on forage grains, the larvae contain in their body 38% of protein, 39% of fat, 5% of chitin. The *Hermetia illucens* larvae are featured with high nutritive value and comprise a rich amino acid content. Saturated acids, 49% of which is lauric acid, prevail in the composition. A physiological experiment has demonstrated the possibility of effective replacement of 5% fish meal with 7% dried *Hermetia illucens* larvae in animals' ration. That resulted in pig's average daily living weight growth of 6.7% as well as in decrease of mixed fodder consumption by 8% per 1 kg of weight gain. The possibility of use of insect micro dosage in mixed fodders for pigs, as a complex probiotic preparation component, has been also demonstrated. Adding the component in the amount of 0.5 kg/t of complex preparation enhanced the daily pigs' weight growth by 14% at lowering the fodder consumption by 12%.

17.1 INTRODUCTION

At present, the world community actively discusses the prospects of using the flour of edible insects, as a food component, and as alternative to fish meal. Considerable consumption of fish meal and fish oil accounts for swine breeding (up to 24.0%) and in aquaculture (up to 46.0%). It is noted that in the period from 2004 till 2008 double increasing of prices for fish meal took place [8], and from this time point a continuous price increase for this product can be observed. At the same time, some research works have been carried out as for the efficiency of the use of uncommon protein sources in animals' feed: algae, bacteria, yeasts, slaughter-house wastes, food molasses, grain wastes, beet pulp, bakery wastes, and others [5]. The disadvantage of vegetable protein sources is a low percentage content of crude protein and high percentage content of carbohydrates. As noted by the FAO (UN Food and Agriculture Organization), "one of basic and accessible sources of nutritive and protein-rich food are insects" (http://prodmagazin.ru/2013/05/13/fao-nasekomyie-otlichnyiy-istochnik-belkov-i-vitaminov/). As the most promising species *Hermetia illucens* fly (or "black soldier" fly) could be considered, whose larvae comprise up to 40% protein, and can be used in feed for animals. The insect belongs to few species of invertebrates able to grow all year round in a pure culture, in closed space, in artificial conditions, and due to that *Hermetia illucens* can be also used for biotechnological purposes. A new direction of insect biomass application is the introduction of bioactive components of insects in probiotic feed, as a growth factor, for the stimulation of probiotic bacteria and for the enhancement of the physiological activity of the preparations.

The purpose of the present work is researching the nutritive value of *Hermetia illucens* larvae raised on forage wheat, as a protein alternative to fish meal, at introducing them for young pigs' fodder at growing period, as well as the estimation of the biological activity of the complex probiotic preparation, which contains *Bacillus* type microorganisms on a phyto-carrier, combined with *Hermetia illucens* larvae biomass, in the ration of young pigs.

17.2 MATERIALS AND METHODS

Under the conditions of a physiological yard of the Federal State Budgetary Scientific Institution "All-Russian Research Institute of Animal husbandry after academy member L.K. Ernst," a physiological experiment on a hybrid (F-1: Large White x Landrace) young pigs stock of 12 animals, the age of 61–87 days, in the growing period, having the start living weight about 17.1 kg, has been carried out. The trial duration was 26 days. 3 groups of young pigs have been formed according to the analog principle (with consideration of origin, age, and living weight). The arrangement of the experiment is shown in the Table 17.1. Compositions and nutritive value of experimental lots of mixed fodder are shown in the Table 17.2. Housing conditions (temperature, humidity, light conditions, and gas composition of air in the room) were the same in the limits of veterinary hygiene norms. After the feeding period a digestion trial for researching the digestibility of mixed fodder nutrients, using nitrogen, calcium and phosphorus, as per the general methodology, has been carried out [4].

To define the influence of larvae on the feed eatability daily individual records of the fodder consumed and their wastes for all the records period, have been performed. After finishing the experiment average samples of

TABLE 17.1 Arrangement of the Experiment

Group	Number of animals	Feeding characteristics
Preliminary period – 19 days		
1 – control group	3	Full ration start mixed fodder – with addition of 5.0% fish meal (FM)
2 – trial group	3	Start mixed fodder with 7.0% of black soldier fly larvae (FL)
3 – trial group	3	Start mixed fodder with 5.0% FM, with introduction of "ProStor" bioactive additive with the larvae (FL) at the dosage 0.5 kg/t

TABLE 17.2 Composition and Nutritive Value of the trial fodder groups

Component, %	Control no. 1	Trial	
		no. 2	no. 3
Larvae meal	—	7.0	—
Fish meal, CP 61.0%	5.0	—	5.0
Wheat	41.8	41.8	41.8
Barley	10.0	10.0	10.0
Corn	8.0	8.0	8.0
Wheat bran	5.0	3.0	5.0
Sunflower oil cake, CP 32%	16.0	16.0	16.0
Sunflower oil	4.0	4.0	4.0
Fodder yeast, CP 34%	7.0	7.0	7.0
Common salt	0.2	0.2	0.2
Tricalcium phosphate	2.0	2.0	2.0
Premix	1.0	1.0	1.0
TOTAL	100.0	100.0	100.0
Metabolic energy, MJ	11.7	12.5	12.2
Dry matter, kg	0.892	0.895	0.892
Crude protein, g	178.7	178.7	178.7
Crude ash, g	57.4	53.2	57.4
Free-nitrogen extracts, g	529.1	527.0	529.1
Starch, g	354.2	346.1	354.2
Sugar, g	27.9	26.9	27.9

TABLE 17.2 Continued

Component, %	Control no. 1	Trial	
		no. 2	no. 3
Lysine, g	7.8	5.3	7.8
Methionine+cystine, g	5.9	4.7	5.9
Threonine, g	6.2	4.9	6.2
Crude fat, g	61.8	67.9	61.8
Crude fiber, g	54.6	59.1	54.6
Calcium, g	10.2	8.2	10.2
Phosphorus, g	8.5	7.4	8.5
Mg, g	1.9	1.8	1.9
S, g	1.0	0.8	1.0
K, g	5.4	5.0	5.4
Na, g	1.6	1.1	1.6
NaCl, g	4.1	2.8	4.1
Vitamin A, IU/kg	20.00	20.00	20.00
Vitamin D_3, ICU/kg	2.00	2.00	2.00
Vitamin E, ppm	20.00	20.00	20.00
Fe, ppm	80.00	80.00	80.00
Cu, ppm	10.00	10.00	10.00
Zn, ppm	60.00	60.00	60.00
Mn, ppm	40.00	40.00	40.00
Co, ppm	0.30	0.30	0.30
I, ppm	0.60	0.60	0.60
Se, ppm	0.20	0.20	0.20

feed, feces and urine have been chemically analyzed in the laboratory of chemical and analytical research at the All-Russian Research Institute of Animal husbandry after academy member L.K. Ernst, in accordance with common analysis procedures. The consumption of feed per unit of living weight growth rate of pigs has been also defined.

The amino acid composition has been defined in the lab of the Scientific Centre "Feed and Metabolism" at the Stavropol State Agricultural University, on the amino acid analyzer AAA-400 (Czech Republic).

The composition of fatty acids has been determined in All-Russian Research Institute of Animal husbandry after academy member L.K. Ernst biochemical/chemical and analytical department, on the Shimadzu GC-2010 gas chromatograph (Japan). Fat separation was carried out after grinding the larvae via a method of extraction of lipids with chloroform-methanol mixture (according to Folch). Obtaining of fatty acid methyl esters (GOST P 51486-99).

All the data material obtained has been statistically processed by means of the Student's variation statistics method with using the Microsoft Excel program in the limits of following significance levels: $*P < 0.05$, $**P < 0.01$.

17.3 RESULTS AND DISCUSSION

Nutritive value of *Hermetia illucens* larvae obtained through their reared on forage grain, is quite high (Table 17.3). At the same time, crude protein contents in the product tested is 22.4% lower than in fish meal, and the amino acid contents is a bit different: contents of lysine (by 5.19%), and of methionine (by 1.70%) is lower, the contents of proline (by 1.13%), alanine (by 0.25%), isoleucine (by 0.7%), phenylanaline (by 0.31%), histidine (by 0.96%) are insignificantly higher, and leucine and tyrosine contents are much higher (by 4.80 and 2.63%, respectively), what is important for growing young animals. The data obtained confirm that the investigation of *Hermetia illucens* larvae as an alternative to fish meal, at their corresponding adding levels to feed, is of indisputable interest. The contents of amino acids, as calculated on 100% of crude protein, are shown in the Figure 17.1.

One of the main components of exoskeleton of insects is chitin, high molecular hard digestible polysaccharide having 18–25 radicals of N-acetylglycoseamine. According to our data, chitin contents in the larval stage of *Hermetia illucens* is insignificant – 5.2%. Taking into consideration that nitrogen of amino acids that is 98.0% of total nitrogen, just a very little portion of total nitrogen can be theoretically inaccessible for pigs. Nevertheless, immune simulating properties of chitin and of its derivate, chitosan, are well known, so that can be positive in enhancing the nonspecific immunity at the use of insect chitin in feed.

TABLE 17.3 Comparative Analysis of Nutritive Values of *Hermetia illucens* Larvae and Fish Meal

Indices	Measuring units	Samples			
		Hermetia illucens larvae		Fish meal*	
Metabolic energy (pigs)	MJ/kg	15.28		15.1	
Dry matter	%	92.55		90.0	
Crude protein	%	37.57		60.0	
Crude fat	%	38.29		8.9	
Chitin	%	5.19		—	
Crude ash	%	3.62		11.2	
Calcium	%	0.41		3.7	
Phosphorus	%	0.32		1.8	
Free-nitrogen extracts	%	14.08		9.5	
Contents of amino acids in crude protein		37.57%	100.0%	60.0%	100.0%
Aspartic acid	%	2.96	7.88	6.10	10.17
Threonine	%	1.42	3.78	2.74	4.57
Serine	%	1.64	4.37	2.61	4.35
Glutamic acid	%	459	12.22	8.61	14.35
Proline	%	2.20	5.86	2.84	4.73
Glycine	%	2.00	5.32	4.00	6.67
Alanine	%	2.61	6.95	4.02	6.70
Valine	%	1.83	4.87	3.19	5.32
Methionine	%	0.64	1.70	1.88	3.13
Cystine	%	0.29	0.77	0.60	1.00
Isoleucine	%	1.91	5.08	2.63	4.38
Leucine	%	4.78	12.72	4.75	7.92
Tyrosine	%	2.36	6.28	2.19	3.65
Phenylalanine	%	1.72	4.58	2.56	4.27
Histidine	%	1.55	4.13	1.90	3.17
Lysine	%	1.95	5.19	5.22	8.70
Arginine	%	1.54	4.10	3.89	6.48
Amino acids total contents	%	35.99	95.80	59.73	99.56

* Table data.

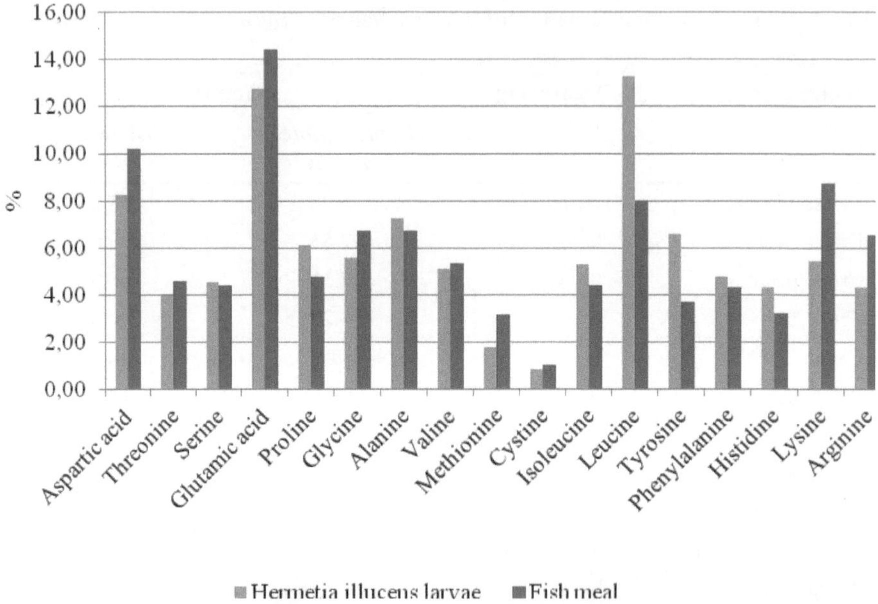

FIGURE 17.1 Comparative contents of amino acids (calculated as 100% of crude protein) in fish meal and in *Hermetia illucens* larvae.

The inhibiting factor of the use of *Hermetia illucens*, especially in swine growing, is relatively high contents of crude fat (upto 39% and more depending on feed substrate), in comparison with fish meal (2.3–11.3%). There is an opinion as for negative influence of quality fat composition on meat products. It is believed that the fat comprises mainly polyunsaturated fatty acids, belonging to ω-3 and ω-6 class acids, and that brings with more active oxidation of the product and softening of the fat tissue structure. But, on other hand, polyunsaturated fatty acids possess richer chemical structure, and that is vitally important for normal body functioning, and saturated fatty acids fix calcium to insoluble salts.

Fatty acid contents in *Hermetia illucens* larvae are illustrated in Figure 17.2. The fatty acid composition shows, that as for percentage contents fatty unsaturated acids prevail: lauric – 47.06; palmitic – 15.17; myristic – 10.48; stearic – 3.47; margaric – 0.21%. Lauric acid is featured with antimicrobial activity, what can be an important factor in prophylactics of intestinal infections and in enhancing the total animal resistance. Polyunsaturated acids constitute only about 3.0% of total fat contents. They are available as eicosapentaenoic acid – 1.7; docosahexaenoic acid – 0.6;

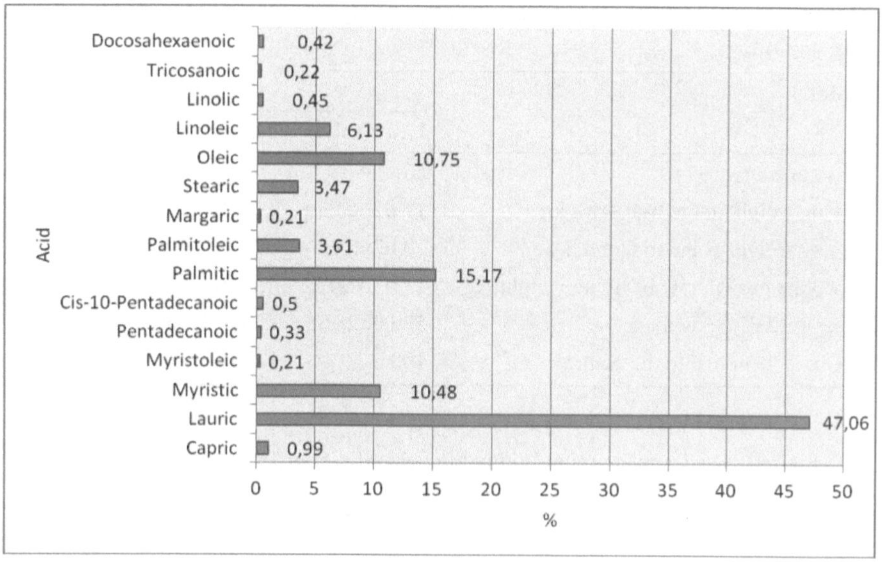

FIGURE 17.2 *Fat/acid* composition of *Hermetia illucens* larvae, %.

α-linoleic acid – 0.7%. Following unsaturated fatty acids are present too: ω-6 – linoleic, ω-9 – oleinic acid. The presence of ω – 3 fatty acids results in enhancing the nutritive value of the product.

The results indicating the growth intensity of pigs during the experiment period, and the feed consumption per unit of the product obtained, are shown in Tables 17.4 and 17.5. Although living weight data as well as absolute growth rate in trial animal groups are statistically unreliable in comparison with control ones, a trend to the living weight growth of young pigs is evident. The maximum growth rate of experimental animal groups was 6.7–14.4% over the values of the control group. Thus, the uncommon feed component – *Hermetia illucens* larvae – did not affect the living weight growth rate of young pigs, and can be considered as an alternative to fish meal, and additional introducing the ProStor probiotic, together with the larvae, even increased the productivity of animals.

As for mixed fodder consumption for a product unit, the indices in trial groups were 0.2–0.3 kg, or corr. 8.0–12.0% lower than in the control group. The expenditures of metabolic energy for 1 kg of weight growth were 0.5–2.19 MJ, or 1.7–7.5% lower.

The data obtained give evidence, that the trial group animals better used feed nutritive substances for the growth of living weight.

TABLE 17.4 Dynamics of the Growth of Test Pigs (on average, in the group, M±m, n=3)

Index	Group		
	1-control	2-trial	3-trial
Days of trial	26		
Living weight at the trial start, kg	17.67±0.29	17.47±0.38	16.80±0.34
Living weight at the trial end, kg	30.67±6.17	31.33±2.70	31.67±0.20
Absolute growth rate of living weight, kg	13.0±0.90	13.87±0.47	14.87±0.24
Average daily growth, g	500.0±28.36	533.3±14.66	571.8±7.57
Idem, in % in real to the control	100.0	106.7	114.4

TABLE 17.5 Feed Consumption Within the Scientific Trial Time (on average for 1 animal, M±m, n=3)

Index	Group		
	1-control	2-trial	3-trial
Contents of metabolic energy in 1 kg of mixed fodder, MJ	11.7	12.5	12.3
Feed consumption within the period, kg	32.5	32.5	32.5
Daily consumption:			
– mixed fodder, kg	1.25	1.25	1.25
– metabolic energy, MJ	14.6	15.6	15.4
Absolute growth rate for the period, kg	13.0±0.90	13.87±0.47	14.87±0.24
in % in rel. to the control	100.0	106.7	114.4
Consumption of mixed fodder for 1 kg of growth, kg	2.5	2.3	2.2
Idem, in % to the control	100.0	92.0	88.0
Consumption metabolic energy for 1 kg of growth, MJ	29.25	28.75	27.06
Idem, in % to the control	100.0	98.3	92.5

Note: metabolic energy in digestible nutrients.

17.3.1 RESULTS OF A DIGESTION TRIAL

Investigation of digestibility indices of nutrients in the feed (Table 17.6) have shown a real difference ($P < 0.05$) while comparing the digestibility

TABLE 17.6 Indices of Digestibility of Nutrients, % (M±m, n=3)

Nutrient	Group		
	1-control	2-trial	3-trial
Dry matter	74.85±1.17	80.4±1.61*	81.28±1.04
Organic substance	76.85±1.02	81.82±1.48	82.63±0.97*
Protein	79.82±1.32	82.45±1.72	82.85±0.67
Fat	53.71±1.78	64.05±6.45	55.37±1.98
Fiber	29.6±6.73	41.39±4.20	51.09±2.15*
Free-nitrogen extracts	81.35±1.06	86.79±0.91*	87.02±1.12*

*The difference is significant at P <0.05.

indices of feed nutrients y animals of the trial groups and the control group: dry matter increase (in limits of 5.0%), organic matter increase (4.7–5.8%), fiber (21.5%), free-nitrogen extracts (5.3–5.7%) in favor to the first groups. As for protein, no real data are available, but a trend can be traced for better digestion of the trial pig groups, in limits of 3.0%, and fat digestibility indices were the same for all the pigs. That confirms that the test pigs used better feed nutrients, and that is proved with better average daily growth rate, which was 14.4% higher in the third trial group.

17.3.2 THE BALANCE AND THE USE OF NITROGEN

While performing the digestion trial for studying the protein metabolism the balance and the use of nitrogen by young pigs, have been calculated (Table 17.7).

The consumption of nitrogen were the same, equaling 45.5 g, and its retention in animals of the trial groups was different, higher in the third trial group by 2.84 g (in comparison with control group analogs). While comparing nitrogen (as taken in and digested) an increase trend is evident, that is especially remarkable at comparing the control animals and animals of the third group.

The analysis of the data obtained has shown that the introduction of a new uncommon component – *Hermetia illucens* larvae – to the feed, made possible better use and retention of nitrogen and, consequently, higher living weight growth rates (at the level of 533.3 g against 500.0 g).

And additional introduction of the ProStor probiotic together with *Hermetia illucens* larvae (the third trial group) yielded living weight growth rates of 571.8 g, for example, 14.4% higher in comparison with control animals.

17.3.3 THE BALANCE OF CALCIUM AND PHOSPHORUS

While performing physiological tests, the balance and the use of calcium and phosphorus in the body of test animals, have been studied, table. 17.8 and 17.9.

TABLE 17.7 Average Daily Balance and Nitrogen Use of Fodders (on average in the group, M±m, n=3)

Index	Group		
	1-control	**2-trial**	**3-trial**
Taken in with food, g	45.5	45.0	45.5
Extracted in feces, g	10.56±0.70	8.70±0.84	8.21±0.31*
Extracted in urine, g	18.19±0.86	17.73±0.34	17.70±3.20
Deposited in the body, g	16.75±1.50	18.57±0.77	19.59±3.34
Used in %:			
taken in	36.81±2.86	41.27±1.63	43.05±7.43
digested	47.94±2.46	51.16±0.90	52.53±8.75

* The difference is significant at P <0.05.

TABLE 17.8 Average Daily Balance and Use of Calcium (at the average in the group, M±m, n=3)

Index	Group		
	1-control	**2-trial**	**3-trial**
Taken in with food, g	14.80	15.48	12.85
Extracted in feces, g	6.46±0.34	3.97±0.67*	3.45±0.80*
Extracted in urine, g	005±0.01	0.12±0.01**	0.08±0.01
Deposited in the body, g	8.22±0.38	11.40±0.45**	9.32±0.65
Used of taken in, %	55.55±2.54	73.61±2.90**	72.53±5.09*

* The difference is *real* at P <0.05.

TABLE 17.9 Average Balance and Use of Phosphorus (on average for the group, M±m, n=3)

Index	Group		
	1-control	**2-trial**	**3-trial**
Taken in with food, g	8.32	7.46	7.57
Extracted in feces, g	3.78±0.25	3.09±0.42	2.49±0.05**
Extracted in urine, g	1.10±0.34	0.45±0.06	1.48±0.11
Deposited in the body, g	3.44±0.51	3.92±0.43	3.61±0.16
Used of taken in, %	41.38±6.22	52.53±5.79	47.60±2.05

* The difference is real at P <0.05.

The data in the groups illustrate that calcium was administered to animals with feed at the same level. The quantity with microelements isolated with feces was twice lower in the second trial food (2.49 g, or 38.5%, and 3.01 g, or 46.6%, at P<0.05, with urine – higher by 0.07 g, or 140.0% and, correspondingly, the retention and use better by 3.18 g, or 38.7%, and 18.1% at P<0.01 (in comparison with the second trial group and the control group), and in the third trial group – 16.98%, at P<0.05.

At the analysis of data of using phosphorus we can observe a real difference in data of the third trial group and the control group: the isolation with feces is 1.29 g, or 34.1%, lower (P<0.01).

The addition of *Hermetia illucens* larvae to food ration of test growing pigs, considered as an alternative food and micro supplement, yielded better use of calcium by animals of the second and the third trial group.

The results as for digestibility of nutrients and balance, as well as for the use of mineral substances, have shown, that the use in food a new uncommon food supplement – *Hermetia illucens* larvae – to the well balanced full ration mixed fodders, both at dosage of 7.0%, and as a component of complex probiotic preparation, did not affect the physiological processes in pigs' bodies, and their productivity.

That gives evidence that the *Hermetia illucens* larvae can be considered as a quality ration alternative for common food (fish meal), which satisfies the needs of young pigs food in the growing period. At the same time it is reasonable to use a micro dosage of the larvae to the animals' ration in the composition of complex probiotic preparation.

KEYWORDS

- **edible insects**
- **fish meal**
- **food additive**
- ***Hermetia illucens***
- **mixed fodder**
- **pigs**
- **probiotic**

REFERENCES

1. Galochkin, V. A., Agafonova, A. V., Krapivina, E. V., Galochkina, V. P., Use of adptogenes for acceleration of isolation of heavy metals and radionuclides at fattening of bull-calves. Biology problems of production animals J. 1, 82–96 (2014). (in Russian).
2. Maximov, V. I., Rovonon, V. E., Voskun, S. E., et al., Preparations on base of chitosan "Solikhit" for treatment of intestine dysbacteriosis. New prospects in research of chitin and chitosan. The Works of the Fifth International Conference. VNIRO, Moscow, 1999, 164–168. (in Russian).
3. Shkatov, M., Alternative mixed fodder ingredients. Husbandry in Russia J. 2, 64–65 (2013). (in Russian).
4. Diener, S., Valorization organic solid waste using the black soldier fly, Hermetia illucens, in low and middle income countries. Dissertation submitted to Doctor of Sciences, 2010, 77 p.
5. Newton, G. L., Booram, C. V., Barker, R. W., Hale, O. M., Dried Hermetia Illucens Larvae meal as a supplement for swine. JAS. 44, 395–400 (1977).
6. http://prodmagazin.ru/2013/05/13/fao-nasekomyie-otlichnyiy-istochnik-belkov-i-vitaminov
7. http://www.wageningenur.nl/en/show/Insects-to-feed-the-world.htm

CHAPTER 18

PEPTIDES OF A PLANT ORIGIN EXERTING HEPATOPROTECTIVE PROPERTIES

O. G. KULIKOVA,[1] D. I. MALTSEV,[1] A. P. ILYINA,[1]
A. O. ROSHCHIN,[1] V. P. YAMSKOVA,[2] and I. A. YAMSKOV[1]

[1]*Nesmeyanov Institute of Organoelement Compounds, Russian Academy of Sciences, ul. Vavilova 28, Moscow, 119991 Russia*

[2]*Koltsov Institute of Developmental Biology, Russian Academy of Sciences, ul. Vavilova 26, Moscow, 119334 Russia, E-mail: yamskova-vp@yandex.ru*

CONTENTS

ABSTRACT

The purpose of the present study was investigation of hepatoprotective properties of peptides, which are considered as a compound of membranotropic homeostatic tissue specific bioregulators (MHTBs) obtained form herbal raw material. Procedure and results of physicochemical properties investigations for peptides isolated from medicinal plants are described and provided. Using the model of *Pleurodeles Waltl* newt liver roller organotypic culturing in vitro, it was shown that examined peptides are responsible for hepatoprotective properties occurring in obtaining source plants. This character of activity is demonstrated by peptides only when applying at ultralow doses (10^{-8}–10^{-15} mg/mL).

18.1 AIM AND BACKGROUND

The aim of a present study was to find and to identify new biologically active substances in medicinal plant tissues.

In previous investigations [1], it was shown that proteins isolated from medicinal plant are responsible for its pharmaceutical activity. According to numerous reports and reviews [2–6], different sorts and kinds of herbs are examined at the moment. Among these plants, such well-known objects as dill (*Anethum graveolens* L.), celandine (*Chelidonium majus* L.), absinth sage (*Artemisia absinthium* L.) and amber (*Hypericum perforatum* L.) should be noted. Hepatoprotective effects is a common defining feature for these herbs.

Dill (Anethum graveolens L.) is a hardy annual, native of the Mediterranean and the Black Sea regions, smaller than common fennel. Ordinarily the plants grow 2–2½ feet tall. The glaucous, smooth, hollow, branching stems bear very threadlike leaves and in midsummer compound umbels with numerous yellow flowers, whose small petals are rolled inward. Very flat, pungent, bitter seeds are freely produced, and unless gathered early are sure to stock the garden with volunteer seedlings for the following year. Under fair storage conditions, the seeds continue viable for three years [7]. It is known as a plant exhibiting a number of medicinal effects. Thus, investigation of hepatoprotective effect for *Anethum graveolens* L. seed oil was performed [4]. In the course of this experiment, hepatotoxicity in rats was caused by using carbon tetrachloride. Experiment length equaled four weeks. As a result, the hepatotoxicity produced by CCl_4 administration was found to be inhibited by dill (*Anethum graveolens* L.) oil with evidence of significant ($p<0.05$) decrease levels of serum AST (aspartate aminotransferase) and ALT (Alanine transaminase). In addition, significant increasing ($p<0.05$) level of serum total protein and albumin should be noted. According to this, it is likely that the mechanism of hepatoprotection of dill oil is due to its antioxidant effect. Dill has a potent hepatoprotective action against CCl_4 induced liver toxicity in rats.

Absinth sage (*Artemisia absinthium* L.) is a strongly aromatic plant, bearing numerous oil-producing glands on leaves, stems, and flowering branches. Its essential oil and other forms of herbal preparations are traditionally used in ethnopharmacology and ethnomedicine [8, 9]. According to Herbal Medicinal Product Committee (HMPC), the herb of A. absinthium is accepted as a traditional herbal medicinal product, which is used in case of temporary loss of appetite and mild dyspeptic/gastrointestinal disorders [10]. In report [6], in vivo hepatoprotective activity of the aqueous extract of *Artemisia absinthium* L., which has been used for the treatment of liver disorders in traditional medicine. Aqueous extract (50, 100 or 200 mg/kg body weight/day) was administered orally to experimental mice. Liver injury was induced chemically, by a single CCl_4 administration (0.1% in olive oil, 10 mL/kg, i.v.), or immunologically, by injection of endotoxin (LPS, 10 μg, i.v.) in BCG-primed mice. The levels of aspartate aminotransferase (AST), alanine aminotransferase (ALT), tumor necrosis factor – α (TNF-α) and inter leukin-1 (IL-1) in mouse sera, as well as superoxide dismutase (SOD), glutathione peroxidase (GPx) and malondialdehyde (MDA) in mouse liver tissues were measured. Obtained results demonstrated that the pretreatment with AEAA significantly ($p < 0.001$) and dose-dependently prevented chemically or immunologically induced increase in serum levels of hepatic enzymes. Aqueous extract of *Artemisia absinthium* L. significantly ($p < 0.05$) reduced the lipid peroxidation in the liver tissue and restored activities of defense antioxidant enzymes SOD and GPx towards normal levels.

Chelidonium majus L. (Papaveraceae) is a biennial or perennial plant native to Europe, North America and Western Asia and commonly known as swallow wort, rock poppy or greater celandine. *Chelidonium majus* L. has been used in traditional Chinese medicine, western phytotherapy, homeopathy and anthroposophy. The German Commission E indicated Chelidonii herba for the use in spastic dis-comfort of bile ducts and gastro-intestinal tract. The fresh latex was externally used in the treatment of warts but also for other skin complaints such as corns, tinea infections, eczema and tumors of the skin. According to the German Commission D monograph on *Chelidonium majus* L., the herbal substance is used in homeopathy for different disorders of the liver and the gallbladder, inflammation of the respiratory organs and the pleura and in rheumatism. According to the Chinese medicine, *Chelidonium majus* L. is mainly used to treat blood stasis, to relieve pain, to promote diuresis in oedema and ascites, to treat jaundice and to relieve cough [3]. The ethanolic extract of *Chelidonium majus* L. exerted marked hepatoprotection against carbon tetrachloride toxicity in two studies on rats, indicated by a reduction in the number of necrotic cells, a prevention of fibrotic changes, and decreased activities of transaminases and bilirubin [11, 12]. The efficacy of an alcohol extract of *Chelidonium majus* L. as an antihepatotoxic agent was tested on rats with CCl_4-induced hepatic injury. CCl_4 (1 mL/kg; twice a week) and extract (125 mg/kg/day) were administered simultaneously. The parameters studied to assess liver damage were plasma ALT, AST. AIP, LDH, bilirubin, cholesterol and liver histology. Significant protection toward CCl_{4-} induced elevation of plasma enzymes, changes in bilirubin, cholesterol and microscopic liver damage was observed. These results indicate the hepatoprotective action of the extract. Fractions of *Chelidonium majus* L. were also efficient in combating p-dimethylaminoazobenzene-induced hepatocarcinogenesis in mice [13].

Hypericum perforatum L. (St. John's wort) is a five-petalled, yellow-flowered perennial weed common to the western United States, Europe, and Asia [14]. Close examination of the flowers reveals small black dots that, when rubbed between the fingers, produce a red stain. This red pigment contains the constituent hypericin. Held up to light, the leaves of the plant display a number of bright, translucent dots. This perforated look led to the species name perforatum. The plant is currently cultivated in Europe, North and South America, Australia, and China [15]. The aerial parts of the plant are harvested during the flowering season and used in modern, standardized extracts. It was, and continues to be, used as a topical treatment for wounds and burns. It has also been used as a folk remedy for kidney, stomach, and lung ailments [16]. Efficiency of *Hypericum perforatum* L. role on hepatic ischemia–reperfusion (I/R) injury in rats also was represented. Hence, albino rats were subjected to 45 min of hepatic ischemia followed by 60 min of reperfusion period. *Hypericum perforatum* L. extract (HPE) at the dose of 50 mg/kg body weight (HPE50) was intraperitoneally injected as a single dose, 15 min prior to ischemia. Rats were sacrificed at the end of reperfusion period and then, biochemical investigations were made in serum and liver tissue. Liver tissue homogenates were used for the measurement of malondialdehyde (MDA), catalase (CAT) and glutathione peroxidase (GPx) levels. At the same time alanine aminotransferase (ALT), aspartate aminotransferase (AST) and lactate dehydrogenase (LDH) were assayed in serum samples and compared statistically. While the ALT, AST, LDH activities and MDA levels were significantly increased, CAT and GPx activities significantly decreased in only I/R induced control rats compared to normal control rats ($p<0.05$). Treatment with HPE50 significantly decreased the ALT, AST, LDH activities and MDA levels, and markedly increased activities of CAT and GPx in tissue homogenates compared to I/R-induced rats without treatment–control group ($p<0.05$). In oxidative stress generated by hepatic ischemia–reperfusion, *Hypericum perforatum* L. as an antioxidant agent contributes an alteration in the delicate balance between the scavenging capacity of antioxidant defense systems and free radicals in favor of the antioxidant defense systems in the body [2].

18.2 INTRODUCTION

At the time it was shown that membranotropic homeostatic tissue specific bioregulators (MHTBs) are observed in different tissues of a mammal and plant origin [17–22]. Bioregulators, which are belong to this group, are localized in intercellular space and exhibit membranotropic activity – influence on viscoelastic properties and cell membrane permeability [23–29]. MHTBs stimulate reactivating and reparation processes in vivo, demonstrate protective activity on in vitro cultivated cells (increase cell viability, conduce the integrity of a tissue structure). These effects of MHTB's as it was observed are characterized by lack of species specificity and by occurring tissue specificity. This type of activity exhibited by bioregulators when applying ultra-low doses (ULD) [30]. It is necessary to notice that MHTBs are characterized by a complex structure, which contains membranotropic-active peptides and proteins with high molecular weights, modulating peptides activity. Our work objective was investigation hepatoprotective properties of ultralow dose applied peptides isolated from tissues of a plant origin, which are belong to medicinal plants MHTBs.

As a study subjects dill (*Anethum graveolens* L.), celandine (*Chelidonium majus* L.), absinth sage (*Artemisia absinthium* L.) and amber (*Hypericum perforatum* L.) were chosen. A great number of medical effects appropriated for these herbal plants determines this choice. Among these effects, it should be noted hepatoprotective activity [31].

18.3 MATERIALS AND METHODS

18.3.1 BIOREGULATOR ISOLATION AND PURIFICATION

For extraction, we used fresh herbal raw material cut into small (1–1.5 cm) fragments, which were incubated at 8–10°C for 5–6 h in an extraction solution containing 2.06×10^{-2} M NH_4NO_3, 1.88×10^{-2} M KNO_3, 3×10^{-3} M $CaCl_2 \cdot 2H_2O$, 1.5×10^{-3} M $MgSO_4 \cdot 7H_2O$, 1.25×10^{-3} M KH_2PO_4. Thus obtained extracts were filtered through several layers of cheesecloth and centrifuged at 3000 g for 30 min; the pellet was discarded. Then, dry ammonium sulfate was added to the plant extracts under stirring to a concentration of 780 g/L until a saturated salt solution formed; pH was adjusted at 7.5–8.0 with ammonium hydroxide. The obtained protein mixtures were incubated at 4°C for 95–100 h. The precipitated proteins were separated by centrifugation at 25,000 g and 4–8°C for 30 min). Thus obtained supernatant and pellet were dialyzed against water until complete removal of ammonium sulfate and then concentrated at 37–40°C in a vacuum rotor evaporator [1].

18.3.2 SDS-ELECTROPHORESIS OF PROTEINS IN POLYACRYLAMIDE GEL

SDS-electrophoresis in 12,5% polyacrylamide gel was performed according to Laemmli [32] under denaturative conditions. The mixture containing molecular-weight protein markers was used. Coumassie R-250 staining was used.

18.3.3 REVERSED-PHASE HIGH PERFORMANCE LIQUID CHROMATOGRAPHY

Supernatants obtained from study plants extracts were analyzed by high-pressure chromatograph ("Kontron," USA) when using hydrophobic HPLC column (4.6×250 мм) Kromasil C18 (Russia). Elution was performed in acetonitrile concentration gradient (4–70%) in 0.1% trifluoroacetic acid (pH 2.2) with 0.5 ml per minute flowing rate in 70 minutes. Detection was performed at 280 nm wavelength.

18.3.4 MASS SPECTROMETRICAL ANALYSIS

The peptides were studied using matrix-assisted laser desorption/ionization time-of-flight (MALDI-TOF) technique in an UltraFlex 2 mass spectrometer ("Bruker Daltonics," Germany) equipped with a nitrogen laser of 337 nm with a pulse frequency of 20 Hz. All measurements were performed in linear and reflection regimens for an estimation of positive ions. Samples for mass-spectrometric analysis were obtained by evaporation until dry with subsequent dilution in 70% acetonitrile containing 0.1% trifluoroacetic acid. The mass spectra were recorded, processed, and analyzed using flexControl 2.4 (Build 38) and flexAnalysis 2.4 (Build 11) software ("Bruker Daltonics," Germany). The accuracy of the mass measurement was ±2 Da. The matrix consisted of a saturated solution of α-cyano-4-hydroxycinnamic acid ("Sigma_Aldrich," Germany) in a mixture of 50% acetonitrile and 2.5% trifluoroacetic acid. All chemicals and water used were of analytical or special mass spectrometry grade of purity [17].

18.3.5 ORGANOTYPIC CULTURING OF NEWT LIVER TISSUES

For obtaining organotypic liver tissue cultures, newt were decapitated under urethane anesthesia, the liver was removed, 3×3×3-mm fragments were taken from the central lobe, transferred into a nutrient medium for amphibian, and cultured at 21–23°C in dark glass vials. In each experiment, liver fragments from one newt were used for different experimental series and control. For roller culturing, tissue fragments (n=5) were placed on a roller and rotation velocity was set at 35 rpm. For culturing on a substrate, the tissue fragments were placed on nitrocellulose filters into vials and incubated stationary without medium stirring. Native tissue was taken without cultivation. Control group was cultivated under the same conditions as experimental groups without any treatment. After 3-day culturing, the tissues were fixed in Bouin fixative, embedded in paraffin blocks, and 3–4-μ histological sections were prepared. The sections were stained with hematoxylin and eosin; the state of the tissue was evaluated by light microscopy, the area of pigmented cell clusters was measured morphometrically. Mitoses in connective tissue cells of the cortical layer were counted. For each experimental point, 150 sections were performed. At least 500 cells in each section were analyzed. The data were processed by methods of variation statistics. The means and errors of the means were calculated. The significance of differences between the means was evaluated using Student t test [33].

18.4 RESULTS AND DISCUSSIONS

According to the approach which was previously applied for MHTBs isolation, extracts for all four subjects were obtained. Subsequently, obtained extracts were fractionated by a 100% ammonium sulfate desalting.

Hereafter formation of two fractions: precipitate and supernatant occurred. For the further investigations fractions of supernatants were used.

The course of investigation will be considered through the example of one subject – Anethum graveolens (dill). All subjects were examined by using SDS polyacrylamide gel electrophoresis (PAGE) method, in all study supernatants the low molecular compounds were found. Particularly, on electrophoregram for dill supernatant (Figure 18.1) following below, presence of a low molecular compound with molecular weight less than 14,200 Da is observable.

Hereafter, all of the supernatant fractions were examined by reversed phase high performance liquid chromatography method. When analyzing

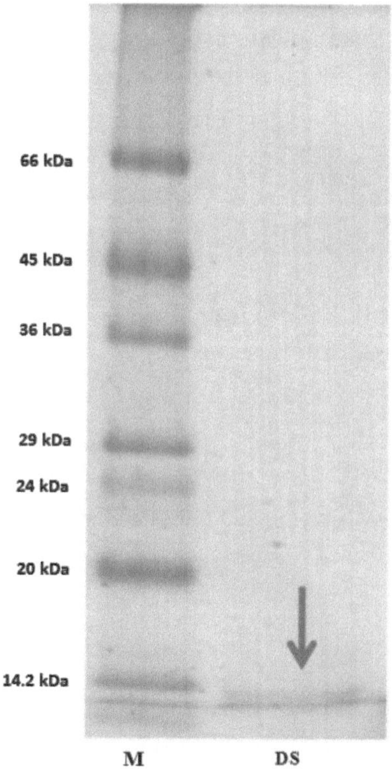

FIGURE 18.1 SDS-electrophoresis in 12.5% polyacrylamide gel of the dill (*Anethum graveolens* L.) supernatant (DS) (As markers (M) we used: α-lactalbumin – 14.2 kDa, soybean trypsin inhibitor – 20.0 kDa trypsinogen from bovine pancreas – 24.0 kDa, carbonic anhydrase – 29.0 kDa, glyceraldehyde – 3 – fosfat dehidrogenaza – 36.0 kDa, ovalbumin – 45.0 kDa, albumin – 66.0 kDa).

dill supernatant using this approach, the following (Figure 18.2) chromatogram was obtained.

Fractions obtained as the result of RP-HPLC were examined by matrix-assisted laser desorption/ionization time-of-flight (MALDI-TOF) mass-spectrometry method. When using this approach, presence of low molecular compounds with molecular weight from 250 to 9000 Da were observed. Molecular weights for peptides, which are included as compounds of dill supernatant, were 267, 1471, 1531, 1845, 1988, 2105, 2763, 3173, 4063, 8607 ± 2 Da.

In order to investigate hepatoprotective properties of earlier obtained peptide-containing fractions, the model of *Pleurodeles Waltl* newt liver roller organotypic culturing in vitro was performed. This model has been chosen due to the fact that it allows estimating the hepatoprotective activity of study subjects. Assessment of the newt liver tissue estate was made by

FIGURE 18.2 DS separation with HPLC. Hydrophobic HPLC column (4.6×250 MM) Kromasil C18 (Russia). Elution was performed in acetonitrile concentration gradient (4–70%) in 0.1% trifluoroacetic acid (pH 2.2) with 0.5 mL per minute flowing rate in 70 minutes. Detection was performed at 280 nm.

measuring pigmented cells surface. According to the literature reports, this characteristics relays amphibian liver tissue ability to respond the injurious influence [21]. After the three days newt liver tissue cultivating procedure, when applying peptides of the studied plants origin at ultralow doses (10^{-8} mg/mL) (Table 18.1), clusters of pigmented cells were detected.

The surface of these cells was higher than those both in control and in native groups. Tissue integrity keeping (i.e., prolongation of an intercellular cooperation, visible activity of a hematogenesis areas, pigmented cells are cooperated into large clusters) in experimental series also should be noted. Histological sections, corresponding to this experiment are represented below (Figure 18.3).

In accordance to foregoing, plant origin peptides capacity at ultralow doses only to influence the activation of defense mechanisms in newt liver tissue was demonstrated.

Moreover, using the same model, investigation of peptide activity at high concentrations (10^{-2} mg/mL) was performed (Table 18.2). In previous publications [34], it was shown that extract supernatant obtained from rat liver tissue demonstrates hepatoprotective effect. Consequently, as a positive control rat liver tissue supernatant was taken for comparative investigation.

When applying peptides isolated from celandine (*Chelidonium majus* L.) at high concentrations, processes of tissue degradation were observed and there was statistically significant difference for pigmented cells surface in experimental and control groups. Vice versa, tissue state in case of peptide treatment at ultralow doses was characterized by a groups of pigmented cells and integrity keeping. Histological sections, corresponding to this experiment are represented below (Figure 18.4).

TABLE 18.1 Hepatoprotective Effect of Studied Plants at Ultralow Doses

Experiment 1	***Surface occupied by pigmented cells (%)***
Native newt liver tissue	7.1±1.4
Control newt liver tissue	5.8±1.3
Peptides from *Chelidonium majus* L.	6.7±1.4 ($p<0.05$)
Peptides from *Artemisia absinthium* L.	7.5±1.3 ($p<0.05$)
Peptides from *Anethum graveolens* L.	9.0±3.7 ($p<0.05$)
Peptides from *Hypericum perforatum* L.	8.5±2.6 ($p<0.05$)

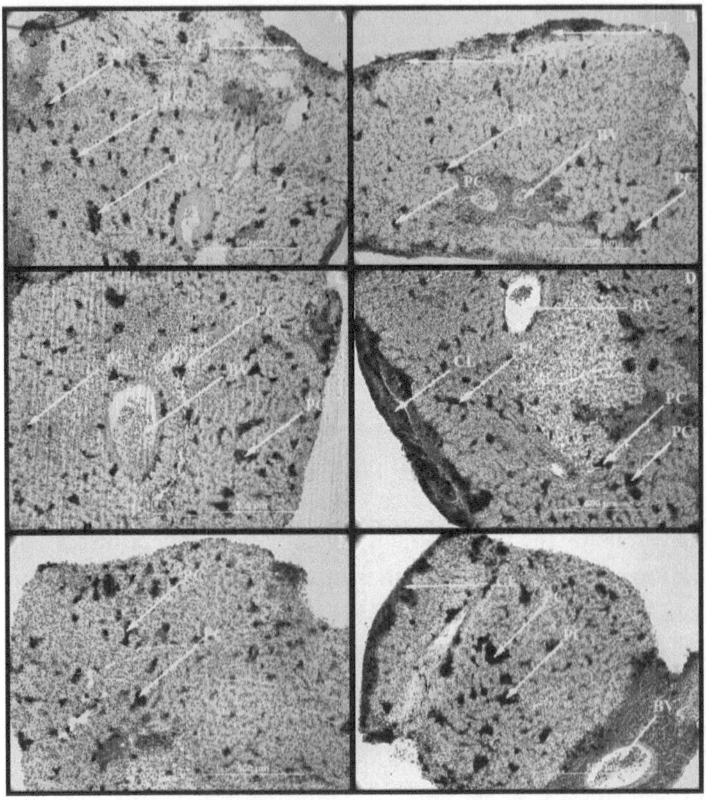

FIGURE 18.3 Histological sections of newt liver tissues for experiment № 1 (A – native newt liver tissue; B – control newt liver tissue; C – peptides from *Chelidonium majus* L.; D – peptides from *Artemisia absinthium* L.; E – peptides from *Hypericum perforatum* L.; F – peptides from *Anethum graveolens* L.; Arrows: PC – pigmented cells; BV – blood vessel; CL – cortical layer).

TABLE 18.2 Comparative Investigation of Dose Corresponded Hepatoprotective Effects

Experiment 2	*Surface occupied by pigmented cells (%)*
Native newt liver tissue	5.5±1.2
Control newt liver tissue	6.1±1.2
Peptides from *Chelidonium majus* L. (10^{-8} mg/mL)	7.4±3.6 ($p<0.05$)
Peptides from Chelidonium majus L. (10^{-2} mg/mL)	4.5±1.1 ($p<0.05$)
Rat liver tissue supernatant	7.6±1.6 ($p<0.05$)

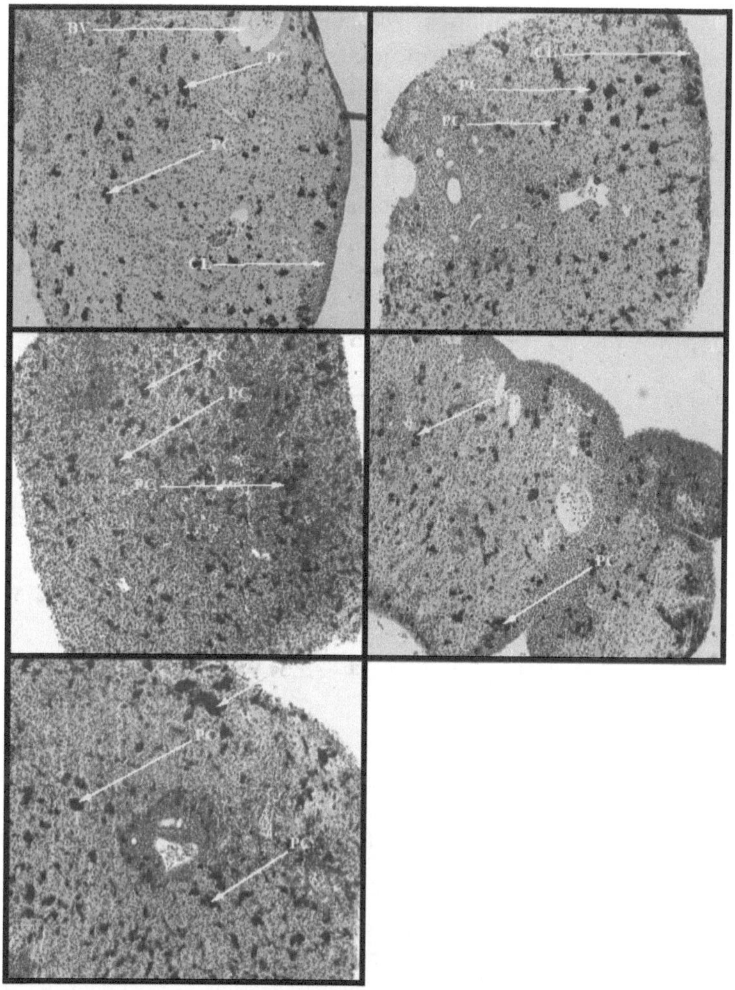

FIGURE 18.4 Histological sections of newt liver tissues for experiment №2 (A – native newt liver tissue; B – control newt liver tissue; C – peptides from *Chelidonium majus* L. (10^{-8} mg/mL); D – peptides from *Chelidonium majus* L. (10^{-2} mg/mL); E – rat liver tissue supernatant. Arrows: PC – pigmented cells; BV – blood vessel; CL – cortical layer).

18.5 CONCLUSIONS

As a result of our study from a number of a medicine plants biologically active at ultralow doses peptide compounds with molecular weights about

1000–9000 Da were obtained. It was shown that peptides of the studied plants origin demonstrate their biological activity only when applying at ultralow doses (10^{-8}–10^{-15} mg/mL). It was stated that peptides of the studied plants origin extracted are responsible for occurrence of hepatoprotective properties in obtaining source plants. It was stated that peptides of the studied plants origin have exactly the same hepatoprotective properties as medicinal plants they were extracted from.

KEYWORDS

- **hepatoprotective effects**
- **medicinal plants**
- **peptides**
- **proteins**

REFERENCES

1. Krasnov, M. S., Yamskova, V. P., Margasyuk, D. V., Kulikova, O. G., Ilina, A. P., Rybakova, E. Yu. Yamskov, I. A., Study of a new group of bioregulators isolated from the greater plantain (*Plantago major* L.). Applied Biochemistry and Microbiology, 2011, Vol. 47, No. 2, pp. 128–135.
2. Bayramoglu, G., Bayramoglu, A., Engur, S., Senturk, H., Ozturk, N., Colak, S., The hepatoprotective effects of *Hypericum perforatum* L. on hepatic ischemia/reperfusion injury in rats. Cytotechnology, Epub 2013 Jun 23
3. Gilca, M., Gaman, L., Panait, E., Stoian, I., Atanasiu, V., *Chelidonium majus* L. – an integrative review: traditional knowledge versus modern findings. Res. Complement. Med., 2010, Vol. 17, No. 5, pp. 241–248.
4. Naeem, M. R., Alaa, O. A., Hepatoprotective effect of dill (*Anethum graveolens* L.) and Fennel (*Foeniculum vulgare* Mill.) oil on hepatotoxic rats. Pakistan Journal of Nutrition, 2014, Vol. 13, No. 6, pp. 303–309.
5. Gilani, A. H., Janbaz, K. H. Preventive and curative effects of *Artemisia absinthium* L. on acetaminophen and CCl_4 -induced hepatotoxicity. General Pharmacology, 1995, Vol. 26, No. 2, pp. 309–315.
6. Amat, N., Upur, H., Blazeković, B., *In vivo* hepatoprotective activity of the aqueous extract of *Artemisia absinthium* L. against chemically and immunologically induced liver injuries in mice. Journal of Ethnopharmacology, 2010, Vol. 131, No. 2, pp. 478–484.

7. Kains, M. G., Book "Culinary Herbs: Their Cultivation Harvesting Curing and Uses." New York, Orange Judd company, 1912, p. 87.
8. Wichtl, M., Herbal Drugs and Phytopharmaceuticals, Boca Raton, FL: CRC Press, 1994, pp. 273–275.
9. Blumenthal, M., Goldberg, A., Brinckmann, J., eds. Herbal Medicine: Expanded Commission E Monographs. Newton, MA: Integrative Medicine Communications, 2000, pp. 359–366.
10. Culpeper N. The English Physician or an Astrologo-physical Discourse on the Vulgar Herbs of this Nation. London, England: Nathanial Brook; 1652.
11. Mitra, S., Gole, K., Samajdar, K., Sur, R. K., Chakraborty, B. N., Antihepatotoxic activity of *Chelidonium majus* L. Int J Pharmacognosy, 1992, Vol. 30, pp. 125–128.
12. Mitra, S., Sur, R. K., Roy, A., Mukherjee, A. S., Effect of *Chelidonium majus* L. on experimental hepatic tissue injury. Phytother Res, 1996, Vol. 10, pp. 354–356
13. Biswas, J., Bhattacharjee, N., Khuda-Bukhsh, A. R., Efficacy of a plant extract (*Chelidonium majus* L.) in combating induced hepatocarcinogenesis in mice. Food Chem Toxicol, 2008, Vol. 46, No. 5, pp. 1474–1487.
14. Abad, M. J., Bedoya, L. M., Apaza, L., Bermejo, P., The Artemisia L. genus: a review of bioactive essential oils. Molecules, 2012, Vol. 17, pp. 2542–2566.
15. Bora KS, Sharma A. The genus Artemisia: a comprehensive review. Pharm Biol 2011; 49: 101–109
16. Herbal Medicinal Product Committee [HMPC]. Assessment report on *Artemisia absinthium* L., herba. London: European Medicines Agency; 2008 (Doc. Ref.: EMEA/HMPC/234444/2008
17. Ilyina, A. P., Kulikova, O. G., Maltsev, D. I., Krasnov, M. S., Rybakov, E. J., Skripnikova, B. C., Kuznetsova, E. S., Buriak, A. K., Yamskova, V. P., Yamskov, I. A., MALDI-TOF mass spectrometric identification of novel intercellular space peptides. Applied Biochemistry and Microbiology, 2011, Vol. 47, No. 2, pp. 118–122.
18. Kulikova, O. G., Yamskova, V. P., A.P. Il'ina, Margasyuk, D. V., Molyavka, A. A., Yamskov, I. A., Identification of a new bioregulator acting in ultralow doses in bulb onion (*Allium cepa* L.). Applied Biochemistry and Microbiology, 2011, Vol. 47, No. 4, pp. 356–360.
19. Yamskova, V. P., Modyanova, E. A., Leventhal, V. I., Lankovskaya, T. P., Bocharova, O. K., Malenkov, A. G., Tissue-specific macromolecular factors from the liver and lung: separation and influence on mechanical strength of the tissue and cells. Biophysics, 1977, Vol. 22, pp. 168–174.
20. Yamskova, V. P., Tumanova, N. B., Loginov, A. S., Comparative study of the effect of liver tissue extracts obtained from mice C57BL and CBA on hepatocytes adhesion. Bulletin of Experimental Biology and Medicine, 1990, Vol. 109, No. 3, pp. 397–399.
21. Yamskova, V. P., Reznikova, M. M., Polypeptide of a low molecular weight obtained from serum of warm-blooded: effects on cell adhesion and proliferation. Journal. general biology, 1991, Vol. 52. No. 2, pp. 181–191.
22. Yamskova, V. P., The role of calcium ions in stabilizing of the rat liver adhesive factor. Biophysics, 1978, Vol. 23. pp. 428–432.
23. Krasnov, M. S., Grigoryan, E. N., Yamskova, V. P., Boguslavskiy, D. V., Yamskov, I A., Regulatory proteins obtained from vertebrates eye tissues. Radiation Biology and Radioecology, 2003, No. 3, pp. 265–268.

24. Margasyuk, D. V., Krasnov, M. S., Blagodatskikh, I. V., Grigoryan, E. N., Yamskova, V. P., Yamskov, I. A., Regulatory Protein from Bovine Cornea: Localization and Biological Activity, pp. 47–59. Book "Biochemical Physics Frontal Research," Ed. by Varfolomeev, S. D., Burlakova, E. B., Popov, A. A., Zaikov, G. E., Hauppauge, N. Y., Nova Science Publishers Inc., 2007, pp. 47–60.
25. Borisenko, A. V., Yamskova, V. P., Krasnov, M. S., Blagodatskikh, I. V., Vecherkin, V. V., Yamskov, I. A., Regulatory proteins from the mammalian liver that display biological activity at ultra low doses, pp. 35–45. Book "Biochemical Physics Frontal Research," Ed. by Varfolomeev, S. D., Burlakova, E. B., Popov, A. A., Zaikov, G. E., Hauppauge, N. Y., Nova Science Publishers Inc., 2007, pp. 35–46.
26. Krasnov, M. S., Gurmizov, E. P., Yamskova, V. P., Yamskov, I. A., Analysis of a Regulatory Peptide from the Bovine Eye Lens: Physicochemical Properties and Effect on Cataract Development in vitro and in vivo. Book "Biochemical Physics Frontal Research," Ed. by Varfolomeev, S. D., Burlakova, E. B., Popov, A. A., Zaikov, G. E., Hauppauge, N. Y., Nova Science Publishers Inc., 2007, pp. 21–34.
27. Yamskova, V. P., Krasnov, M. S., Rybakova, E. Y., Vecherkin, V. V., Borisenko, A. V., Yamskov, I. A., Analysis of regulatory proteins from bovine blood serum that display biological activity at ultra low doses: 2. Tissue localization and role in wound healing," pp. 68–77. Book "Biochemical Physics Frontal Research," Ed. by Varfolomeev, S. D., Burlakova, E. B., Popov, A. A., Zaikov, G. E., Hauppauge, N. Y., Nova Science Publishers Inc., 2007, P. 160.
28. Gundorova, R. A., Khoroshilova-Maslova, I. P., Chentsova, E. V., Ilatovskaya, L. V., Yamskova, V. P., Romanov, I. Y., Adgelon application in the treatment of penetrating wounds of the cornea in the experiment. Questions of Ophthalmology, 1997, Vol. 113, No. 2, pp. 12–15.
29. Nazarova, P. A., Yamskova, V. P., Krasnov, M. S., Filatova, A. G., Yamskov, I. A., Regulatory proteins biologically active in ultralow doses from mammalian glands and their secretions, pp. 78–88. Book "Biochemical Physics Frontal Research," Ed. by Varfolomeev, S. D., Burlakova, E. B., Popov, A. A., Zaikov, G. E., Hauppauge, N. Y., Nova Science Publishers Inc., 2007, pp. 73–82.
30. Yamskova, V. P., Krasnov, M. S., Yamskov, I. A., New experimental and theoretical aspects in bioregulation. The mechanism of action of tissue-specific homeostatic membranotrop nyh bioregulators. Saarbrucken: Lambert Academic Publishing, 2012, p. 136.
31. Chukhno, V. P., Rozhko N. Illustrated Encyclopedic Dictionary "Medicinal Plants." Eksmo, 2007, p. 360.
32. Laemmli, U. K., Nature, 1970.
33. Yamskova, P. V., Borisenko, V. A., Krasnov, S. M., Ilina, P. A., Rybakova, Yu. E., Malcev, I. D., Yamskov, A. I., On Mechanisms Underlying Regeneration and Reparation Processes in Tissues. Bulletin of Experimental Biology and Medicine, 2010, Vol. 149, No. 1, pp. 140–143.
34. Yamskova, V. P., Borisenko, A. V., Krasnov, M. S., Il'yina, A. P., Rybakova, E. Y., Maltsev, D. I., Yamskov, I. A., Effect of Bioregulators Isolated from the Liver, Blood Serum, and Bile of Mammals on the State of Newt Liver Tissue in Organotypic Culture. Bulletin of Experimental Biology and Medicine, 2010, Vol. 150, No. 1, pp. 140–148.

CHAPTER 19

INTERCALATION AND RELEASE OF DINUCLEOTIDES FROM A NANODIMENSIONAL LAYERED DOUBLE HYDROXIDE

A. S. SHCHOKOLOVA, A. N. RYMKO, D. V. BURKO, S. V. KVACH, and A. I. ZINCHENKO

Institute of Microbiology, National Academy of Sciences, Kuprevich Str. 2, 220141, Minsk, Belarus; Fax: +375(17)267-47-66; E-mail: zinch@mbio.bas-net.by

CONTENTS

ABSTRACT

Diadenosine-5′,5‴-P[1], P[4]-tetraphosphate (Ap_4A) and bis-(3′-5′)-cyclic dimeric deoxyguanosine monophosphate (c-di-GMP) intercalated into nanosized (200–300 nm) Al, Mg layered double hydroxides (LDHs) were originally prepared. Binding of both dinucleotides with inorganic carrier ranged from 0.8 to 1 μmol/mg LDH. Immobilization process was shown not to alter chemical structure of the dinucleotides. Parameters of Ap_4A and c-di-GMP elution from nanoparticles were studied. It was found that rate of dinucleotide release from LDH nanocomplexes (LDH/Ap_4A and LDH/c-di-GMP) varied depending on pH of the media (higher at acid pH and lower at neutral pH values). The obtained results evidence in favor of using LDH/Ap_4A and LDH/c-di-GMP nanocomposites as pH-controlled passive system for delivery of pharmaceutically valuable dinucleotides into target cells.

19.1 INTRODUCTION

Delivery of pharmaceuticals to affected tissues envisages handling problems related to preservation of their structure and activity along the transit route from introduction zone to the target cells and further translocation through cell membrane. Moreover, the drug supplied by conventional means tends to be evenly distributed in the body, reaching not only the target cells where it is expected to display therapeutic effect but also other vital systems causing potentially adverse impact. It appears natural therefore that currently the keen interest of researchers has been focused on elaboration of methods for passive or precise transportation of medicines into cells using organic and inorganic nanovehicles [1–3].

Application of nanoparticles as drug carriers enables to reduce considerably systemic intoxication of the organism as compared to molecular form, to protect active substances from premature decomposition in blood stream, to modify their pharmacokinetic parameters and to promote their penetration through plasmatic membrane and intracellular partitioning [4].

Vivid interest around the world has been aroused by systems of drug delivery engaging nanodimensional inorganic supports [5]. It was provoked by discovery of special wall characteristic in vessels feeding tissues exposed to cancerogenesis, inflammation, infarction, hypoxia, etc.

denoted as hyper transparent capacity, and with respect to tumor cells – enhanced permeability and retention of inflowing substances (termed EPR effect) [6, 7]. Unlike normal, such blood vessels are distinguished by loose structure passing relatively large molecules (about 500 nm). The described phenomenon lays the ground for engineering drug-nano-carrier composites as constituents of passive delivery system.

A whole spectrum of materials was proposed for construction of such drug-courier nanocomplexes – calcium phosphate, gold, carbonaceous compounds, silica, ferric oxide, and layered double hydroxides (LDHs) [8, 9].

LDHs, also called "anionic clays" represent a family of materials that have attracted increasing attention in recent years due to their technological importance in catalysis, separation technology, optics, medical science, and nanocomposite engineering. LDHs consist of positively charged metal hydroxide layers and interlayer gallery, in which the anions (along with water) are stabilized in order to compensate the positive layer charges. The general chemical formula of LDH clays is expressed as $[M^{II}_{1-x}M^{III}_{x}(OH)_2]^{x+}(A^{n-})_{x/n}\cdot yH_2O$, where M^{II} is a divalent metal ion, such as Mg^{2+}, Ca^{2+}, Zn^{2+}, etc., M^{III} is a trivalent metal ion, such as Al^{3+}, Cr^{3+}, Fe^{3+}, Co^{3+}, etc. and A^{n-} is an anion of any type with charge n, such as Cl^-, CO_3^{2-}, NO_3^-, etc. [10]. The electrostatic interactions, hydrogen bonds between layers and contents of the gallery hold the layers together, making a three-dimensional structure (Figure 19.1). There are several combinations of divalent and trivalent cations that can form LDHs. For these ions, the only requirement is that their radii are not too different from those of Mg^{2+} and Al^{3+}. The anions occupy the interlayer gallery region

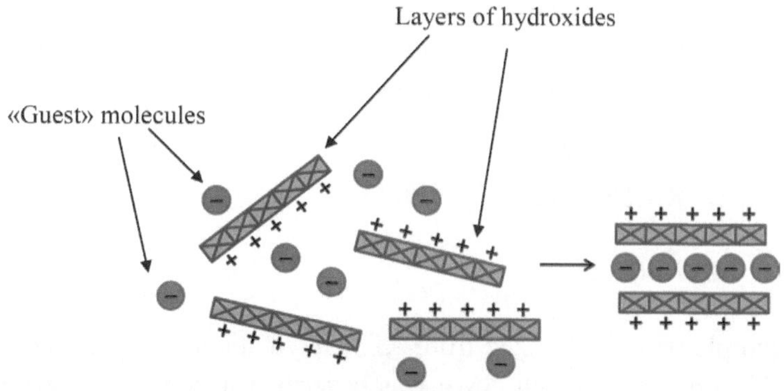

FIGURE 19.1 Schematic structure of layered double hydroxides.

of these crystalline materials. The remarkable behavior of LDHs is their high reactivity toward various organic anions, which can exchange as much as 80–100% of the interlayer molecules.

LDHs have historically been of interest as catalysts, ceramic precursors, traps for anionic pollutants, catalyst supports, ion exchangers, and additives for polymers. More recently, the successful synthesis of LDH materials on the nano scale has paved the way for many novel applications particularly in the field of nanomedicine. For example, LDH materials are increasingly explored as controlled release systems [8, 11, 12]. Up to now many kinds of anions were intercalated in the LDH interlayer gallery, such as common inorganic anions, organic anions, polymeric anions, complex anions, macrocyclic ligands and their metal complexes, polyoxalates, and biochemical anions (amino acids, CMP, AMP, ATP and nucleic acid fragments) [13, 14].

LDHs possess a series of advantages as transporters over other nanocarriers. They display several beneficial properties, including primarily ready bioavailability and low toxicity. According to literature reports, synthesis of LDHs differing in qualitative and quantitative composition of metal cations yields under definite conditions nanoparticles ranging in size from 50 to 500 nm [15]. The substance embedded in interlayer spacing is safely protected from deteriorating impact of damaging factors [16].

LDHs carry a minor positive charge facilitating efficient intercalation of anions and sufficient electrostatic attraction to negatively charged cell surface [8]. In addition, LDH particles rapidly decay in media with low pH values to relatively non-toxic kidney-evacuated metal ions (e.g., Mg^{2+} and Al^{3+}). It is vital that LDHs retain stability at neutral pH values.

In the present study, we prepared artificially novel bio-inorganic nanohybrids of LDH and pharmaceutically valuable biomolecules Ap_4A and c-di-GMP (Figure 19.2) and investigated the possibility of using intercalated LDH as a drug delivery system.

19.2 MATERIALS AND METHODS

19.2.1 MATERIALS

Ap_4A sample was synthesized from ATP enzymatically as previously reported [17]. Synthesis of c-di-GMP was described in our paper earlier [18].

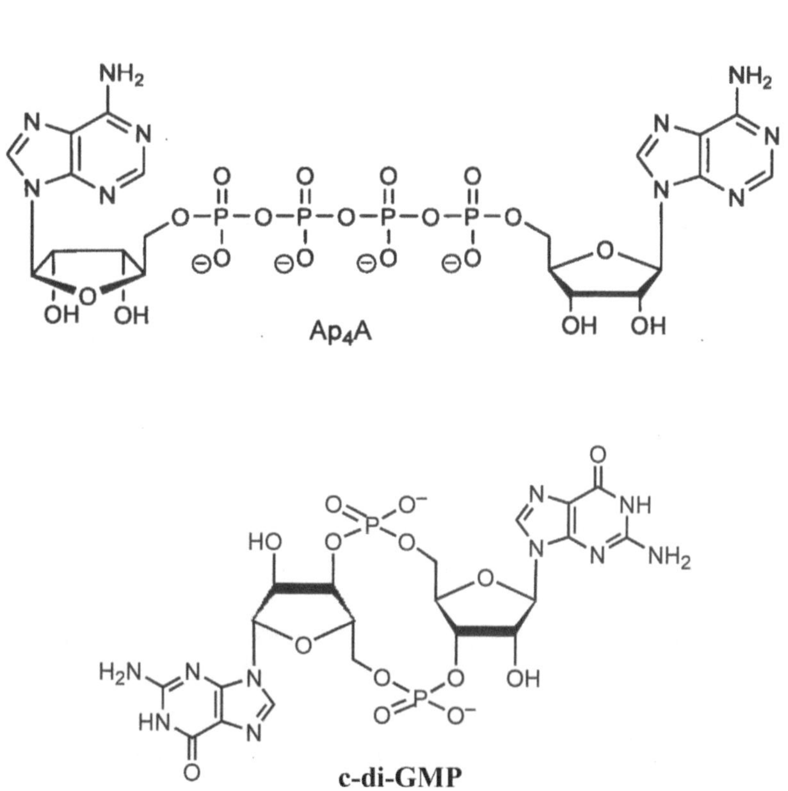

FIGURE 19.2 Structure of the dinucleotides, Ap_4A and c-di-GMP.

ATP, AMP, adenosine and adenine were purchased from Sigma-Aldrich (USA). Twice-distilled water from which carbon dioxide was removed by boiling was used in all experiments.

19.2.2 PREPARATION OF LDH/AP$_4$A AND LDH/C-DI-GMP NANOCOMPOSITE PARTICLES

The nanocomposites with Mg:Al ratio 2:1 were synthesized by coprecipitation method according to [19], with subsequent crystallization at 75°C in the presence of ammonium hydroxide. Briefly, 10 mL salt solution containing 0.02 M $Mg(NO_3)_2$ and 0.01 M $Al(NO_3)_3$ was added drop-wise with vigorous stirring at 25°C to 10 mL of 10 mM NH_4OH solution comprising 10 mM Ap_4A or c-di-GMP. The reaction slurry was aged at 75°C for 12 h.

The resulting white precipitate was collected by centrifugation (20 000 g for 10 min), washed five times with water, and finally, with acetone. All the samples were air-dried at 60°C for 5–6 h.

19.2.3 CHARACTERIZATION OF LDH/AP$_4$A NANOHYBRIDS

The obtained LDH nanoparticles were imaged using a JSM-2010 (JEOL, Japan) transmission electron microscope (TEM) at the acceleration voltage 200 kV. The loading capacity of LDH/Ap$_4$A and LDH/c-di-GMP nanoparticles and the efficiency of this process under different experimental conditions were evaluated using the following method. A known amount of the nanocomposites was placed in a 10 mL volumetric flask and LDH layers were dissolved in 0.02 M HCl solution. Concentration of Ap$_4$A or c-di-GMP in the solution was determined by monitoring the absorbance at λ=260 nm for Ap$_4$A (ε=30800 $M^{-1}\cdot cm^{-1}$) or λ=253 nm for c-di-GMP (ε=23700 $M^{-1}\cdot cm^{-1}$) with Shimadzu Corporation model UV-1202 spectrophotometer.

19.2.4 RELEASE OF DINUCLEOTIDES FROM LDH/AP$_4$A OR LDH/C-DI-GMP NANOHYBRIDS

To measure the amount of Ap$_4$A and c-di-GMP released from the nanohybrids, the in vitro drug release tests were performed at 25°C by stirring powdered LDH/Ap$_4$A or LDH/c-di-GMP nanohybrids (0.05 g) in 20 mL of 0.05 M phosphate-citrate buffer solution (pH 4.4 or 7.5, respectively) [20]. Aliquots (1 ml) of the suspension were sampled at desired time intervals, centrifuged and the Ap$_4$A and c-di-GMP contents of supernatant were determined by reading UV absorbance at λ=260 nm for Ap$_4$A or λ=253 nm for c-di-GMP to calculate the amount of dinucleotides released from the nanohybrids. The percentage released at each time point was expressed as a fraction of the total amount of the dinucleotides.

19.3 RESULTS AND DISCUSSION

LDHs are currently produced, as a rule, by one of the following methods: (i) coprecipitation [21]; (ii) ion exchange [22]. Co-precipitation regarded as preferential technique envisages adding alkaline solution to the mixture containing divalent and trivalent metal cations. As to particulate dimensions, it should be stated that the size and morphology of LDH nanoparticles depend on conditions of synthesis and nature of gallery (guest) molecules, while precise correlation of morphological changes with LDH composition remains to be established [23].

The LDHs, containing Ap_4A (LDH/Ap_4A) or c-di-GMP (LDH/c-di-GMP) were prepared by coprecipitation from a mixed aqueous solution containing Mg^{2+}, Al^{3+} and the dinucleotides. In order to choose the optimal conditions for intercalation, various reaction parameters, like temperature, time, concentration of the components and pH, were examined. Based on these experiments, it was concluded that the optimal intercalation conditions for Ap_4A are as follows: Mg:Al molar ratio 2:1, room temperature, pH 10.0, and 12 h aging. The maximum amount of Ap_4A in the intercalated compound was around 0.83 mmol per 1 g of LDH-nanocomposite.

The similar c-di-GMP immobilization procedure allowed to achieve drug loading approximately 1 mmol per 1 g of LDH-nanohybrid (Table 19.1). For comparison, the Table 19.1 provides data on immobilization of additional nucleic acid compounds – ATP, AMP, adenosine and adenine.

It is clear from Table 19.1 that the number of guest molecules intercalated into LDHs is a function of several factors. The key parameter is molecular charge. For instance, adenine and adenosine lacking negative charge are not intercalated into LDH structure. On the other hand, elevated negative charge results in reduced number of attracted molecules, probably due to deficiency of available positive charges in LDHs [24]. It is well illustrated by comparative amounts of intercalated AMP and ATP (Ap_4A) – 1.44 versus

TABLE 19.1 The Loading of the Guest Molecules in LDH-Nanocomposites

Guest molecule	Guest (mmol/g)	Guest (g/g)	Loading percentage (w/w)
C-di-GMP	1.1±0,05	0.81±0,04	81
Ap_4A	0.83±0,03	0.69±0,03	69
ATP	0.79±0,03	0.38±0,02	38
AMP	1.44±0,07	0.50±0,02	50
Adenosine	–	–	–
Adenine	–	–	–

0.79 (0.83) mmol/g LDH. The other factors governing incorporation of guest compounds into LDH structure are the form and size of the molecule, its hydrophobicity, solubility of LDH/guest complex [24].

The morphology and dimensions of LDH/Ap_4A nanohybrids have been estimated by electron microscopy. TEM image of LDH/Ap_4A nanoparticles is presented in Figure 19.3. It should be noted that presence of small hexagonal platelets having a diameter of about 200 nm together with some larger platelets (about 350 nm), could be explained by formation of aggregates. Microphotographs of LDH/c-di-GMP nanoparticles resembled images of LDH/Ap_4A (data not provided).

The intercalated dinucleotides were quantitatively recovered from the host lattices by treating LDH/Ap_4A and LDH/c-di-GMP nanoparticles with 0.02 M HCl [24]. This technique coupled to high performance liquid chromatography (LKB, Sweden; column Ultropac TSK ODS-120T) of guest molecules before and after elution from LDH complex (Figure 19.4) led us to assume that Ap_4A and c-di-GMP compounds were intercalated into LDH in intact form.

In order to investigate the possibility of using intercalated LDH nanohybrid as a drug delivery system, deintercalation of Ap_4A and c-di-GMP was examined. Typical kinetic curves of Ap_4A and c-di-GMP release from the nanocomposites at different pH values are shown in Figure 19.5. Noteworthy,

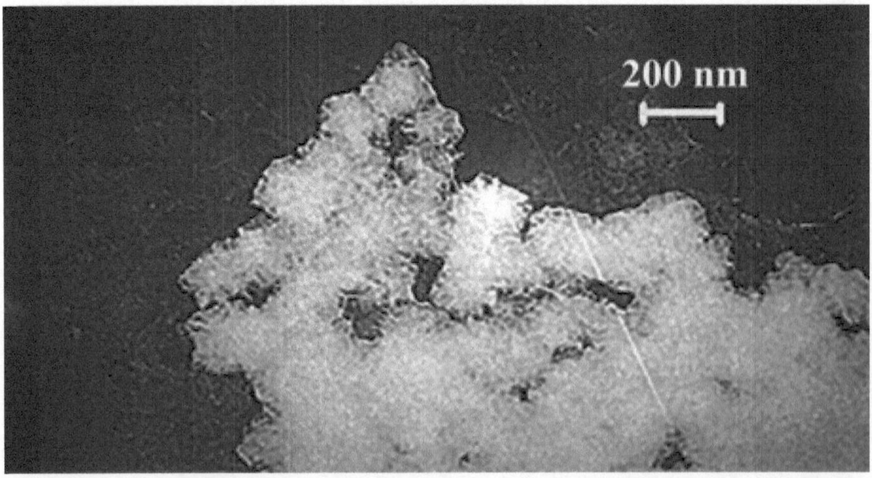

FIGURE 19.3 Transmission electron microscopy image of LDH/Ap_4A nanoparticles.

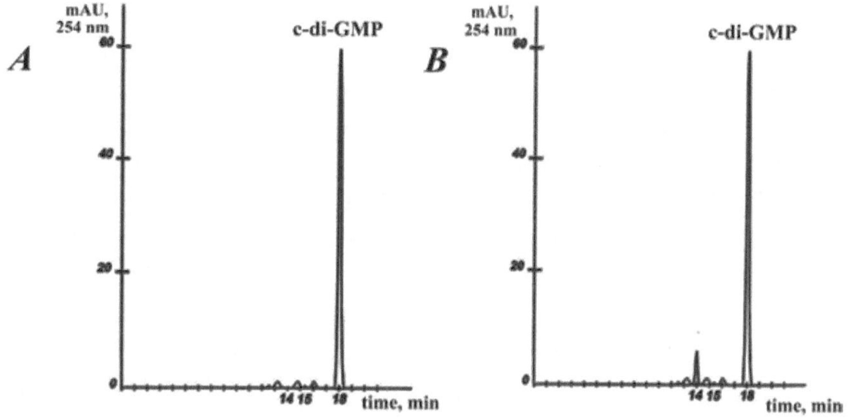

FIGURE 19.4 Purity of c-di-GMP tested by HPLC. (A) HPLC profile of c-di-GMP before intercalation into LDH. (B) HPLC profile of c-di-GMP after elution from LDH/dinucleotide nanocomposite.

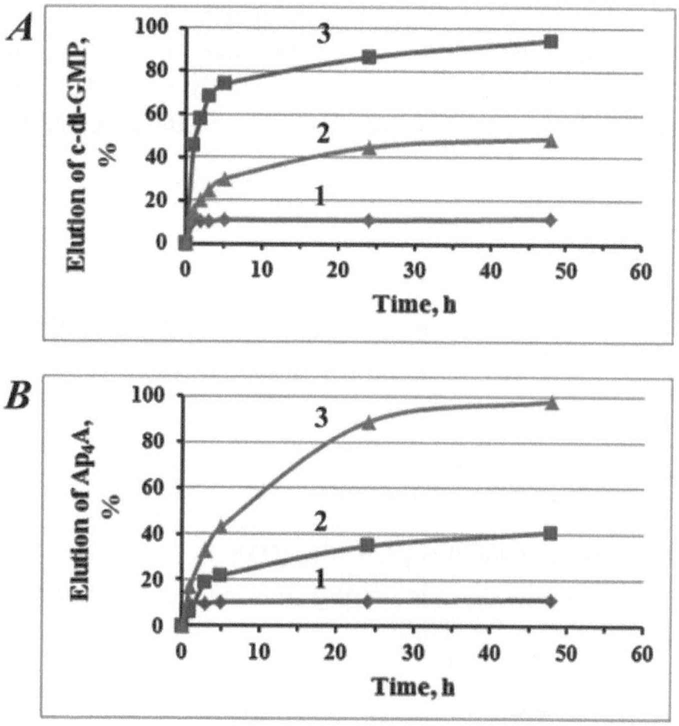

FIGURE 19.5 Release of c-di-GMP (A) and Ap_4A (B) from LDH/dinucleotide nanocomposites in H_2O (1) and 0.05 M phosphate-citrate buffer solution at pH 7.5 (2) and pH 4.4 (3).

that the release of Ap_4A and c-di-GMP from the nanohybrids is evidently pH dependent, and its rate at pH 7.5 is remarkably lower than that at pH 4.4.

Such a discrepancy in the release rate at pH 4.4 and pH 7.5 may be due to a possible difference in mechanism responsible for the dissociation of the guest molecules from the nanohybrid [25]. At acidic pH, LDHs begin to dissolve. This would indicate that release of an interlayer molecule should occur mainly through the removal of inorganic host. At pH above 7.0, LDHs are likely to be more stable, and as a result, release may be attributed to the restricted motion of guest molecules arising from steric effect of LDHs and the electrostatic interaction between the anions and positively charged LDHs layers. That is to say, the mechanism of release in the pH 4.4 environment should engage both dissolution of LDH layers and ion exchange, while at pH 7.5 the mechanism should include mainly ion exchange of gallery and the buffer solution [25]. The lower detachment rate of Ap_4A and c-di-GMP from the nanohybrids at pH 7.5 indicates that LDH nanocomposites are indeed an excellent potential drug delivery system.

Summing up, this study proved the first successful demonstration that Ap_4A and c-di-GMP may be reversibly intercalated into LDH. Dinucleotide-LDH nanohybrid appears to be an attractive carrier for passive drug delivery to the target tissues.

ACKNOWLEDGEMENTS

The work was supported by the grant 3.07 from the Belarus State Research Program "Basics of Biotechnologies."

KEYWORDS

- **Al, Mg layered double hydroxides (LDHs)**
- **bis-(3′-5′)-cyclic dimeric deoxyguanosine monophosphate (c-di-GMP)**
- **diadenosine-5′,5′′′-P1**
- **nanocomposites**
- **P4-tetraphosphate (Ap_4A)**

REFERENCES

1. Xu, Z. P., Zeng, Q. H., Lu, G. Q., Yu, A. B., Chem. Eng. Sci., 61 (3), 1027 (2006).
2. Liu, C. X., Nano Biomed. Eng., 3 (2), 73 (2011).
3. Hanafi-Bojd, M. Y., Jaafari, M. R., Ramezanian, N., Xue, M., Amin, M., Shahtahmassebi, N., Malaekeh-Nikouei, B., Eur. J. Pharm. Biopharm., 89, 248 (2015).
4. Irache, J. M., Esparza, I., Gamazo, C., Agueros, M., Espuelas, S., Vet. Parasitol., 180 (1–2), 47 (2011).
5. Yang, J. H., Han, Y. S., Park, M., Park, T., Hwang, S. J., Choy, J. H., Chem. Mater., 19 (10), 2679 (2007).
6. Torchilin, V., Adv. Drug Del. Rev., 63 (3), 131 (2011).
7. Fang, J., Nakamura, H., Maeda, H., Fang, J., Adv. Drug Del. Rev., 63 (3), 136 (2011).
8. Xu, Z. P., Lu, G. Q., Pure Appl. Chem., 78, 1771 (2006).
9. Chen, Y., Chen, H., Shi, J., *Mol. Pharmaceutics.* 11 (8), 2495 (2014).
10. Cavani, F., Trifiro, F., Vaccari, A., Catal. Today. 11, 173 (1991).
11. Choy, J. H., Park, M., Clay Sci., 12 (Suppl. 1), 52 (2005).
12. Hussein, M. Z., S. H. Al Ali, Zainal, Z., Hakim, M. N., Int. J. Nanomed., 6, 1373 (2011).
13. Barahuie, F., Hussein, M. Z., Fakurazi, S., Zainal, Z., Int. J. Mol. Sci., 15, 7750 (2014).
14. Pang, X., Cheng, J., Chen, L., Li, D., Appl. Clay Sci., 104, 128 (2015).
15. Manzi-Nshuti, C., Chen, D., Su, S. P., Wilkie, C. A., Thermochim. Acta. 495, 63 (2009).
16. Xu, Z. P., Walker, T. L., Liu, K. L., Cooper, H. M., Lu, G. Q., Bartlett, P. F., Int. J. Nanomed., 2 (2), 163 (2007).
17. Burko, D. V., Kvach, S. V., Zinchenko, A. I., Biotechnology: State Of The Art And Prospects For Development (Ed. G.E. Zaikov), Nova Science Publishers, Inc., New York, 31 (2008).
18. Korovashkina, A. S., Rymko, A. N., Kvach, S. V., Zinchenko, A. I., J. Biotechnol., 164 (2), 276 (2012).
19. Olanrewaju, J., Newalkar, B. L., Mancino, C., Komarneni, S., Mater. Lett., 45 (6), 307 (2000).
20. Liu, C., Hou, W., Li, L., Li, Y., Liu, S., J. Solid State Chem., 181, 1792 (2008).
21. Olfs, H. W., Torres-Dorante, L. O., Eckelt, R., Kosslick, H., Appl. Clay Sci., 43 (3–4), 459 (2009).
22. Aisawa, S., Ohnuma, Y., Hirose, K., Takahashi, S., Hirahara, H., Narita, E., Appl. Clay Sci., 28 (1–4), 137 (2005).
23. Liu, C. X., Hou, W. G., J. Dispers. Sci. Technol., 30 (2), 174 (2009).
24. Choy, J. H., Oh, J. M., Park, M., Sohn, K. M., Kim, J. W., Adv. Mater., 16 (14), 1181 (2004).
25. Tyner, K. M., Schiffman, S. R., Giannelis, E. P., J. Control Rel., 95, 501 (2004).

INDEX

E

F

G

H

I

K

L

M

N

O

T

U